Probability and Statistics

Probability and Statistics

E. Rukmangadachari

Professor of Mathematics
Department of Humanities and Sciences
Malla Reddy Engineering College
Secunderabad

Pearson

ISBN 9788131761366
First Impression, 2012
Current Impression, 2026

Published by Pearson India Education Services Pvt. Ltd, CIN: U72200TN2005PTC057128.

Head Office: 1st Floor, Berger Tower, Plot No. C-001A/2, Sector 16B, Noida - 201 301, Uttar Pradesh, India.
Registered Office: 6th Floor, Tower A, Unit A and B,International Tech Park, CapitaLand Chennai,

200 Feet Radial Road, Zamin Pallavaram, Old Pallavaram,Chennai – 600117, Tamil Nadu, India

Website: in.pearson.com; Email: companysecretary.india@pearson.com

Printed in India at: Manipal Technologies Limited, Manipal

To

My Beloved Parents

Enikapati Krishnamachari,

Rajamma, Ademma

About the Author

E. Rukmangadachari is former head of Computer Science and Engineering as well as Humanities and Sciences at Malla Reddy Engineering College, Secunderabad. Earlier, he was a reader in mathematics (PG course) at Government College, Rajahmundry. He is an M.A. from Osmania University, Hyderabad, and an M.Phil. and Ph.D. degree holder from Sri. Venkateswara University, Tirupathi.

A recent recipient of the Andhra Pradesh State Meritorius Teacher's award in 1981, Professor Rukmangadachari has published over 40 research papers in national and international journals. With a rich repertoire of over 45 years' experience in teaching mathematics to undergraduate, postgraduate and engineering students, he is currently the vice president of the Andhra Pradesh Society of Mathematical Sciences, Hyderabad. An ace planner with fine managerial skills, he was the organizing secretary for the conduct of the 17th Congress of the Andhra Pradesh Society for Mathematical Sciences, Hyderabad.

Contents

Preface

I am pleased to present this edition of *Probability and Statistics* to the B.Tech. students.

The topics have been dealt with in a coherent manner, supported by illustrations for better comprehension. Each chapter is replete with examples and exercises, along with solutions and hints, wherever necessary.

The book also has numerous Mutiple Choice Questions and Fill in the Blanks at the end of each chapter, thus providing the student with an abundant repository of exam specific problems.

Suggestions for the improvement of the book are welcome and will be gratefully acknowledged.

Acknowledgements

I express my deep sense of gratitude to Sri. Ch. Malla Reddy, Chairman, and Sri. Ch. Mahender Reddy, Secretary, Malla Reddy Group of Institutions (MGRI), whose patronage has given me the opportunity to write a book.

I am thankful to Prof. Madan Mohan, Director (Academics) and Col. G. Ram Reddy, Director (Administration), MRGI; and Dr R. K. Murthy, Principal, Malla Reddy Engineering College, Secunderabad, for their kindness, guidance and encouragement.

E. Rukmangadachari

List of Symbols

$\mathbb{R}$	Set of real numbers
S	Sample space
s	Sample point
A, B and C	Sets
$(\bar{\ })$, $()'$ and $()^c$	Complement of a set
$\cup$	Union of sets
$\cap$	Intersection of sets
$\sum$	Summation symbol
X, Y and Z	Random variables
x	Value of random variable X
$P()$	Probability
$P(X = x)$	Probability of the event $X = x$
p	Probability value or probability of success
E	Event/maximum error of estimate
F	Cumulative probability distribution
$f(x)$	Probability density function
μ	Mean
σ^2	Variance
σ	Standard deviation ($= \sqrt{\text{Variance}}$)
ϕ	Empty set or impossible event
q	Probability of failure (Bionomial Distribution)
$b(x; n, p)$	Probability function of bionomial Distribution
$B(x; n, p)$	Cumulative probability distribution of bionomial distribution
λ	Parameter (Poisson Distribution)
$f(x, \lambda)$	Probability function of Poisson distribution
$F(x, \lambda)$	Cumulative probability distribution of Poisson distribution
$f(x; \mu, \sigma^2)$	Probability function of normal distributionwith parameters μ
Z	Standard normal variable
z	Value of Z
$f(Z)$	Probability function of Z
$F(Z)$	Cumulative distribution function of Z
α	Level of significance
$N(\mu, \sigma^2)$	Normal distribution with μ and σ^2
$\in$	Small positive integer
N	Population size
$\bar{x}$ or $\bar{X}$	Sample mean
$\mu_{\bar{x}}$	Mean of sampling distribution of the mean $\bar{x}$
$\sigma^2_{\bar{x}}$	Variance of sampling distributions of the mean $\bar{x}$
χ	χ^2 - distribution
s^2	Sample variance
ν	Number of degrees of freedom
$F_\alpha(\nu_1, \nu_2)$	F-distribution
$t_{\alpha,\nu}$	t-distribution
$\hat{\theta}$	Unbiased estimator of θ
θ	Population parameter

Z_α	Value of Z-distribution for specified α
t_α	Value of Student's t-distribution for specified α
L and U	Lower and upper bounds respectively
H_0	Null hypothesis
H_1	Alternate hypothesis
δ	Difference of μ_1 and μ_2 or P_1 and P_2
p	Proportion
o_{ij} and e_{ij}	Observed and expected cell frequency respectively
$\hat{y} = a + bx$	Estimated regression line equation
a and b	Estimate of α and β respectively

ρ	Karl Pearson (population) coefficient of correlation
$\rho\left(\frac{\sigma_y}{\sigma_x}\right)$ and $\rho\left(\frac{\sigma_y}{\sigma_x}\right)$	Coefficient regression
s_e^2	Estimate of variance σ^2
Cov(X, Y)	Covariance between X and Y
r or $\hat{\rho}$	Estimate of sample correlation coefficient ρ
Z_f	Fisher Z
μ_Z	Mean of Z_f
σ_Z^2	Variance of Z_f
d_i	Rank of x_i minus rank of y_i
r_Δ	Spearson's rank correlation coefficient
$\bar{x}$ and $\bar{y}$	Mean of x and y values respectively

1 Probability

1.1 Introduction

The beginning of probability theory dates back to the mid-seventeenth century when Pascal[1] and Fermat[2] independently found solution to a problem faced by a gambler, though the first book on the subject *Book on Games of Chance* written by Cardan[3] was published in 1663. Outstanding contributions to probability theory were made by Huygens,[4] Bernoulli,[5] Demoivre,[6] Laplace,[7] Gauss,[8] Chebyshev,[9] Markov,[10] and Kolmogorov[11]—the last being credited with the development of the axiomatic theory of probability.

The early workers recognized the significance of investigation of laws governing random events. The increasing interest in natural sciences made it necessary to extend the theory of probability beyond the games of chance. Probability theory today is connected with many other branches of mathematics and many fields of natural science, engineering technology and economics.

Probability was developed to analyse the games of chance. It is a mathematical modelling of the phenomenon of chance or randomness. The measure of chance or likelihood is called the probability of statement. Closely related to probability is statistics which is the science of handling, assembling, analysing, characterizing and interpreting data, and drawing conclusion. It can be regarded as a branch of applied probability.

[1]Pascal, Blaise (1623–62) is a French philosopher, mathematician, great geometer, probabilist, combinotorist and physicst. He and Fermat independently founded probability theory.

[2]Fermat, Pierre de (1601–65) is a brilliant versatile amateur, French mathematician and an unexcelled number theorist.

[3]Cardan, Jerome (1501–76) is an Italian physician and mathematician.

[4]Huygens, Christiaan (1629–95) is a Dutch physical scientist, astronomer and mathematician. He did pioneering work on continued fractions, tautochrone, probability and analysis toward the invention of calculus.

[5]Bernoulli, Jacob (1654–1704) is a Swiss mathematician, physicist, analyst, combinatorist, probabilist and statistician.

[6]de Moivre, Abraham de (1667–1754) is French mathematician, statistician, probabilist and analyst. He was born in France, studied in Belgium and settled in England.

[7]Laplace, Pierre Simon de (1749–1827) French mathematician analyst probabilist, astronomer and physicist known for his work in celestial mechanics and probability theory.

[8]Gauss, Carl Friedrich (1777–1855) is a German mathematician. He is considered along with Archimedes and Newton made contributions to algebra, analysis, geometry, number theory, probability, etc.

[9]Chebyshev, Pafnuty Lvovich (1821–94) is a Russian mathematician worked in algebra, analysis, geometry, number theory, probability, etc.

[10]Markov, Andrei Andreyevich (1856–1922) is a Russian mathematician, probabilist, algorist, algebraist and topologist.

[11]Kolmogorov, Andrei Nikolaevich (1903–87) is a Russian analyst, probabilist, topologist. He laid set theoretic foundation for probability theory in 1933.

Modern mathematical statistics has been applied in a wide range of fields. In engineering, for instance, it is applied in testing materials, in robotics, automation in general, and in the control of production processes. In other fields, like agriculture, biology, computer science, demography, ecogeography, management of natural and human resources, medicine, meteorology, politics, psychology, sociology and traffic control.

Probability theory provides mode of probability distribution to be tested by statistical tests and will furnish the mathematical foundation of those tests and other methods. It finds its application in many areas of engineering. As an example, we may mention reliability engineering as it is important to estimate if a system is likely to fail in a time interval. It is vital if the failure of the system results in the probability of injury or loss of life.

It is widely used in production engineering particularly in quality control.

In communication engineering, noise control using models based on probability theory is very important.

1.2 Sets and Set Operations

Sets

A set is a collection of objects—concrete or abstract, finite or infinite.

Examples

1. The collection of subjects of probability and statistics:

   {Probability, Statistics}
2. The collection of binary digits:

   $\{0, 1\}$
3. The collection of natural numbers:

   $\{0, 1, 2, 3, \ldots\}$
4. The collection of the prime numbers:

   $\{2, 3, 5, 7, 11, 13, \ldots\}$
5. The collection of the sons of Rama:

   {Kusha, Lava}

When we conceive of a set A and consider an object as there arise two situations. The objects x may be in the set A in which case we write $x \in A$. It means that x is a member (element) of the set A. Otherwise, x may not belong to the set A which is denoted by $x \notin A$.

The method of representing a set by listing its members in some order (as in Example, 1.5) is called the roster method or tabular form of representing a set.

Another method of representing a set is by mentioning the property governing the elements of the set. Example 3 above can be written as $\{x \mid x \text{ is a natural number}\}$.

There is a set which contains no elements and it is defined by $\{x \mid x \neq x\}$. It is unique and is denoted by ϕ, it is called the empty set or the null set.

Example $\{x \mid x^2 = 1, x \text{ is even}\}$

Subset Suppose A and B are sets. If $x \in A \Rightarrow x \in B$ then A is a subset of B. It is represented as $A \subseteq B$. Here, B is called the superset of A.

Example Let $A = \{2, 3\}$ and $B = \{x \mid x \in I, 0 < x < 10\}$. Since 2 and 3 in A also belong to B, so $A \subseteq B$.

Equality of Sets Two sets A and B are said to be equal, denoted by $A = B$, if both contain the same elements in some order.

Note that $A = B$ if and only if $A \subseteq B$ and $B \subseteq A$.

Remarks

1. The null set is a subset of every set A.
2. Let $A \subseteq B$ and $A \neq B$ then B contains at least one element which is not in A.

Comparable Sets Two sets A and B are comparable if $A \subseteq B$ or $B \subseteq A$, otherwise A and B are incomparable.

Equivalent Sets The number of elements m in a finite set A is called its cardinal number, and is denoted by $n(A)$ or $|A|$.

Two finite sets A and B are said to be equivalent, denoted $A \sim B$, if their cardinal numbers are equal. The sets of Examples 1, 2 and 5 are equivalent, since each has cardinal number 2.

Two infinite sets are said to be equivalent, if there exists a one–to–one correspondence between their elements.

The sets of even and odd integers are equivalent since $m \leftrightarrow 2m$ for each integer.

Universal Set A set containing all the objects of study or consideration is called a universal set.

Power Set Power set of a set A is the set of all subsets of A. It is denoted by $P(A)$.

Example If $A = \{a, b, c\}$, then $P(A) = \{\phi, \{a\}, \{b\}, \{c\}, \{a, b\}, \{a, c\}, \{b, c\}\}$. Note that $n(P(A)) = 2^3 = 8$.

Note If $n(A) = m$ then $n(P(A)) = 2^m$.

Set Operations

Union of *A* and *B* (Figure 1.1)

$A \cup B = \{x \mid x \in A \text{ or } x \in B \text{ or both}\}$

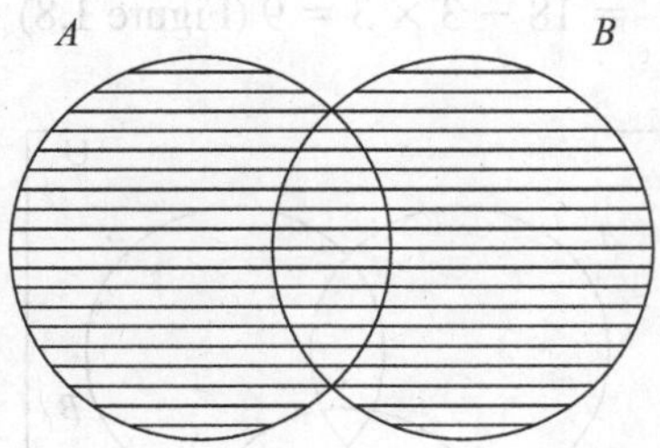

Figure 1.1 Union of A and B.

Intersection of *A* and *B* (Figure 1.2)

$A \cap B = \{x \mid x \in A \text{ and } x \in B\}$

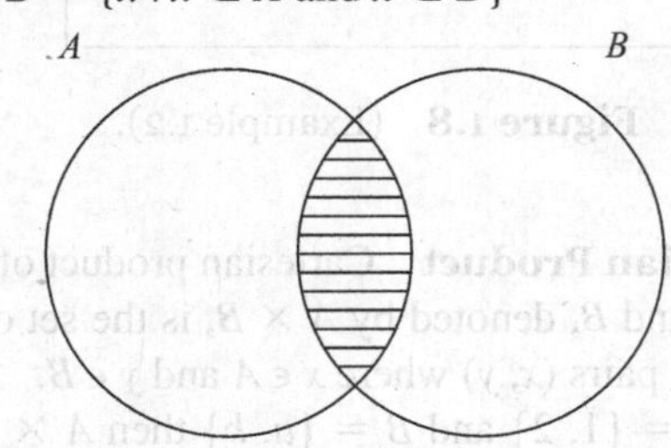

Figure 1.2 Intersection of A and B.

Disjoint Sets (Figure 1.3)

A and B are disjoint if $A \cap B = \phi$.

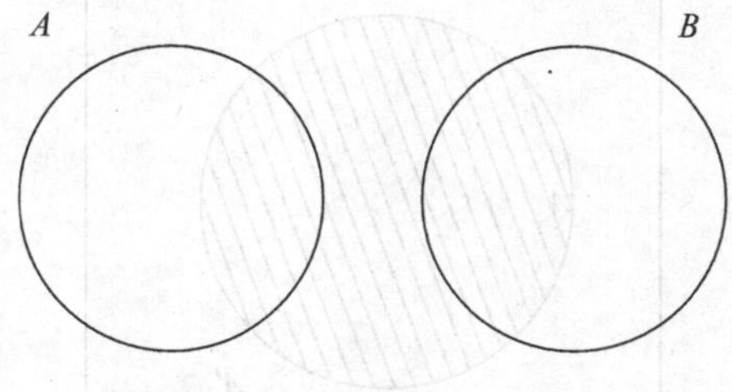

Figure 1.3 Disjoint sets.

Relative Complement of *A* in *B* or Difference of Sets (Figure 1.4)

$B - A = \{x \mid x \in B, x \notin A\}$

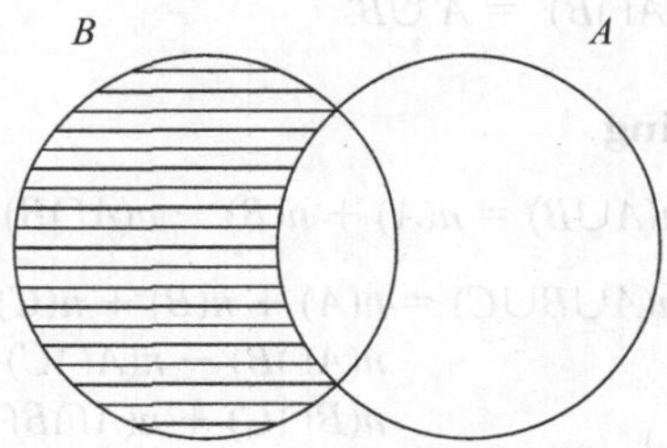

Figure 1.4 Difference of sets.

Symmetric Difference (Figure 1.5)

$A \Delta B = (A - B) \cup (B - A)$

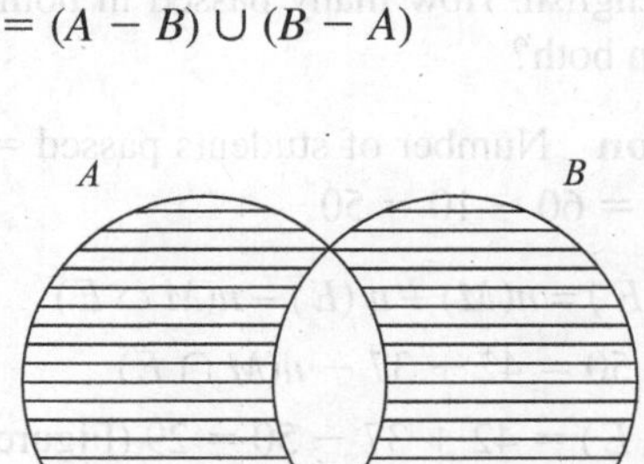

Figure 1.5 Symmetric difference.

Complement

Complement of A relative to the universal set U, denoted by $\bar{A}$ or A^c or A', is the set $\{x \in U \mid x \notin A\}$.

Inverse Law (Figure 1.6)

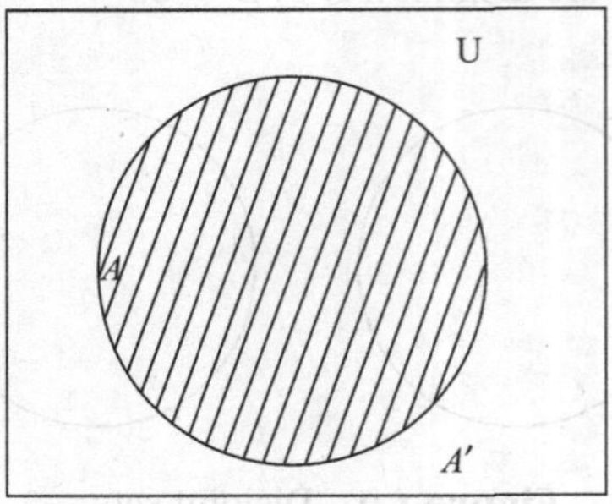

Figure 1.6 Inverse law.

De Morgan's Laws

(a) $(A \cup B)' = A' \cap B'$

(b) $(A \cap B)' = A' \cup B'$

Counting

(a) $n(A \cup B) = n(A) + n(B) - n(A \cap B)$

(b) $n(A \cup B \cup C) = n(A) + n(B) + n(C) - n(A \cap B) - n(A \cap C) - n(B \cap C) + n(A \cap B \cap C)$

Example 1.1

In a class of 60 students, 42 passed in Maths and 37 in English. How many passed in both, if 10 failed in both?

Solution Number of students passed $= n(U) - n(F) = 60 - 10 = 50$

$$n(M \cup E) = n(M) + n(E) - n(M \cap E)$$

$$50 = 42 + 37 - n(M \cap E)$$

$n(M \cap E) = 42 + 37 - 50 = 29$ (Figure 1.7)

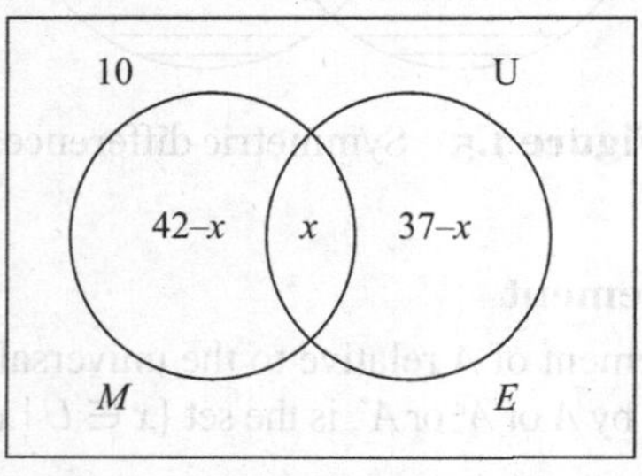

Figure 1.7 (Example 1.1).

Example 1.2

A college awarded 38 medals in football, 15 in basketball and 20 in cricket. If these medals went to a total of 58 students and only 3 men got medals in all the three sports, how many received medals in exactly two of the three sports?

Solution Let F, B, and C denote the sets of students who received medals in football, basketball and cricket respectively.

Then $n(F) = 38$

$n(B) = 15$

$n(C) = 20$

$n(F \cup B \cup C) = 58$ and $n(A \cap B \cap C) = 3$

We have

$n(F \cup B \cup C) = n(F) + n(B) + n(C) - n(F \cap B) - n(F \cap C) - n(B \cap C) + n(F \cap B \cap C)$

Therefore, $n(F \cap B) + n(F \cap C) + n(B \cap C) = 38 + 15 + 20 + 3 - 58 = 18$

From this we have to subtract $3n(F \cap B \cap C)$ to get the required number

$= 18 - 3 \times 3 = 9$ (Figure 1.8)

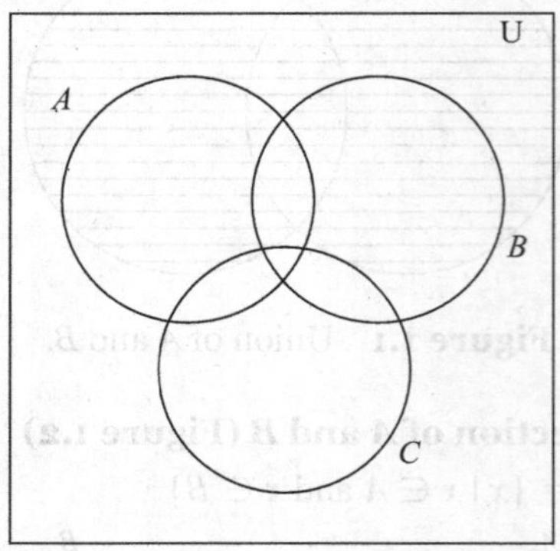

Figure 1.8 (Example 1.2).

Cartesian Product Cartesian product of two sets A and B, denoted by $A \times B$, is the set of all ordered pairs (x, y) where $x \in A$ and $y \in B$.

If $A = \{1, 2\}$ and $B = \{a, b\}$ then $A \times B = \{(1, a), (1, b), (2, a), (2, b)\}$

If $n(A) = m$ and $n(B) = n$ then $n(A \times B) = m \times n$.

1.3 Principle of Counting

Sum Rule Let a task be performed in n_1 ways and a second task in n_2 ways, and an *m*th task in n_m ways. Suppose the tasks are such that they cannot be performed simultaneously. Then the number of ways of performing these *m* tasks is $n_1 + n_2 \ldots + n_m$.

Example 1.3

Let a class consist of 42 boys and 18 girls. The number of ways in which we may select one student either a boy or a girl is $42 + 18 = 60$.

Product Rule If a task can be performed in n_1 different ways and after doing it in any one way, a second task can be done in n_2 different ways and finally *m*th task can be done in n_m different ways then all these *m* tasks can be done in specified order in succession in $n_1 \cdot n_2 \cdot \ldots n_m$ different ways.

Example 1.4

Let there be 3 ways—road (*B*), rail (*T*) and air (*P*) to reach Mumbai from Hyderabad; and 2 ways—air (*P*) and sea (*S*) to reach New York from Mumbai. Then the number of ways of reaching New York from Hyderabad is $3 \times 2 = 6$ ways.

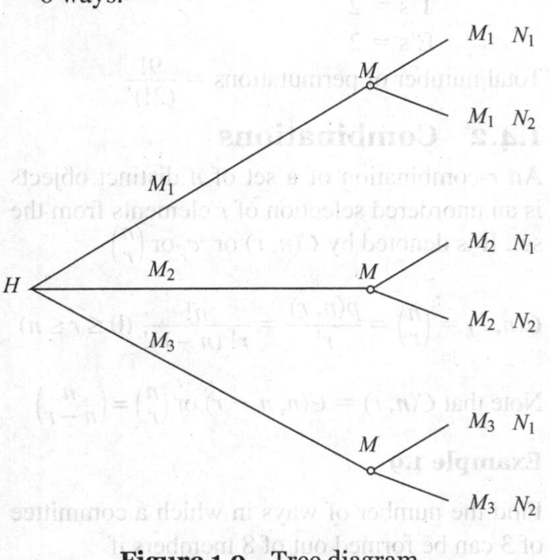

Figure 1.9 Tree diagram.

Tree Diagram Let M_1, M_2, and M_3 be the 3 ways in which one can travel from Hyderabad to Mumbai and let N_1 and N_2 be the 2 ways of reaching New York. Having chosen one of these 3 modes of travel one has 2 ways of reaching New York. The different ways of reaching New York could be graphically and poignantly represented through a tree diagram as shown in Figure 1.9.

1.4 Permutations and Combinations

1.4.1 Permutations

A permutation of a set of *n* distinct objects is an ordered arrangement of these *n* objects.

An r-permutation is an ordered arrangement of *r* elements taken from *n* given objects.

Example 1.5

Let $A = \{1, 2, 3\}$ then 123, 132 and 312, are permutations of *A*. The total number of permutations with the elements of *A* is 6. It is same as filling up 3 places with these 3 numbers. The first place can be filled with any one of the 3 numbers. Having filled the first place by one of these, the second place can be filled by any one of the remaining 2 numbers and the third place by the remaining one, i.e., $3 \times 2 \times 1 = 6$:

1 2 3	1 3 2
2 3 1	2 1 3
3 1 2	3 2 1

The 2-permutations of *A* are 12, 13; 21, 23; and 31, 32.

Factorial Function In this context, we introduce the *factorial function* which is basic. By definition, $0! = 1$ and $n! = (n - 1)! \times n$.

For large *n*, an approximation to its value is given by Stirling's[12] formula:

$$n! \sim \sqrt{2\pi n}(n/e)^n$$

Here ~ means asymptotically equal. It implies that the ratio of the LHS and the RHS terms approach unity as $n \to \infty$.

---

[12]Stirling, James (1692–1770) is a Scottish mathematician.

Number of *r*-permutations The number of *r*-permutations of a set with *n* objects, denoted by ^np_r or $P(n, r)$, is

$$P(n, r) = n(n-1)(n-r+1) \ r \text{ factors}$$

$$= \frac{n!}{(n-r)!} \quad 0 \le r \le n$$

Corollary

When $r = n$, we get $P(n, n) = n!$

$r = 0$, we get $P(n, 0) = 1$

$P(n, r)$ is the number of linear arrangements of *n* objects taken *r* at a time *when repetition* is not allowed.

When repetition is allowed the number of arrangements is n^r.

Example 1.6

How many three-digit numbers can be formed with the digits 1–6:

(a) When repetition is not allowed
(b) When repetition is allowed

Solution Here $n = 6, r = 3$

(a) $P(6, 3) = \frac{6!}{(6-3)!} = 6 \times 5 \times 4 = 120$

It is same as filling up 3 blanks with 6 objects. The first place can be filled in 6 ways. Having filled the first place by any one of the 6 objects, the second place can be filled in 5 ways with any one of the remaining 5 objects. Similarly, the third place can be filled in 4 ways with any one of the remaining 4 objects.

Total number of ways of filling 3 places with 6 objects when repetition is not allowed is $6 \times 5 \times 4 = 120$ ways:

(b) When repetition is allowed, the first place can be filled in 6 ways by any one of the 6 objects.

Having filled the first place, the second and the third places also can be filled in 6 ways since 6 objects are available each time.

Total number of ways $= 6 \times 6 \times 6 = 6^3 = 216$.

Example 1.7

In how many ways can 3 boys and 5 girls can be seated in a row if

(a) No two boys sit together
(b) All the girls sit together

Solution In any arrangement of 5 girls in a row, there are 6 places which 3 boys have to occupy. This can be done in $P(6, 3)$ ways.

The 5 girls can be arranged among themselves in 5! ways:

$$\text{Total number of arrangements} = 5! \times P(6, 3) = 14{,}400$$

(a) Consider 5 girls as one unit since they have to be together. So 3 boys and 1 unit of girls can be arranged in 4! ways:

Total number of arrangements $= 4! \times 5! = 2880$,

since the 5 girls can be arranged among themselves in 5! ways.

Example 1.8

Find the number of different permutations of the letters of the word COMMITTEE.

Solution There are 9 letters of which

M's = 2
O, I, and C each = 1
T's = 2
E's = 2

$$\text{Total number of permutations} = \frac{9!}{(2!)^3}$$

1.4.2 Combinations

An *r*-combination of a set of *n* distinct objects is an unordered selection of *r* elements from the set. It is denoted by $C(n, r)$ or ^nc_r or $\binom{n}{r}$

$$C(n, r) = \binom{n}{r} = \frac{p(n, r)}{r!} = \frac{n!}{r!\,(n-r)!} \ (0 \le r \le n)$$

Note that $C(n, r) = C(n, n-r)$ or $\binom{n}{r} = \binom{n}{n-r}$

Example 1.9

Find the number of ways in which a committee of 3 can be formed out of 8 members if

(a) A specified member is always to be included

(b) A specified member is always to be excluded

(c) Either A or B must be included

Solution (a) Since a specified member is to be included, we have to select 2 out of the remaining 7.

This is done in $\binom{7}{2} = \frac{7 \times 6}{1 \times 2} = 21$ ways.

(b) Since a specified member is always to be excluded, we have to select 3 members out of the remaining 7.

This is done in $\binom{7}{3} = \frac{7 \times 6 \times 5}{1 \times 2 \times 3} = 35$ ways.

(c) Since A or B is to be included, we have to select 2 out of the remaining 6.

This can be done in $\binom{6}{2} = \frac{6 \times 5}{1 \times 2} = 15$ ways.

Number of ways of selecting one from A and B= 2 ways.

Total number of ways of making selection = 2 × 15 = 30 ways.

Combination with Repetition The number of combinations of n objects taken r at a time with repetition is $\binom{n+r-1}{r} = \frac{(n+r-1)!}{r!\,(n-r)!}$

1.5 Binomial Expansion

$(x + a)^n = \binom{n}{0} x^n + \binom{n}{1} x^{n-1}a^1 + \dots + \binom{n}{r} x^{n-r}a^r + \dots + \binom{n}{n} a^n.$

Here $\binom{n}{r} = \frac{n(n-1)\dots(n-r+1)}{r!}$, $r = 0, 1, 2, \dots$, n are called the binomial coefficients.

1.6 Introduction to Probability

Consider that a machine produces certain components that have to meet certain specifications. The quality control department takes samples of components and checks how many of them meet the specifications. Suppose on average 90 out of 100 components meet the specifications. A component is selected at random and let A be the outcome that the component meets the specifications and B the outcome that the component fails to meet the specifications.

Then we say the *probability* of A occurring is 90/100 = 0.90

And the probability of B occurring is 10/100 = 0.10

We write $P(A)$ = probability of A occurring = 0.9

$P(B)$ = probability of B occurring = 0.1

We note that the sum of all probabilities of all possible outcomes is unity.

The process of selecting a component is called an *experiment* or *trial*. The possible outcome is called an *event*. In the above example, there are only two possible events A and B.

The set of all possible outcomes is called the sample space S, then $B = A'$ is the event which is the complement of A (Figure 1.10).

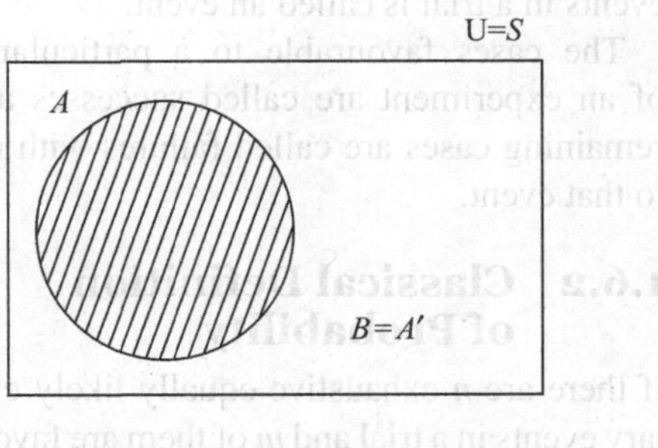

Figure 1.10 Sample space.

1.6.1 Random Experiment

An experiment which can be repeated any number of times under essentially identical conditions and which can be associated with a set of known results is called a *random experiment* or *trial* if the result of any single repetition of the experiment is not certain and cannot be predicted with certainty.

Examples

1. Tossing of a coin
2. Rolling a die
3. Drawing a card from a pack of 52 cards

Elementary or Simple Event The result of any single repetition of a random experiment is called an elementary or simple event.

Examples

1. In the random experiment of tossing a coin, getting a head (H) or getting a tail (T) is an elementary event.
2. In the random experiment of rolling a die, getting any of the numbers 1−6 is an elementary event.
3. In the random experiment of drawing a card, getting a king or an ace etc. is an elementary event.

The collection of all elementary events in a trial is called the set of exhaustive events. In the above examples 1 and 2, {H, T} and {1, 2, 3, 4, 5, 6} are the set of exhaustive events respectively.

Any combination of one or more elementary events in a trial is called an event.

The cases favourable to a particular event of an experiment are called successes and the remaining cases are called failures with respect to that event.

1.6.2 Classical Definition of Probability

If there are n exhaustive equally likely elementary events in a trial and m of them are favourable to an event A, then m/n is called the probability of A:

$$P(A) = m/n$$

1.6.3 Von Mises Statistical Definition of Probability

If a trial is conducted n times and m of them are favourable to an event A, then m/n is called the relative frequency of A and is denoted by $R(A)$. If $\lim_{n\to\infty} R(A)$ exists, then the limit is called probability of A.

1.6.4 Mathematical Definition of Probability

The set of all possible outcomes (results) in a trial is called the sample space for the trial and is denoted by S. The elements of S are called sample points.

Examples

1. Tossing of two coins: {(H, H) (H, T) (T, H) (T, T)} which, for the sake of simplicity, is written as {HH, HT, TH, TT}
2. Rolling of a die: {1, 2, 3, 4, 5, 6}

Let S be a sample space of a random experiment. Every subset of S is called an event. In particular, ø and S are two events in S. ø is called the impossible event and S the certain event in S. An event which contains only one sample point is called a simple event.

Simple events in a sample space are precisely the elementary events in the random experiment.

1.7 Axioms of Probability

Probability Definition 2

Let S be a finite sample space and P be a probability function on S. If A is an event in S then $P(A)$, the image of A, is called probability of A. A real valued function $P: P(S) \to R$ is said to be a probability function on S if

1. $P(A) \ge 0, \forall A \in P(S)$.
2. $P(S) = 1$.
3. A, B $[P(S), A \cap B = \emptyset] \Rightarrow P(A \cup B) = P(A) + P(B)$.

If $A_1, A_2, \ldots, A_n$ are n mutually exclusive events in a sample space S, then

$$P(A_1 \cup A_2 \cup \ldots \cup A_n) = P(A_1) + P(A_2) + \cdots + P(A_n).$$

1.8 Basic Theorems

Complementation Note The complement of an event A in a sample space S is unique. If A' is the complement of A then

$$P(A') = 1 - P(A)$$

We have $S = A \cup A'$ and $A \cup A' = \emptyset$

Hence by axioms 2 and 3, we have

$$1 = P(S) = P(A) + P(A') \Rightarrow P(A')$$
$$\Rightarrow 1 - P(A)$$

Example 1.10 Coin Tossing

Three fair coins are tossed simultaneously. Find the probability of the event A where A is the event in which at least one head turns up.

Solution The sample space S consists of $n(S) = 2^3 = 8$ outcomes. Since the coins are fair, we may assign the same probability to each outcome.

The event A′ is no head turning up if it consists of only 1 outcome.

Hence $P(A') = 1/8$. The answer is $P(A) = 1 - P(A') = 7/8$.

Mutually Exclusive Events Events are said to be mutually exclusive if the happening of any one of them precludes the happening of all the others. That is, if no two or more events happen at the same time.

Example

1. In tossing of a coin the events H (turning up of a head) and T (turning up of tail) are mutually exclusive.
2. In throwing a die the events of showing up of {1} {2} ... and {6} are mutually exclusive.

Theorem 1: Addition Rule for Mutually Exclusive Events

If $A_1, A_2, \ldots, A_m$ are m mutually exclusive events in a sample space S, then

$$P(A_1 \cup A_2 \cup \ldots \cup A_m) = P(A_1) + P(A_2) + \ldots + P(A_m) = \sum_{r=1}^{m} P(A_r).$$

Example 1.11

If $P(A) = 1/5$, $P(B) = 2/3$ and $P(A \cap B) = 1/15$, find the following:

(a) $P(A \cup B)$, (b) $P(A' \cap B)$, (c) $P(A \cap B')$, (d) $P(A' \cap B')$ and (e) $P(A' \cup B')$.

Solution

(a) $P(A \cup B) = P(A) + P(B) - P(A \cap B) = 1/5 + 2/3 - 1/15 = 4/5$

(b) $P(A' \cap B) = P(B) - P(A \cap B) = 2/3 - 1/15 = 3/5$

(c) $P(A \cap B') = P(A) - P(A \cap B) = 1/5 - 1/15 = 2/15$

(De Morgan's Laws)

(d) $P(A' \cap B') = P((A \cup B)') = 1 - P(A \cup B) = 1 - 4/5 = 1/5$

(e) $P(A' \cup B') = P((A \cap B)') = 1 - P(A \cap B) = 1 - 1/15 = 14/15$

Theorem 2: Addition Rule for Arbitrary Events

If A and B are any two events in a sample space S
$P(A \cup B) = P(A) + P(B) - P(A \cap B)$

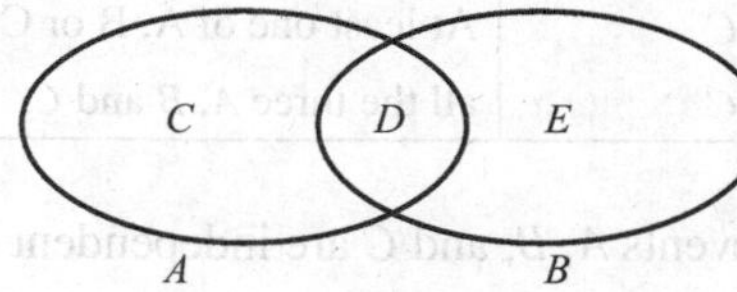

Figure 1.11 By axiom 3 and disjointedness.

Proof In Figure 1.11, sets C, D and E are mutually exclusive (disjoint). By Theorem 1,

$$P(C) + P(D) = P(A) \quad (1)$$

and
$$P(E) = P(B) - P(D) = P(B) - P(A \cap B) \quad (2)$$

From (1) + (2), we get

$$P(A \cup B) = P(A) + P(B) - P(A \cap B)$$

Example 1.12 Union of Arbitrary Events

Find the probability of getting an odd number or a number less than 4.

Solution Let A be the event of getting an odd number and B be the event of getting a number less than 4, then $A \cap B$ is the event of getting an odd number less than 4 = {1,3}:

$P(A) = 3/6$, $P(B) = 3/6$
and $P(A \cap B) = 2/6$

$$P(A \cup B) = P(A) + P(B) - P(A \cap B) = 3/6 + 3/6 - 2/6 = 4/6 = 2/3$$

The following combinations of events arise in probability theory and their meanings are explained in Table 1.1.

Table 1.1 Combination of events and their meanings.

Combination	Meaning
$A \cup B$	either A or B or both
$A \cap B$	both A and B
A' or $\bar{A}$	not A
$A \cap B = \phi$	A and B are mutually exclusive
$A' \cap B'$ or $(A \cup B)'$	neither A nor B
$A \cap B'$	Only A
$A' \cap B$	Only B
$(A \cap B) \cup (A' \cup B)$	Exactly one of A and B
$A \cup B \cup C$	At least one of A, B or C
$A \cap B \cap C$	all the three A, B and C

Three events A, B, and C are independent if

$P(A \cap B) = P(A) \times P(B)$,
$P(B \cap C) = P(B) \times P(C)$,
$P(C \cap A) = P(C) \times P(A)$
and $P(A \cap B \cap C) = P(A) \times P(B) \times P(C)$.

Sampling Consider drawing objects randomly at a time from a given set of objects. This is called sampling from a population. There are two ways of sampling: with replacement and without replacement.

1. In sampling with replacement, the object that was drawn at random is placed back into the given set and the set is mixed thoroughly. Then we draw the next object at random.
2. In sampling without replacement, the object that was drawn in is put aside.

Example 1.13 Sampling with and without Replacement

A box contains 10 balls, 3 of which are red and the others white. Two balls are drawn at random. Find the probability that none of the 2 balls is red.

Solution Let A and B be the two events:

A: First drawn ball is white.
B: Second drawn ball is white.

Since, we are sampling at random, each of the 10 balls has the probability 1/10 of being picked.

Also, 7 out of 10 balls are white.

The probability of a ball that is picked is white is

$$P(A) = 7/10.$$

If we sample with replacement, the situation before the second drawing is the same as at the beginning.

$$P(B) = 7/10$$

Since the events are independent

$P(A \cap B) = P(A) \times P(B) = 7/10 \times 7/10 = 49/100 = 0.49$ or 49%

If we sample without replacement then $P(A) = 7/10$ as before.

If A has occurred, then there are 9 balls left in the box, 3 of which are red:

$P(B/A) = 6/9 = 2/3$
$P(A \cap B) = P(A) \times P(B/A) = 7/10 \times 2/3 = 7/15 = 46.6\%$

1.9 Conditional Probability and Independent Events

1.9.1 Conditional Probability

If A and B are two events in a sample space s and $P(A) \neq 0$, then the probability of B after the event A has occurred is called conditional probability of B given A. It is denoted by $P(B/A)$:

$$P\left(\frac{B}{A}\right) = \frac{P(A \cap B)}{P(A)} \qquad P(A) \neq 0$$

Similarly, the conditional probability of A given B is

$$P\left(\frac{A}{B}\right) = \frac{P(A \cap B)}{P(B)} \qquad P(B) \neq 0$$

Solving the two equations, we obtain the multiplication theorem of probability:

1. $P(A \cap B) = P(A) \times P(B/A)$
2. $P(A \cap B) = P(B) \times P(A/B)$

1.9.2 Independent Events

Two events A and B in a sample space s are independent if and only if

$$P(A \cap B) = P(A) \times P(B)$$

In this case, if $P(A) \neq 0$ and $P(B) \neq 0$ then

$$P(A/B) = P(A) \text{ and } P(B/A) = P(B)$$

This means that the probability of A does not depend on the occurrence of B. This justifies the term 'independent'.

Example 1.14

Three coins are tossed simultaneously. Find the probability of getting one head and at least one head.

Solution

S = {HHH, HHT, HTH, THH, TTH, THT, HTT, TTT}

$n(S) = 8$
$P(S) = 8/8 = 1$

If A is the event of getting one head, A = {TTH, THT, HTT}

$n(A) = 3$
$P(A) = 3/8$

If B denotes the event of getting at least one head and C denotes the event of getting no head,

$$P(B) = P(S) - P(C) = 1 - \frac{1}{8} = 7/8$$

Example 1.15

Find the probability of getting 3 of hearts if we draw 2 cards from a pack of 52 cards.

Solution Total number of cards in a pack = 52

Two cards can be chosen from 52 cards:

$$\text{Number of exhaustive cases} = {}^{52}c_2 = \frac{52 \times 51}{1 \times 2} = 1326$$

There are 13 hearts out of which we have to select 3 cards and can be done in ${}^{13}c_3$ ways

$$= \frac{13 \times 12 \times 11}{1 \times 2 \times 3} = 286$$

$$\frac{13 \times 12 \times 11}{1 \times 2 \times 3} \times \frac{1 \times 2}{52 \times 51} = \frac{286}{1326} = \frac{11}{51}$$

Example 1.16

What is the probability of drawing an ace from a well-shuffled deck of 52 playing cards?

Solution

Number of exhaustive cases $n(A) = {}^{52}c_1 = 52$

Number of favourable cases $n(B) = {}^{4}c_1 = 4$

$$\text{The required probability} = \frac{n(A)}{n(B)} = \frac{4}{52} = \frac{2}{13}$$

Example 1.17

What is the probability of (a) a non-leap year having 53 Sundays and (b) a leap year having 53 Sundays?

Solution

(a) Let A denote the event of having 53 Sundays in a non-leap year. A non-leap year contains 365 days = 52 complete weeks + 1 day extra. The extra day can be any one of the 7 days S, M, T, W, TH, F or Sat. Out of the 7 possibilities there is only one favourable case:

$$\therefore P(A) = \frac{1}{7}$$

(b) A leap year contains 366 days so that 366 = 52 weeks + 2 days extra. There are 7 possible outcomes: SM, MT, TW, WTH, THF, FSat, and SatS, of which the first and the last two are favourable:

$$\therefore P(B) = \frac{2}{7}$$

Mutually Exclusive Events

Example 1.18

Two fair dice are thrown. Find the probability of getting doubles (both the dice showing the same number) or the sum of 7.

Solution

Sample space S = $\{(x, y) \mid \{1 \le x, y \le 6\}$
$n(S) = 6^2 = 36$

Let A be the event of getting doubles and B be the event of getting the sum of 7. Then

$A = \{(1, 1)\ (2, 2)\ (3, 3)\ (4, 4)\ (5, 5)\ (6, 6)\}$
$n(A) = 6$

$B = \{(1, 6)\ (2, 5)\ (3, 4)\ (4, 3)\ (5, 2)\ (6, 1)\}$
$n(B) = 6$

Therefore, A and B are mutually exclusive events:

$P(A \text{ or } B) = P(A \cup B) = P(A) + P(B)$

$= \frac{1}{6} + \frac{1}{6} = \frac{1}{3}$

Example 1.19

Let A and B be the two mutually exclusive events of an experiment. If $P(\text{not } A) = 0.65$, $P(A \cup B) = 0.65$ and $P(B) = \text{P}$ then find P.

Solution We are given that $P(\text{not } A) = P(A') = 0.65$ and $P(A \cup B) = 0.65$

Then $P(A) = 1 - P(A') = 1 - 0.65 = 0.35$

Since A and B are mutually exclusive

$P(A \cup B) = P(A) + P(B)$

$0.65 = 0.35 + P(B)$

$P(B) = P = 0.65 - 0.35 = 0.30$

Conditional Probability

Example 1.20

A die is tossed and the number that turns up is odd. What is the probability that it is prime?

Solution Out of 1, 2, 3, 4, 5 and 6, the numbers 2, 4 and 6 are even and 1, 3, and 5 are odd. Out of three odd numbers, two 3 and 5 are prime.

Conditional probability is with respect to the reduced sample space of odd numbers only.

Number of favourable cases = 2

Total number of cases = 3

$P(\text{prime/odd}) = \frac{2}{3}$

Example 1.21

If A and B are events with $P(A) = \frac{1}{3}$, $P(B) = \frac{1}{4}$ and $P(A \cup B) = \frac{1}{2}$, find (a) $P(A/B)$, (b) $P(B/A)$, (c) $P(A \cap B')$ and (d) $P(A/B')$.

Solution We have $P(A \cup B) = P(A) + P(B) - P(A \cap B)$

$$\frac{1}{2} = \frac{1}{3} + \frac{1}{4} - P(A \cap B)$$

$$P(A \cap B) = \frac{1}{3} + \frac{1}{4} - \frac{1}{2} = \frac{4 + 3 - 6}{12} = \frac{1}{12}$$

$$\text{(a) } P(A/B) = \frac{P(A \cap B)}{P(B)} = \frac{\frac{1}{12}}{\frac{1}{4}} = \frac{1}{3}$$

$$\text{(b) } P(B/A) = \frac{P(A \cap B)}{P(A)} = \frac{\frac{1}{12}}{\frac{1}{3}} = \frac{1}{4}$$

$$\text{(c) } P(A \cap B') = P(A) - P(A \cap B) = \frac{1}{3} - \frac{1}{12} = \frac{1}{4}$$

$$P(B') = 1 - P(B) = 1 - \frac{1}{4} = \frac{3}{4}$$

$$\text{(d) } P(A/B') = \frac{P(A \cap B')}{P(B')} = \frac{\frac{1}{4}}{\frac{3}{4}} = \frac{1}{3}$$

Example 1.22

If the probability that a communication system will have high fidelity is 0.81 and the probability that it will have high fidelity and the selectivity is 0.18.What is the probability that a system with high fidelity will also have high selectivity.

Solution Let A be the event of the communication system having high fidelity and B be the event of the communication system having high selectivity. We are given that $P(A) = 0.81$ and $P(A \cap B) = 0.18$.

By the definition of conditional probability

$$P(B/A) = \frac{P(A \cap B)}{P(A)} = \frac{0.18}{0.81} = \frac{2}{9}$$

Example 1.23

Two dice are thrown and their sum is 7. Find the probability that at least one of the dice shows up 2.

Solution If A denotes the event that sum 7 is obtained then

$$A = \{(1,6)\ (2,5)\ (3,4)\ (4,3)\ (5,2)\ (6,1)\}$$

$$n(A) = 6$$

If B denotes the event that at least one die shows up 2

$$B = \{(2, y) \mid \{1 \le y \le 6\} \cup \{(x, 2) \mid 1 \le x \le 6\}$$

$$n(B) = 12$$

$$A \cap B = \{(2, 5)\ (5, 2)\}$$

By the definition of conditional probability

$$P(B/A) = \frac{P(A \cap B)}{P(A)} = \frac{2}{6} = \frac{1}{3}$$

Example 1.24

Weather records show that the probability of high barometric pressure is 0.82 and the probability of rain and high barometric pressure is 0.20. Find the probability of rain, given high barometric pressure.

Solution Let A denote the event of rain and B high barometric pressure.

$$\text{Then } P(A/B) = \frac{P(A \cap B)}{P(B)} = \frac{0.20}{0.82} = 0.2446$$

Independent Events

Example 1.25

Two friends A and B appear for an interview for two posts. The probability of A's selection is $\frac{1}{6}$ and that of B's selection is $\frac{2}{5}$. What is the probability that (a) both are selected, (b) only one is selected, (c) none is selected and (d) at least one is selected.

Solution Let A and B denote the event that A is selected and B is selected respectively. It is given that

$$P(A) = \frac{1}{6} \text{ and } P(B) = \frac{2}{5}$$

There are two posts. So the selection of one does not affect that of the other. Therefore, A and B are independent events.

(a) $P(A \cap B) = P(A) \times P(B) = \frac{1}{6} \times \frac{2}{5} = \frac{1}{15}$, since A and B are independent

(b) $A \cap B'$ and $A' \cap B$ are mutually exclusive.

$P((A \cap B') \cup (A' \cap B)) = P(A \cap B') + P(A' \cap B)$

$= P(A) \times P(B') + P(A') \times P(B)$ ($A \cap B'$ and $A' \cap B$ are mutually exclusive; A and B are independent)

$= P(A)\,[1 - P(B)] + [1 - P(A)] \times P(B)$

$= \frac{1}{6}\left(1 - \frac{2}{5}\right) + \left(1 - \frac{1}{6}\right)\frac{2}{5} = \frac{13}{30}$

(c) $P(A' \cap B') = P(A') \times P(B')$ (A and B are independent)

$= [1 - P(A)][1 - P(B)] = \left(1 - \frac{1}{6}\right)\left(1 - \frac{2}{5}\right)$

$= \frac{5}{6} \times \frac{3}{5} = \frac{1}{2}$

(d) $P(A \cup B) = P(A) + P(B) - P(A \cap B)$

$= \frac{1}{6} + \frac{2}{5} - \frac{1}{15} = \frac{5 + 12 - 2}{30} = \frac{15}{30} = \frac{1}{2}$

Example 1.26

The chances of three students A, B and C solving a problem given in mathematics Olympiad are $\frac{1}{2}, \frac{1}{3}$ and $\frac{1}{4}$ respectively. What is the probability of the problem being solved?

Solution Let A, B and C be the events of the students A, B, and C solving the problem respectively. It is given that $P(A) = \frac{1}{2}$, $P(B) = \frac{1}{3}$ and $P(C) = \frac{1}{4}$.

The probability of solving the problem

$= P(A \cup B \cup C) = 1 - P[(A \cup B \cup C)']$

$= 1 - P(A' \cap B' \cap C')$

$= 1 - [P(A') \times P(B') \times P(C')]$

$= 1 - \left[\left(1 - \frac{1}{2}\right)\left(1 - \frac{1}{3}\right)\left(1 - \frac{1}{4}\right)\right]$

$= 1 - \left(\frac{1}{2} \times \frac{2}{3} \times \frac{3}{4}\right) = \frac{3}{4}$

1.10 Theorem of Total Probability (or the Rule of Elimination)

Theorem Let $B_1, B_2, \ldots, B_k$ be a partition of the sample space S with $P(B_i) \neq 0$ for $i = 1, 2, \ldots, k$. Then for any event A of S,

$$P(A) = \sum_{i=1}^{k} P(B_i \cap A) = \sum_{i=1}^{k} P(B_i)\, P\left(\frac{A}{B_i}\right)$$

Proof It is given that $B_1, B_2, \ldots, B_k$ are a partition of S, therefore $S = \cup_i^k B_i$ and $B_i \cap B_j = \phi$ for any i and j, i.e. their union is S and the B_i's are mutually disjoint sets (Figure 1.12).

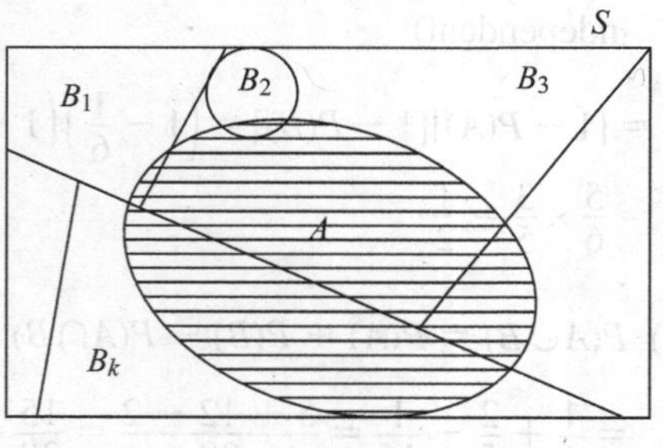

Figure 1.12 Total probability theorem.

$A = A \cap S = A \cap \left(\cup_{i=1}^{k} B_i\right)$

$= A \cap (B_1 \cup B_2 \cup \ldots \cup B_k)$

$= (A \cap B_1) \cup (A \cap B_2) \cup \ldots (A \cap B_k)$

$A \cap B_1, A \cap B_2, \ldots, A \cap B_k$ are all mutually disjoint sets. Applying total probability or the rule of elimination, we have

$P(A) = P[(A \cap B_1) \cup (A \cap B_2) \cup \ldots \cup (A \cap B_k)]$

$= P(B_1 \cap A) + P(B_2 \cap A) + \ldots + P(B_k \cap A)$

Finally applying the multiplicative rule

$$P(A) = \sum_{i=1}^{k} P(B_i \cap A) = \sum_{i=1}^{k} P(B_i)\, P\left(\frac{A}{B_i}\right)$$

Theorem of Total Probability

Example 1.27

Coloured balls are distributed in three indistinguishable boxes B_1, B_2 and B_3 as shown below:

	B_1	B_2	B_3
Red	2	4	3
White	3	1	4
Blue	5	3	3
Total	10	8	10

A box is selected at random from which a ball is selected at random. What is the probability that the ball's colour is (a) red, (b) white or (c) blue?

Solution Let A be the colour of the ball (red, white or blue)

By the theorem on total probability,

$$P(R) = P(B_1) \times P\left(\frac{A}{B_1}\right) + P(B_2) \times P\left(\frac{A}{B_2}\right) + P(B_3) \times P\left(\frac{A}{B_3}\right)$$

The three boxes B_1, B_2 and B_3 are indistinguishable.

∴ The probability of selecting one at random

$P(B_1) = P(B_2) = P(B_3) = \frac{1}{3}$

$$P\left(\frac{R}{B_1}\right) = \frac{2}{10},\ P\left(\frac{R}{B_2}\right) = \frac{4}{8} \text{ and } P\left(\frac{R}{B_3}\right) = \frac{3}{10}$$

$$P(R) = \frac{1}{3} \times \frac{2}{10} + \frac{1}{3} \times \frac{4}{8} + \frac{1}{3} \times \frac{3}{10} = \frac{1}{3}$$

Similarly

$$P(W) = \frac{1}{3} \times \frac{3}{10} + \frac{1}{3} \times \frac{1}{8} + \frac{1}{3} \times \frac{4}{10} = \frac{11}{40}$$

$$P(B) = \frac{1}{3} \times \frac{5}{10} + \frac{1}{3} \times \frac{3}{8} + \frac{1}{3} \times \frac{3}{10} = \frac{47}{120}$$

Example 1.28

A building contractor has undertaken a construction job. The probability that there will be strike is 0.65, that the construction job will be completed on time if there is no strike is 0.80 and that the work will be completed on time if there is a strike is 0.32. Find the probability that the construction job will be completed on time.

Solution Let A be the event that the construction job will be completed on time and B be the event that there will be a strike. We have to find $P(A)$.

$P(B) = 0.65$

$\Rightarrow P(\text{no strike}) = P(B') = 1 - P(B)$

$= 1 - 0.65 = 0.35$

$P(A/B) = 0.32$ and $P(A/B') = 0.80$

The events B and B' form a partition of the sample space S. By the theorem on total probability, we have

$$P(A) = P(B) \times P(A/B) + P(B') \times P(A/B')$$
$$= 0.65 \times 0.32 + 0.35 \times 0.8$$
$$= 0.208 \times 0.28 = 0.488$$

1.11 Bayes' Theorem or Rule

Theorem Let the sample space S be partitioned into k subsets $B_1, B_2, \ldots, B_k$ with $P(B_i) \neq 0$ for $i = 1, 2, \ldots, k$. Then for any arbitrary event A in S with $P(A) \neq 0$, we have, for $r = 1, 2, \ldots, k$,

$$P\left(\frac{B_r}{A}\right) = P(B_r \cap A) \Big/ \sum_{i=1}^{k} P(B_i \cap A)$$

$$= \frac{P(B_r)\, P\left(\frac{A}{B_r}\right)}{\sum_{i=1}^{k} P(B_i)\, P\left(\frac{A}{B_i}\right)} \quad (1)$$

Proof By the definition of conditional probability

$$P\left(\frac{B_r}{A}\right) = \frac{P(B_r \cap A)}{P(A)} \quad (2)$$

Also, by the theorem of total probability

$$P(A) = \sum_{i=1}^{k} P(B_i \cap A) = \sum_{i=1}^{k} P(B_i)\, P\left(\frac{A}{B_i}\right) \quad (3)$$

Finally, by the multiplication rule

$$P(B_r \cap A) = P(B_r) \times P\left(\frac{A}{B_r}\right) \quad (4)$$

Now, we obtain (1) by substituting (3) and (4) in (2).

Note Bayes' theorem is also known as the formula for the probability of 'causes'. It gives the probability of a particular cause B_r given that event A has happened.

$P(B_i)$ is 'a priori probability' known even before the experiment, $P(A/B_i)$ is the 'likelihoods' and $P(B_i/A)$ is 'a posteriori probability' which depends often on the result of the experiment.

Example 1.29

In a bolt manufacturing company, three machines manufacture 20, 30 and 50% of its total output and of these 6, 3 and 2% are found defective respectively. A bolt is drawn at random and is found defective. Find the probability that the defective bolt is manufactured by machine 1, 2 or 3.

Solution Let A, B and C be the events that the defective bolt is manufactured by machine 1, 2 and 3 respectively. It is given that the probabilities of these events are

$P(A) = \frac{20}{100} = \frac{1}{5}$, $P(B) = \frac{30}{100} = \frac{3}{10}$ and $P(C) = \frac{50}{100} = \frac{1}{2}$

Let D denote the event that the bolt drawn at random is defective .Then by the definition of the conditional probability, we have

$P(D/A) = \frac{6}{100} = 0.06$, $P(D/B) = 0.03$ and $P(D/C) = 0.02$

The probability that the defective bolt is from machine 1 is, by Bayes' theorem,

$$P(A/D) = \frac{P(A) \times P(D/A)}{P(A) \times P(D/A) + P(B) \times P(D/B) + P(C) \times P(D/C)}$$

$$= \frac{\frac{1}{5} \times 0.06}{\frac{1}{5} \times 0.06 + \frac{3}{10}\, 0.03 + \frac{1}{2} \times 0.02} = \frac{12}{31}$$

Similarly,

$$P(B/D) = \frac{P(B) \times P(D/B)}{P(A) \times P(D/A) + P(B) \times P(D/B) + P(C) \times P(D/C)}$$

$$= \frac{\frac{3}{10} \times 0.03}{\frac{1}{5} \times 0.06 + \frac{3}{10}\ 0.03 + \frac{1}{2} \times 0.02} = \frac{9}{31}$$

$$P(C/D) = \frac{P(C) \times P(D/C)}{P(A) \times P(D/A) + P(B) \times P(D/B) + P(C) \times P(D/C)}$$

$$= \frac{\frac{1}{2} \times 0.02}{\frac{1}{5} \times 0.06 + \frac{3}{10}\ 0.03 + \frac{1}{2} \times 0.02} = \frac{10}{31}$$

Example 1.30

A university bought 45%, 25% and 30% of computers from HCL, WIPRO and IBM respectively, and 2%, 3% and 1% of these were found to be defective. Find the probability of a computer selected at random is found to be defective.

Solution Let A, B and C be the events of computer purchased from HCL, WIPRO and IBM companies respectively, and D be the event that the computer was found to be defective. Then

$$P(A) = \frac{45}{45 + 25 + 30} = 0.45,$$

$$P(B) = \frac{25}{100} = 0.25 \text{ and } P(C) = \frac{30}{100} = 0.30$$

$$P(D/A) = \frac{2}{100} = 0.02,\ P(D/B) = \frac{3}{100} = 0.03$$

$$\text{and } P(D/C) = \frac{30}{100} = 0.30$$

By Bayes' theorem

$$P(D) = P(A) \times P(D/A) + P(B) \times P(D/B) + P(C) \times P(D/C)$$

$$= 0.45(0.02) + 0.25(0.03) + 0.30(0.01)$$

$$= 0.0090 + 0.0075 + 0.0030 = 0.0195$$

Example 1.31

The following observations were made at a clinic where HIV test was performed:

(a) 15% of the patients at the clinic have HIV.

(b) Among those who have HIV, 95% were tested positive on the ELISA test.

(c) Among those that do not have HIV, 2% tested positive on the ELISA test.

Find the probability that a patient has HIV if the ELISA test was positive.

Solution Let A, B and C be the events that a patient has HIV, does not have HIV and has tested positive respectively.

It is given that $P(A) = 1/15 = 0.15$

$$P(B) = P(A') = 1 - P(A) = 1 - 0.15 = 0.85$$

$$P(C/A) = \frac{95}{100} = 0.95$$

$$\text{and } P(C/B) = \frac{2}{100} = 0.02$$

Now, by Bayes' theorem,

$$P(A/C) = \frac{P(A) \times P\left(\frac{C}{A}\right)}{P(A) \times P\left(\frac{C}{A}\right) \times P(B) \times P\left(\frac{C}{B}\right)}$$

$$= \frac{0.15 \times 0.95}{0.15 \times 0.95 + 0.85 \times 0.02}$$

$$= 0.89$$

Example 1.32

Each of the identical boxes B_1, B_2 and B_3 contains two coins: B_1 contains both gold coins, B_2 both silver coins and B_3 contains one gold and one silver coin. If a box is chosen at random and a coin is picked at random and if the coin is gold, what is the probability that the other coin in the box is also of gold?

Solution Let E_1, E_2, and E_3 be the events that boxes B_1, B_2, and B_3 are chosen respectively.

Then $P(E_1) = P(E_2) = P(E_3) = \frac{1}{3}$

Let G be the event that the coin drawn is of gold.

Then $P(G/E_1) = P(\text{gold coin from } B_1) = \frac{2}{2} = 1$

$P(G/E_2) = P(\text{gold coin from } B_2) = 0$

$P(G/E_3) = P(\text{gold coin from } B_3) = \frac{1}{2}$

Now the probability that the other coin in the box is of gold and the probability that gold coin is drawn from B_1, by Bayes' theorem

$$P(E_1/G) = \frac{P(E_1) \times P(G/E_1)}{P(E_1)\times P(G/E_1)+P(E_2)\times P(G/E_2)+P(E_3)\times P(G/E_3)}$$

$$= \frac{\frac{1}{3}\times 1}{\frac{1}{3}\times 1+\frac{1}{3}\times 0+\frac{1}{3}\times\frac{1}{2}}$$

$$= \frac{1}{2}$$

Example 1.33

A man is known to speak truth 3 out of 4 times. He throws a die and reports that it is six, Find the probability that it is actually a six.

Solution Let E be the event that the man reports that six occurs while throwing the die and let S be the event that six occurs. Then

$P(S)$ = Probability that six occurs $= \frac{1}{6}$

$P(S')$ = Probability that six does not occur $= 1 - \frac{1}{6} = \frac{5}{6}$

$P(E/S)$ = Probability that the man reports that six occurs when six has actually occurred

= Probability that the man speaks the truth $= \frac{3}{4}$

$P(E/S')$ = Probability that the man report that six occurs when six has not actually occurred

= Probability that the man does not speak the truth

$= 1 - \frac{3}{4} = \frac{1}{4}$

By Bayes' theorem

$P(S/E)$ = Probability that the man reports that six occurs when six has actually occurred

$$= \frac{P(S)\,P\left(\frac{E}{S}\right)}{P(S)\times P\left(\frac{E}{S}\right)+P(S')\times P\left(\frac{E}{S'}\right)}$$

$$= \frac{\frac{1}{6}\times\frac{3}{4}}{\frac{1}{6}\times\frac{3}{4}+\frac{5}{6}\times\frac{1}{4}} = \frac{1}{8}\times\frac{24}{8} = \frac{3}{8}$$

EXERCISES

Multiplication Theorem of Probability

1. A urn contains 10 black and 5 white balls. Two balls are drawn from the urn one after the other without replacement. What is the probability that both the drawn balls are black?

Ans: $P(\text{black is I drawn}) = P(A) = \frac{10}{15}$

$P(\text{black is II drawn}) = P\left(\frac{B}{A}\right) P = \frac{9}{14}$

$P(A \cap B) = P(A) \times P\left(\frac{B}{A}\right) = \frac{10}{15} \times \frac{9}{14} = \frac{3}{7}$

2. Three cards are drawn successively without replacement from a pack of 52 well-shuffled cards. What is the probability that the first two cards are kings and the third card drawn is an ace?

Ans: $P(K) = \frac{4}{52}, P\left(\frac{K}{K}\right) = \frac{3}{15}$ and $P\left(\frac{A}{KK}\right) = \frac{4}{50}$

$P(KKA) = P(K) \times P\left(\frac{K}{K}\right) \times P\left(\frac{A}{KK}\right)$

$= \frac{4}{52} \times \frac{3}{51} \times \frac{4}{50} = \frac{2}{5525}$

Independent Events

3. A die is thrown. Let A be the event that the number that appears is a multiple of 3 and B the number that appears is even. Find whether A and B are independent.

Ans: $A \cap B = P\{3, 6\} \cap \{2, 4, 6\} = \{6\}$

$$P(A) = \frac{2}{6}, P(B) = \frac{3}{6}$$

and $P(A \cap B) = \frac{1}{6} = P(A) \times P(B)$

A and B are independent.

4. Three coins are tossed simultaneously. Let A be the event in which three heads or three tails turn up, B getting at least two heads and C getting at the most two heads. Among the pairs (A, B), (A, C) and (B, C) identify the independent and dependent pairs.

Ans: (A, B) are independent and others dependent.

5. A pair of dice is tossed twice. Find the probability of scoring 7 points (a) once, (b) at least once and (c) twice.

Ans: (a) $\frac{5}{36} + \frac{5}{36} = \frac{5}{18}$, (b) $\frac{11}{36}$ and (c) $\frac{1}{36}$

Conditional Probability

6. If $P(A) = \frac{7}{13}$, $P(B) \frac{9}{13}$ = and $P(A \cap B) = \frac{4}{13}$ Find $P\left(\frac{A}{B}\right)$

Ans: $P\left(\frac{A}{B}\right) = \frac{P(A \cap B)}{P(B)} = \frac{4}{9}$

7. Ten cards numbered 1 to 10 are placed in a box, mixed up thoroughly and then one card is drawn at random. If the number on the drawn card is more than 3, what is the probability that it is an even number?

Ans: $P(\text{even}) = P(A) = \frac{5}{10}$

and $P(\geq 3) = P(B) = \frac{7}{10}$

$P(A \cap B) = \frac{4}{10}$ and $P\left(\frac{A}{B}\right) = \frac{4}{7}$

8. Two digits are selected at random from the digits 1 to 9. If the sum is even, find the probability that both the numbers are odd.

Ans: $\frac{10}{16}$ [Hint: even + even = even, ${}^4c_2 = 6$ ways

odd + odd = even, ${}^5c_2 = 10$ ways]

9. A bag contains 10 gold and 8 silver coins. Two successive draws of 4 coins are made such that

(a) Coins are replaced before the second trial

(b) Coins are not replaced before the second trial

Find the probability that the first draw will give four gold and the second four silver coins.

Ans: (a) $P(A \cap B) = P(A) \times P(B)$

$$= \frac{10C_4}{18C_4} \times \frac{8C_4}{18C_4}$$

(b) $P(A \cap B) = P(A) \times P\left(\frac{B}{A}\right)$

$$= \frac{10C_4}{18C_4} \times \frac{8C_4}{18C_4}$$

10. An urn contains 10 white and 3 black balls, while another contains 3 white and 5 black balls. Two balls are drawn from the first urn and placed into the second and then a ball is drawn at random from the second urn. What is the probability that it from the second urn? What is the probability that it is a white ball?

Ans: $P(A) = \frac{10C_2}{13C_2}$, $P(B) = \frac{3C_2}{13C_2}$ and

$$P(C) = \frac{10C_1\frac{3}{4}}{13C_2}$$

After transfer of two balls cases,

(a) 5W 5B, (b) 3W 7B and (c) 4W 6B

P(drawing W from urn 2)

$$= \sum P(A) \times P\left(\frac{W}{A}\right) = \frac{59}{130}$$

Bayes' Theorem

11. Suppose that the reliability of a HIV test is specified as follows: Of the people having HIV, 90% of the test detect the disease but 10% go undetected. Of the people free of HIV, 99% of the test are judged HIV negative but 1% are diagnosed as showing HIV positive. From a large population of which only 0.1% have HIV, one person is selected at random, given the HIV test, and the pathologist reports the person as HIV positive. What is the probability that the person actually has HIV?

Ans: $P(A) = \dfrac{0.1}{100} = 0.001, P(A')$

$= 1 - 0.001 = 0.999,$

$$P(B/A) = \frac{90}{100} = 0.9$$

$$\text{and } P(B/A') = \frac{1}{100} = 0.01$$

$$P(A/B) = \frac{P(A)\,P(B/A)}{P(A) \times P(B/A) + P(A') \times P(B/A')}$$

$$= \frac{0.001 \times 0.9}{0.001 \times 0.9 + 0.999 \times 0.01}$$

$$\frac{90}{1089} = 0.083 \text{ (approx.)}$$

12. In a bolt manufacturing factory, machines A, B and C produce 25, 35 and 40% of the bolts respectively, of which 5, 4 and 2% are found to be defective. A bolt is drawn at random and is found to be defective. What is the probability that the defective bolt is from machine B?

Ans: E, the bolt is defective.

$P(A) = 25\%, P(B) = 35\%$ and $P(C) = 40\%$

$P(E/A) = 5\%, P(E/B) = 4\%$ and

$P(E/C) = 2\% P(B/E) = 28/69$

13. There are three bags: the first containing 1 white, 2 red and 3 green balls; the second containing 2 white, 3 red and 1 green balls; and the third containing 3 white, 1 red and 2 green balls. Two balls are drawn from a bag chosen at random. These are found to be 1 white and 1 red balls. Find the probability that the balls so drawn come from the second bag.

Ans: B_1, B_2, and B_3 are first, second and third bag chosen respectively

A is that the two balls are 1 white and 1 red

$P(B_i) = 1/3$ ($i = 1, 2$ and 3)

$P(A/B_1) = P$(white and red balls are drawn from B_1) =

$$\frac{{}^1c_1 \times {}^2c_1}{{}^6c_2} = \frac{2}{15}$$

$$P\left(\frac{A}{B_2}\right) = \frac{{}^2c_1\,{}^3c_1}{{}^6c_1} = \frac{2}{5}$$

$$\text{and } P\left(\frac{A}{B_2}\right) = \frac{{}^3c_1 \times {}^1c_1}{{}^6c_1} = \frac{1}{5}$$

By Bayes' theorem, $P(B_2/A) = \dfrac{6}{11}$

MULTIPLE CHOICE QUESTIONS

1. The probability that A happens is $\frac{1}{3}$. The odds against the happening of A are

(a) 2:1 (b) 2:3 (c) 3:2 (d) 5:2

Ans: (a)

2. The odds in favour of an event A are 5:4. The probability of success of A is

(a) 4/5 (b) 5/9 (c) 4/9 (d) 3/5

Ans: (b)

3. The probability that A passes a test is 2/3 and B passes a test is 3/5. The probability that only one of them will pass a test is

(a) 2/5 (b) 4/15 (c) 2/15 (d) 7/15

Ans: (d)

4. A buys a lottery ticket in which the chance of winning is 1/10 and B has a ticket in which his chance of winning is 1/20. The chance that at least one of them will win is

(a) 1/200 (b) 29/200

(c) 30/200 (d) 170/200

Ans: (b)

5. The probability that a non-leap year will have 53 Tuesdays is

(a) 3/7 (b) 2/7 (c) 1/7 (d) 4/7

Ans: (c)

6. The probability that a leap year will have 53 Sundays is

(a) 1/7 (b) 2/7 (c) 3/7 (d) 4/7

Ans: (b)

7. The probability of getting 2, 3 or 4 from a throw of a single die is

(a) 1/6 (b) 2/3 (c) 1/3 (d) 1/2

Ans: (d)

8. A single die is tossed once. The probability that a 2 or 5 will turn up is

(a) 1/3 (b) 1/2 (c) 1/6 (d) 2/3

Ans: (a)

9. The probability that a single toss of a die will result in a number less than 4 is

(a) 1/3 (b) 1/2 (c) 2/3 (d) 3/4

Ans: (b)

10. The probability that a single toss of a die will result in an odd number less than 4 is

(a) 1/2 (b) 1/3 (c) 2/3 (d) 3/4

Ans: (c)

11. A ball is drawn at random from a box containing 6 red, 4 white and 5 blue balls. The probability that the ball drawn is red is

(a) 3/5 (b) 1/5 (c) 4/5 (d) 2/5

Ans: (d)

12. In Question 11, the probability of drawing a red or white ball is

(a) 1/5 (b) 2/5 (c) 2/3 (d) 1/3

Ans: (c)

13. In Question 11, the probability of drawing a blue ball is

(a) 1/3 (b) 2/3 (c) 1/6 (d) 3/4

Ans: (a)

14. In Question 11, the probability of drawing a red or a white ball is

(a) 1/5 (b) 2/5 (c) 1/3 (d) 2/3

Ans: (c)

15. A pair of dice is thrown at a time. The probability of getting a sum more than each die is

(a) 3/6 (b) 4/6 (c) 1/6 (d) 1/36

Ans: (d)

16. A pair of dice is rolled. The probability of obtaining an even prime number on each die is

(a) 0 (b) 1/3 (c) 1/2 (d) 1/36

Ans: (d)

17. Two events A and B will be independent if

(a) A and B are mutually exclusive

(b) $P(A' \cap B') = [1 - P(A)] \times [1 - P(B)]$

(c) $P(A) = P(B)$

(d) $P(A) + P(B) = P(B) = 1$

Ans: (b)

18. A bag contains 3 red, 4 white and 7 black balls. The probability of drawing a red or a black ball is

(a) 2/7 (b) 5/7 (c) 3/7 (d) 4/7

Ans: (b)

19. A number is chosen at random from the first 100 natural numbers. The probability that the number chosen is a multiple of 5 or 15 is

(a) 28/5 (b) 1/5 (c) 3/5 (d) 4/5

Ans: (b)

20. If the two outcomes A and B of an experiment are independent where $P(A) = 0.4$ and $P(A \cup B) = 0.7$, then $P(B)$ is

(a) 0.5 (b) 0.3 (c) 4/7 (d) 2/7

Ans: (a)

21. For two given events A and B, $P(A \cap B)$ is

(a) $< P(A) + P(B)$

(b) $> P(A) + P(B)$

(c) $= P(A) + P(B)$

(d) $P(A) + P(B) + P(A \cup B)$

Ans: (a), (b) and (c)

22. If $P(E) = 0.25$, $P(F)=0.5$ and $P(E' \cap F') = 0.14$ then the probability of occurrence of neither E nor F is

(a) 0.61 (b) 0.39 (c) 0.89 (d) 0.08

Ans: (d)

23. A purse contains 4 copper and 3 silver coins. Another purse contains 6 copper and 2 silver coins. The probability that a coin drawn at random from either purse is a copper coin is

(a) 4/7 (b) 3/4 (c) 3/7 (d) 37/56

Ans: (d)

24. Two cards are drawn at random from a pack of 52 cards. The probability that these are aces is

(a) $2/(^{52}c_2)$ (b) 1/221

(c) 1/663 (d) 2/221

Ans: (b)

25. If a card is drawn from a well-shuffled pack of 52 cards, the probability that it is a spade or a queen is

(a) 17/52 (b) 4/13

(c) 13/52 (d) 4/51

Ans: (b)

26. Two balls are drawn from a bag containing 3 white, 4 black and 5 red balls. The probability that the balls drawn are of different colours is

(a) 47/66 (b) 17/52

(c) 5/22 (d) 2/11

Ans: (a)

27. The probability of drawing a card which is at least a spade or a king from a well-shuffled pack of cards is

(a) 1/26 (b) 17/52

(c) 1/52 (d) 4/13

Ans: (d)

28. A person draws a card from a pack of playing cards, replaces it and shuffles the pack. He continues doing this until he gets a spade. The probability that he will fail the first two times is

(a) 9/64 (b) 3/32 (c) 7/64 (d) 5/64

Ans: (a)

29. In a family of 4, children the probability that there will be at least one boy is

(a) 1/16 (b) 3/4 (c) 15/16 (d) 1/4

Ans: (c)

30. An experiment fields 3 mutually exclusive events *A*, *B* and *C*. If $P(A) = 2$ and $P(B) = 3$ then $P(C)$ is

(a) 2/11 (b) 3/11 (c) 6/11 (d) 5/11

Ans: (c)

FILL IN THE BLANKS

1. A card is drawn from an ordinary pack and a gambler bets that it is a spade or an ace. The odds against his winning the bet are ____.

Ans: 9 : 4

2. An integer is chosen at random from the first 200 positive integers. The probability that the integer chosen is divisible by is ____.

Ans: 1/4

3. A problem in statistics is given to five students. Their chances of solving it are 1/2, 1/3, 1/4, 1/4 and 1/5. The probability that the problem will be solved is ____.

Ans: 17/20.

4. Three coins are tossed simultaneously. The probability that at least two tails are attained is ____.

Ans: 1/2

5. In a single throw of two distinct dice, the probability of obtaining a total of 7 is ____.

Ans: 1/6

6. In Question 5, the probability of obtaining a total of 13 is ____.

Ans: 0

7. In Question 5, the probability of obtaining a total as an even number is ____.

Ans: 1/2

8. A and B are two among 10 persons sitting round a circular table. The probability that there are exactly three persons between A and B is ____.

Ans: 2/9

9. The number of four-digit numbers that can be formed with the digits 1–7 if repetition is not allowed is ____.

Ans: 840

10. In Question 9, the number of numbers formed if repetition is allowed is ____.

Ans: 2/9

11 From a box containing 10 balls, the number of ways in which 3 balls can be drawn at random is ____.

Ans: 720

12. The probability of drawing an ace from a well-shuffled pack of 52 playing cards is ____.

Ans: 1/13

13. In a random experiment, *A* and *B* are two events such that $P(A/B) = P(A)$. The events *A* and *B* are ____.

Ans: Independent

14. In a random experiment, E_1 and E_2 are two possible independent events. Then $P(E_2/E_1)$ = ____.

Ans: $P(E_2)$

15. If A and B are two events such that $P(A \cup B) = P(A \cap B)$, then $P(A)$ and $P(B)$ are related such that ____.

Ans: $P(A) = P(B)$

16. If P is a probability function defined as the sample space $s = \{a_1, a_2, a_3, a_4\}$ such that $P(a_1) = 1/2$, $P(a_2) = P$, $P(a_3) = 1/4$ and $P(a_4) = 1/8$, then P = _____.

Ans: 1/8

17. A and B are two events of a sample space. Then $\bar{A} \cap \bar{B}$ is the event of _____.

Ans: A and B not happening

18. A and B are two events of a sample space. Then $A \cup B$ is the event of _____.

Ans: Happening of B or non-happening of A

19. If $P(A) = 1/4$, $P(B) = 1/3$ and $P(A \cap B) = 1/5$ then $P(A \cup B) =$ _____.

Ans: 23/60

20. If $P(A) = 2/3$ and $P(A \cap B) = 1/4$ then $P(A \cap B') =$ _____.

Ans: 5/8

21. If A and B are mutually exclusive events then $P(A' \cap B') =$ _____.

Ans: 1

22. The probability of getting an ace and a king when we draw 2 cards with replacement from a pack of 52 cards is _____.

Ans: 1/169

23. In tossing three coins, the probability of getting a head is _____.

Ans: 7/8

24. If $P(A) = P_1$ and $P(B) = P_2$ and A and B are independent then $P[(A \cup B)'] =$ _____.

Ans: $(1 - P_1)(1 - P_2)$

25. In throwing two dice, if A is the event that the sum is odd and B the sum is 6 then $P(A \cap B) =$ _____.

Ans: 5/72

2 Probability Distribution

2.1 Introduction

Often we have to measure many different variables, e.g., the output voltage of a system, the strength of a team or the cost of a project. Further, we have to know the probability that a variable falls within a given range (0 to 1). The way the probability is distributed across various ranges of values gives rise to the concept of a probability distribution. The probability distribution or simply distribution shows the probabilities of events in an experiment.

2.2 Random Variables

Quantities whose variation has an element of chance are called random variables. Some examples are as follows:

1. The length of time during which a product is in working condition.
2. The force required to stretch a spring for a specified length.
3. The output voltage of a system.
4. The number of units produced in a day from a factory.
5. The number of electrical sockets in a house.

All the above quantities vary. Items 1–3 vary continuously. They are specified by real numbers and are continuous random variables. Items 4 and 5 assume integer values. They are discrete random variables.

It is clear that in a random experiment, the outcomes or results are governed by chance. Also, the sample space of a random experiment consists of all outcomes of the experiment. The elements of S are the sample points which are the outcomes or events. They are often non-numeric, but they can be quantified by assigning a real number to every event of the sample space, e.g., the number of heads in tossing two or three coins and the sum of the points on a pair of dice when they are thrown simultaneously. This rule of assigning a real number to each sample point in S is called a random variable or variate. It is sometimes called a stochastic variable, meaning that it is related to chance.

2.2.1 Random Variable Definition

A random variable X on a sample space S (control) is a function X: $S \to R$, from S to the set of real numbers R which assigns a real number $X(s) = x$ to each sample point (Figure 2.1).

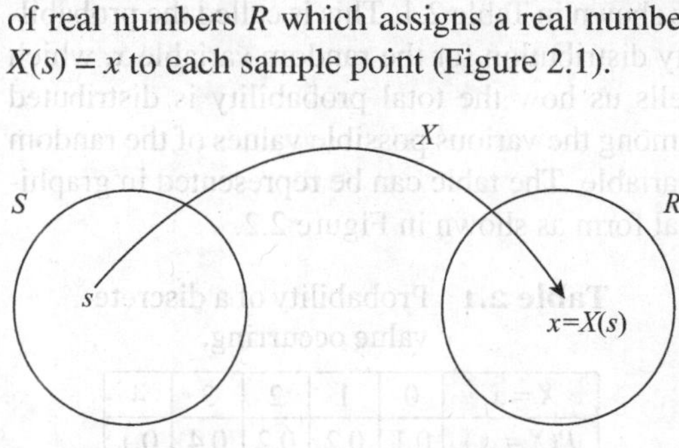

Figure 2.1 Random variable.

The range space $R_x \subseteq R$ is the set of images of points in S and is as such of all possible values x of X.

Though X is called a random variable or variate, it is in fact a single-valued function.

Examples

1. X = the number of defects on a randomly selected piece of furniture
2. X = the number that shows up when a die is thrown at random

Random variables are classified into the following:

1. Discrete random variables
2. Continuous random variables based on the values they assume

2.3 Probability Distribution

2.3.1 Discrete Probability Distributions

Discrete Random Variable The range of values of a variable may not give sufficient information about the variable. There is a need to know which values are likely to occur more frequently and which values less frequently. Suppose, for example, x is a discrete random variable which can take values 0, 1, 2, 3 and 4. Questions such as 'Which value is most likely to occur?', 'Is 3 more likely to occur than 1?', etc. are of interest for us. The information about the probability of occurrence of each of these values is shown in Table 2.1. This is called the probability distribution for the random variable x, which tells us how the total probability is distributed among the various possible values of the random variable. The table can be represented in graphical form as shown in Figure 2.2.

Table 2.1 Probability of a discrete value occurring.

$X = x_i$	0	1	2	3	4
$P(X = x_i)$	0.1	0.2	0.2	0.4	0.1

A random variable is said to be a discrete random variable if the set of all possible outcomes that is the sample space S is countable. In other words S is a finite or an infinite sequence.

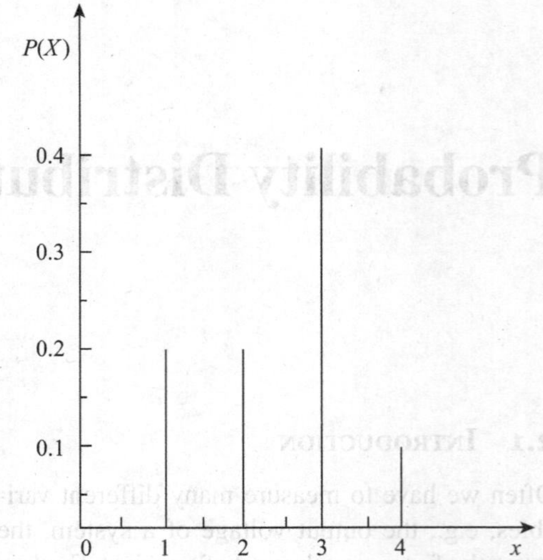

Figure 2.2 Plot of data in Table 2.1.

Example 2.1

Let two fair coins be tossed. Suppose that x denotes the number of heads observed. Find the probability distribution for x.

Solution The simple events for this experiment with the corresponding probabilities are shown in Table 2.2. Since E_1 = HH results in two heads, this simple event results in the value $x = 2$. Similarly, the value $x = 1$ is assigned to E_2 and to E_3 and $x = 0$ to E_4.

Table 2.2 Simple events and probabilities in tossing two coins.

Simple Event	Coin 1	Coin 2	$P(E_i)$	x
E_1	H	H	$\frac{1}{4}$	2
E_2	H	T	$\frac{1}{4}$	1
E_3	T	H	$\frac{1}{4}$	1
E_4	T	T	$\frac{1}{4}$	0

For each value of x, we can calculate $p(x)$ by adding the probabilities of the simple events in that event. For example, when $x = 0$, $p(0) = p(E_4) = \frac{1}{4}$ and when $x = 1$, $p(1) = p(E_2) + p(E_3) = \frac{1}{2}$.

The values of x and the corresponding probabilities $p(x)$ are shown in Table 2.3.

Table 2.3 Probability distribution for x (x = Number of heads).

x	Simple Events in x	$p(x)$
0	E_4	$\frac{1}{4}$
1	E_2 and E_3	$\frac{1}{4}$ and $\frac{1}{4}$
2	E_1	$\frac{1}{4}$

$$\sum p(x) = 1$$

Example 2.2

A coin is tossed six times. Find the probability of obtaining four or more heads.

Solution In a single toss, the probability of getting a head $p = \frac{1}{2}$ and the probability of getting a tail $q = 1 - p = 1 - \frac{1}{2} = \frac{1}{2}$

Also, the number of trails $n = 6$.

The probability of getting 4 heads = $P(4)$

$$= \binom{6}{4}\left(\frac{1}{2}\right)^4\left(\frac{1}{2}\right)^2 = \frac{6 \times 5}{1 \times 2} \times \frac{1}{2} = \frac{15}{64} = 0.234$$

The probability of getting 5 heads = $P(5)$

$$= \binom{6}{5}\left(\frac{1}{2}\right)^5\left(\frac{1}{2}\right)^2 6\left(\frac{1}{2^6}\right) = \frac{6}{64} = 0.095$$

The probability of getting 6 heads = $P(6)$

$$= \binom{6}{6}\left(\frac{1}{2}\right)^6 1 = \frac{1}{64} = 0.016$$

The probability of getting 4 or more heads

$$= P(4) + P(5) + P(6)$$
$$= 0.234 + 0.095 + 0.016 = 0.345$$

Example 2.3

Find the probability that in a family of 4 children there will be (a) at least one boy and (b) at least one boy and at least one girl, assuming that the probability of a male birth is $\frac{1}{2}$.

Solution Here the total number of trials $n = 4$ and $p = \frac{1}{2}$.

The number of successes x varies from $x = 1, 2, 3$ and 4

Thus, we have to find $b(x; n, p)$ for $x = 1, 2, 3$ and 4.

(a) $P(\text{1 boy}) = \binom{4}{1}\left(\frac{1}{2}\right)^1\left(\frac{1}{2}\right)^3 = \frac{1}{4}$

$P(\text{2 boys}) = \binom{4}{2}\left(\frac{1}{2}\right)^2\left(\frac{1}{2}\right)^2 = \frac{3}{8}$

$P(\text{3 boys}) = \binom{4}{3}\left(\frac{1}{2}\right)^3\left(\frac{1}{2}\right)^1 = \frac{1}{4}$

$P(\text{4 boys}) = \binom{4}{4}\left(\frac{1}{2}\right)^4\left(\frac{1}{2}\right)^0 = \frac{1}{16}$

Then,

$P(\text{at least 1 boy}) = P(\text{1 boy}) + P(\text{2 boys}) + P(\text{3 boys}) + P(\text{4 boys})$

$$= \frac{1}{4} + \frac{3}{8} + \frac{1}{4} + \frac{1}{16} = \frac{15}{16}$$

(b) $P(\text{at least 1 boy and at least 1 girl}) = 1 - P(\text{no boy}) - P(\text{no girl})$

$$= 1 - \left(\frac{1}{16}\right) - \left(\frac{15}{16}\right) = \left(\frac{7}{8}\right)$$

Aliter Let X be a random variable denoting the number of boys in families with 4 children. Then,

(a) $P(X \geq 1) = P(X = 1) + P(X = 2) + P(X = 3) + P(X = 4) = \frac{15}{16}$

(b) $P(\text{at least 1 boy and at least 1 girl})$

$$= 1 - 2P(X = 4) = 1 - \left(2 \times \frac{1}{16}\right) = \frac{7}{8}$$

Example 2.4

Out of 24 eggs, 6 are rotten. If 2 eggs are drawn at random, find the probability distribution of the number of rotten eggs that can be drawn.

Solution Let X denote the number of rotten eggs drawn.

Then X takes on the values 0, 1 and 2.

$$P(X = 0) = \frac{\binom{18}{2}}{\binom{24}{2}} = \frac{18 \times 17}{1 \times 2} \times \frac{1 \times 2}{24 \times 23} = \frac{51}{92}$$

$$P(X=1)=\frac{\binom{18}{1}\times\binom{16}{1}}{\binom{24}{2}}=\frac{18\times 16}{24\times 23}\times 1\times 2=\frac{9}{23}$$

$$P(X=2)=\frac{\binom{6}{2}}{\binom{24}{2}}=\frac{6\times 5}{1\times 2}\times\frac{1\times 2}{24\times 23}=\frac{5}{92}$$

The probability distribution of X is tabulated in Table 2.4.

Table 2.4 Probability distribution for X (Example 2.4).

$X = x$	0	1	2
$P(X = x)$	$\frac{51}{92}$	$\frac{9}{23}$	$\frac{5}{92}$

Example 2.5

60% of the riders of two wheelers put on their helmets. Then find the probability that

(a) 4 out of 5 will be using their helmets
(b) At least 4 out of 5 riders will use their helmets
(c) Fewer than 4 out of 5 will be using their helmets.

Solution The problem can be solved by using the binomial distribution $b(x; n, p)$ where $n = 5$, $p = \frac{60}{100} = 0.60$ and x = number of riders using helmet. Note that $q = 1 - p = 0.4$.

(a) $P(X = 4) = b(4;\ 5,\ 0.60) = \binom{5}{4}(0.6)^4 (0.4)^{5-4} = 0.26$

(b) $P(X \geq 4) = P(X = 4) + P(X = 5)$
$= b\,(4; 5, 0.6) + b\,(5; 5, 0.6)$
$= 0.26 + 0.08 = 0.34$

$$b(5; 5, 0.6) = \binom{5}{5}(0.6)^5 (0.4)^0 = 0.08$$

(c) $P(X < 4) = 1 - P(X \geq 4) = 1 - 0.34 = 0.66$

Example 2.6

Find the probability of getting 7 exactly twice in 3 throws with a pair of dice.

Solution In a single throw of a pair of dice, 7 can occur in the following 6 ways:

$$\{(6, 1), (5, 2), (4, 3), (3, 4), (2, 5), (1, 6)\}$$

Out of 6 × 6 = 36 ways, the probability of the occurrence of 7 in a throw is $p = \frac{6}{36} = \frac{1}{6}$ and $q = 1 - \frac{1}{6} = \frac{5}{6}$.

Number of trials $n = 3$
Number of successes $x = 2$
∴ The probability of getting 7 exactly twice in 3 throws

$$= b(x; n, p) = b\left(2; 3, \frac{1}{6}\right) = \binom{3}{2}\left(\left(\frac{1}{6}\right)\right)^2\left(\frac{5}{6}\right)^{3-2}$$

$$= \frac{3}{1}\times\frac{1}{36}\times\frac{5}{6} = \frac{5}{72}$$

Example 2.7

Find the probability distribution of the number of good apples in a box of 6 chosen at random if 90% of the apples are good in a large consignment.

Solution The probability of a good apple $= p = \frac{90}{100} = 0.9$

Its complement $q = 1 - p = 1 - 0.9 = 0.1$

Let X be the number of good apples and $n = 6$.

Table 2.5 Probability distribution for x (Example 2.7).

X	1	2	3	4	5	6
$P(X)$	0.000054	0.001215	0.015000	0.098000	0.354290	0.531100

The required binomial distribution is $b(x; n, p)$

$= \binom{6}{x} (0.9)^x (0.1)^{6-x}$ for $x = 0, 1, 2, \ldots\ldots, 6$

The results are tabulated in Table 2.5

Example 2.8

In tossing a coin two times in succession, the sample space of the experiment is

$$S = \{HH, HT, TH, TT\}$$

If X denotes the number of heads obtained, then X is a random variable. For each outcome, its value is as follows:

$$X(HH) = 2, X(HT) = 1, X(TH) = 1 \text{ and } X(TT) = 0$$

More than one random variable can be defined on the same sample space.

For example, let Y denote the number of heads minus the number of tails, for each outcome of the above sample space S. Then,

$$Y(HH) = 2, Y(HT) = 0, Y(TH) = 0 \text{ and } Y(TT) = -2$$

Thus, X and Y are two different random variables defined on the same sample space S.

Discrete Probability Distributions

Each event in a sample space has certain probability of occurrence. A formula representing all these probabilities which a discrete random variable X assumes is called the discrete probability distribution.

In Example 2.8, the sample space

$$S = \{HH, HT, TH, TT\}$$

We assign uniform probability of $\frac{1}{4}$ to each element of S. If $X: S \to R$ such that X be the discrete random variable that assigns to each element of S the number of heads:

$$X(HH) = 2, X(HT) = 1, X(TH) = 1 \text{ and } X(TT) = 0$$

The sample space of the random variable X is $S = \{0, 1, 2\}$.

Example 2.9

Let two dice be thrown at random. Let X be the discrete random variable that assigns to each point (a, b) the maximum of its numbers, i.e. $X(a, b) = \max(a, b)$.

Then X is a function from the sample space S consisting of 36 ordered pairs $\{(1, 1), (2, 2), \ldots, (6, 6)\}$ to a subset of real numbers $\{1, 2, 3, \ldots, 6\}$.

The event maximum 3 can occur in the following 5 cases: $\{(1, 3), (2, 3), (3, 3), (3, 2), (3, 1)\}$. Thus the random variable X assigns to this event of the sample space a real number 5. The probability of such an event happening is $\frac{5}{36}$ since there are 36 exhaustive cases. This is represented as $P(X = x_i) = p_i = f(x_i)$

So, $P(X = 3) = f(3) = \frac{5}{36}$.

Computing the other probabilities similarly, we have

$$p(1) = P(X = 1) = P\{(1, 1)\} = \frac{1}{36}$$

$$p(2) = P(X = 2) = P\{(2, 1), (1, 2), (2, 2)\} = \frac{3}{36}$$

$$p(3) = P(X = 3) = P\{(1, 3), (2, 3), (3, 3), (3, 2), (3, 1)\} = \frac{5}{36}$$

$$p(4) = P(X = 4) = P\{(1, 4), (2, 4), (3, 4), (4, 4), (4, 3), (4, 2), (4, 1)\} = \frac{7}{36}$$

$$p(5) = P(X = 5) = P\{(1, 5), (2, 5), (3, 5), (4, 5), (5, 5), (5, 4), (5, 3), (5, 2), (5, 1)\} = \frac{9}{36}$$

$$p(6) = P(X = 6) = P\{(1, 6), (2, 6), (3, 6), (4, 6), (5, 6), (6, 6), (6, 5), (6, 4), (6, 3), (6, 2), (6, 1)\} = \frac{11}{36}$$

The distribution of probabilities of this discrete random variable X is displayed in Table 2.6.

Table 2.6 Probability distribution for x (Example 2.9).

$X = x_i$	1	2	3	4	5	6
$P(X = x_i) = f(x_i) = p_i$	$\frac{1}{36}$	$\frac{3}{36}$	$\frac{5}{36}$	$\frac{7}{36}$	$\frac{9}{36}$	$\frac{11}{36}$

The discrete probability distribution, probability function or probability mass function of a discrete random variable X is the function $f(x)$ which satisfies the following conditions:

1. $f(x) \geq 0$
2. $\sum_x f(x) = 1$
3. $P(X = x) = f(x)$

Thus, probability distribution is the set of ordered pairs $(x, f(x))$ where x is the outcome and $f(x)$ is its probability.

Cumulative distribution of a discrete random variable X is $F(x)$ defined by

$$F(x) = P(X \leq x) = \sum f(t) \text{ for } t \leq x, \; -\infty < x < \infty$$

It follows that

$$F(-\infty) = 0 \text{ and } F(+\infty) = 1$$

$$p(x_j) = P(X = x_j) = F(x_j) - F(x_{j-1})$$

2.3.2 Continuous Probability Distributions

We now define certain terms.

Probability Density Function (pdf) or Continuous Random Variable For a continuous random variable X, the function $f(x)$ which satisfies the following conditions is called the probability density function (pdf) or simply a density function of X (Figure 2.3):

1. $f(x) \geq 0$
2. $\int_{-\infty}^{\infty} f(x)dx = 1$
3. $P(a < x < b) = \int_a^b f(x)dx$

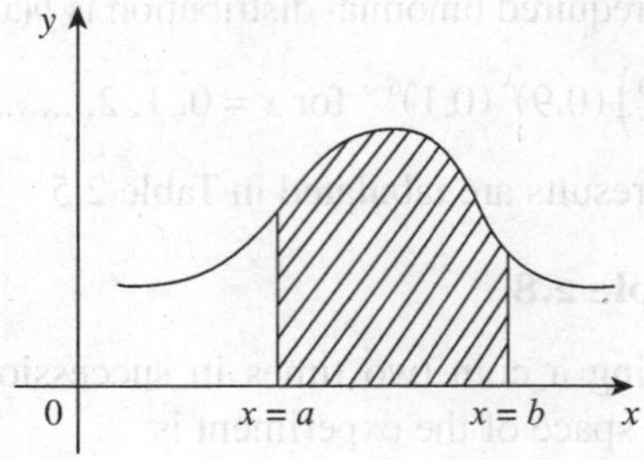

Figure 2.3 Probability density function (pdf): area under the curve $y = f(x)$, between the ordinate $x = a$ and $x = b$.

Notes

1. $P(a < x < b) = P(a \leq x < b) = P(a < x \leq b) = P(a \leq x \leq b)$

   Inclusion or exclusion of the end points a and b does not affect the probability for a continuous distribution, which is not the case in the discrete distribution.
2. Probability at a point 'a':

$$P(x = a) = \int_{a-\delta x}^{a+\delta x} f(x)dx$$

2.3.3 Cumulative Distributions

Let X be a continuous random variable and let $f(x)$ be its pdf. Then the cumulative distribution $f(x)$ of X is defined by

$$F(x) = P(X \leq x) = \int_{-\infty}^{x} f(u)du, \; -\infty < x < \infty$$

$F(x)$ satisfies the following:

$$F(-\infty) = 0 \text{ and } F(+\infty) = 1 \quad 0 \leq F(x) < 1 \text{ for } -\infty < x < \infty$$

$$f(x) = \frac{dF(x)}{dx} = F'(x) \geq 0 \text{ and}$$

$$P(a < x < b) = F(b) - F(a)$$

2.4 Expectation or Mean or Expected Value

The expectation or mean or expected value of a random variable X denoted by $E(X)$ or μ is defined by

$$E(X) = \sum_i x_i f(x_i) \quad \text{if } X \text{ is discrete}$$

$$= \int_{-\infty}^{\infty} x f(x)dx, \quad \text{if } X \text{ is continuous}$$

The mean μ characterizes the central location of the distribution.

Notes

1. x is the median if

$$P(X < x) \leq \frac{1}{2} \text{ and } P(X > x) \leq \frac{1}{2}$$

2. x is the mode if $f(x)$ or $P(x_i)$ attains its maximum at x.

2.5 Variance and Standard Deviation

Variance The variance $\sigma^2 = E(X - \mu)^2 = \sum_x (x_j - \mu)^2 f(x_j)$, if X is discrete

$$= \int_{-\infty}^{\infty} (x - \mu)^2 f(x)dx, \text{ if } X \text{ is continuous}$$

The variance characterizes the spread or variability of the distribution.

Standard Deviation The positive square root of variance σ^2 is called the standard deviation and is denoted by σ.

Results

1. For discrete random variable X, we have

$$\sigma^2 = \sum_x (x - \mu)^2 f(x) = \sum_x (x^2 - 2\mu x + \mu^2) f(x)$$

$$= \sum_x x^2 f(x) - 2\mu \sum_x x f(x) + \mu^2 \sum_x f(x)$$

$$= E(X^2) - 2\mu \times \mu + \mu^2 \times 1 = E(X^2) - \mu^2$$

2. For continuous random variable X, we have

$$\sigma^2 = \int_{-\infty}^{\infty} (x - \mu)^2 f(x)dx = \int_{-\infty}^{\infty} (x^2 - 2\mu x + \mu^2) f(x) dx$$

$$= \int_{-\infty}^{\infty} x^2 f(x)dx - 2\mu \int_{-\infty}^{\infty} x f(x)dx + \mu^2 \int_{-\infty}^{\infty} f(x)dx$$

$$= E(X^2) - 2\mu \times \mu + \mu^2 \times 1 = E(X^2) - \mu^2$$

Mathematical Expectation The mathematical expectation E is computed by the formula $E = \sum_i^k = \sum a_i p_i$ where p_i is the probability of obtaining the amount a_i ($i = 1, 2, ..., k$).

Mean (μ) Let X be a random variable whose possible values $x_1, x_2, x_3, \ldots, x_n$ occur with probabilities $p_1, p_2, p_3, \ldots, p_n$ respectively. The mean of X, denoted by μ, is the number $\sum_{i=1}^{k} x_i p_i$. Thus, the mean of X is the weighted average of all possible values of X, each value being weighted by its probability with which it occurs.

The mean of a random variable X is also called the expectation of X, denoted by $E(X)$.

Thus,

$$E(X) = \mu = \sum_{i=1}^{n} x_i p_i = x_1 p_1 + x_2 p_2 + \cdots + x_n p_n$$

In other words, the mean or expectation of a random variable X is the sum of the products of all possible values of X by their respective probabilities.

Example 2.10

Let a pair of dice be thrown. If X is the sum of the numbers that appear on the two dice, find the mean μ (or expectation $E(X)$) of X.

Solution The sample space of the experiment consists of $6^2 = 36$ simple events $\{(x_i, y_i) | i = 1, 2, \ldots, 6\}$.

The random variable X being the sum of the numbers on the two dice takes on the values 2, 3, 4, …, 12.

Now,

$P(X = 2) = P\{(1,1)\} = \frac{1}{36}$

$P(X = 3) = P\{(1,2), (2,1)\} = \frac{2}{36}$

$P(X = 4) = P\{(1,3), (2,2), (3,1)\} = \frac{3}{36}$

$P(X = 5) = P\{(1,4), (2,3), (3,2), (4,1)\} = \frac{4}{36}$

$P(X = 6) = P\{(1,5), (2,4), (3,3), (4,2), (5,1)\} = \frac{5}{36}$

$P(X = 7) = P\{(1,6), (2,5), (3,4), (4,3), (5,2), (6,1)\} = \frac{6}{36}$

$P(X = 8) = P\{(2,6), (3,5), (4,4), (5,3), (6,2)\} = \frac{5}{36}$

$P(X = 9) = P\{(3,6), (4,5), (5,4), (6,3)\} = \frac{4}{36}$

$P(X = 10) = P\{(4,6), (5,5), (6,4)\} = \frac{3}{36}$

$P(X = 11) = P\{(5,6), (6,5)\} = \frac{2}{36}$

$P(X = 12) = P\{(6,6)\} = \frac{1}{36}$

The probability distribution of X is tabulated in Table 2.7.

Table 2.7 Probability distribution for X (Example 2.10).

X or x_i	2	3	4	5	6	7	8	9	10	11	12
$P(X)$ or p_i	$\frac{1}{36}$	$\frac{2}{36}$	$\frac{3}{36}$	$\frac{4}{36}$	$\frac{5}{36}$	$\frac{6}{36}$	$\frac{5}{36}$	$\frac{4}{36}$	$\frac{3}{36}$	$\frac{2}{36}$	$\frac{1}{36}$

Hence,

$$\mu = E(X) = \sum_{i=1}^{n} x_i p_i = 2 \times \frac{1}{36} + 3 \times \frac{2}{36} + 4 \times \frac{3}{36} + 5 \times \frac{4}{36} + 6 \times \frac{5}{36} + 7 \times \frac{6}{36} + 8 \times \frac{5}{36} + 9 \times \frac{4}{36} + 10 \times \frac{3}{36} + 11 \times \frac{2}{36} + 12 \times \frac{1}{36} = 7.$$

Example 2.11

If the probability density of a random variable is given by

$$f(x) = \begin{cases} K(1 - x^2) & \text{for } 0 < x < 1 \\ 0 & \text{elsewhere} \end{cases}$$

Find the value of K and the probabilities that a random variable will take on a value

(a) Between 0.1 and 0.2
(b) Greater than 0.5

Solution Since the total probability is to be equal to unity, we must have

$$\int_{-\infty}^{\infty} f(x)dx = 1$$

$$\Rightarrow \int_{-\infty}^{0} f(x)dx + \int_{0}^{1} f(x)dx + \int_{1}^{\infty} f(x)dx = 1$$

$$\Rightarrow \int_{-\infty}^{0} 0\, dx + K\int_{0}^{1} (1 - x^2)dx + \int_{0}^{\infty} 0\, dx = 1$$

$$\Rightarrow K\left(x - \frac{1}{3}x^3\right)_0^1 = 1 \quad \Rightarrow \frac{2}{3}K = 1 \text{ or } K = \frac{3}{2}$$

$$\text{(a) } P(0.1 < x < 0.2) = \int_{0.1}^{0.2} f(x)dx$$

$$= \frac{3}{2}\int_{0.1}^{0.2} (1 - x^2)dx = \frac{3}{2}\left(x - \frac{1}{3}x^3\right)_{0.1}^{0.2}$$

$$= \frac{3}{2}\left\{(0.2 - 0.1) - \frac{1}{3}[(0.2)^3 - (0.1)^3]\right\}$$

$$= \frac{3}{2}\left[0.1 - \frac{1}{3}(0.007)\right] = 0.1465$$

(b) $P(x > 0.5) = \frac{3}{2}\int_{0.5}^{1}(1 - x^2)dx$

$$= \frac{3}{2}\left(x - \frac{1}{3}x^3\right)_{0.5}^{1}$$

$$= \frac{3}{2}\left\{(1 - 0.5) - \frac{1}{3}[1^3 - (0.5)^3]\right\}$$

$$= \frac{3}{2}\left(0.5 - \frac{0.875}{3}\right) = \frac{1.5}{2} - \frac{0.875}{2}$$

$$= \frac{0.625}{2} = 0.3125$$

2.6 Probability Density Functions

Let x be a continuous random variable which takes on any value on [0, 1]. Listing of all possible values is clearly impossible, in this case. There are infinitely many values on [0, 1]. So, the probability of any one particular occurring is zero. We can, therefore, divide the interval [0, 1] into number of subintervals and attach probabilities to each subinterval. Thus, there results a probability distribution.

Table 2.8 gives an example. The probability that x will be between 0.4 and 0.6 is 0.35, i.e., $P(0.4 \le x \le 0.6) = 0.35$, $P(0.6 \le x \le 0.8) = 0.25$, etc. Figure 2.4 shows a graphical representation of the data in Table 2.8.

Table 2.8 Probability that x lies in a subinterval of [0, 1].

x	[0.0, 0.2]	[0.2, 0.4]	[0.4, 0.6]	[0.6, 0.8]	[0.8, 1.0]
$p(x)$	0.10	0.20	0.35	0.25	0.10

Note that $\sum p(x) = 1$

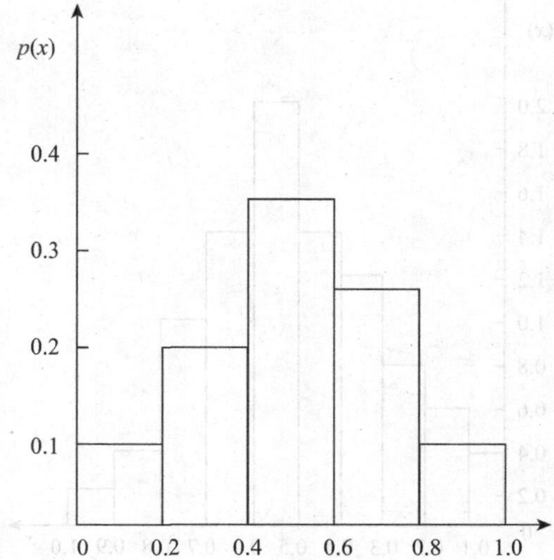

Figure 2.4 Graphical representation of the data in Table 2.8.

By reducing the size of each subinterval, a more refined distribution could be attained. This is shown in Table 2.9 and graphically represented in Figure 2.5.

The probability that x lies in a particular interval is given by the sum of the heights of the rectangles on that interval. For example, the probability that x lies in [0.3, 0.6] is 0.47.

Table 2.9 More refined distribution of probability that x lies in a subinterval of [0, 1].

x	[0.0, 0.1]	[0.1, 0.2]	[0.2, 0.3]	[0.3, 0.4]	[0.4, 0.5]	[0.5, 0.6]	[0.6, 0.7]	[0.7, 0.8]	[0.8, 0.9]	[0.9, 1.0]
$p(x)$	0.04	0.06	0.08	0.12	0.15	0.20	0.15	0.10	0.08	0.02

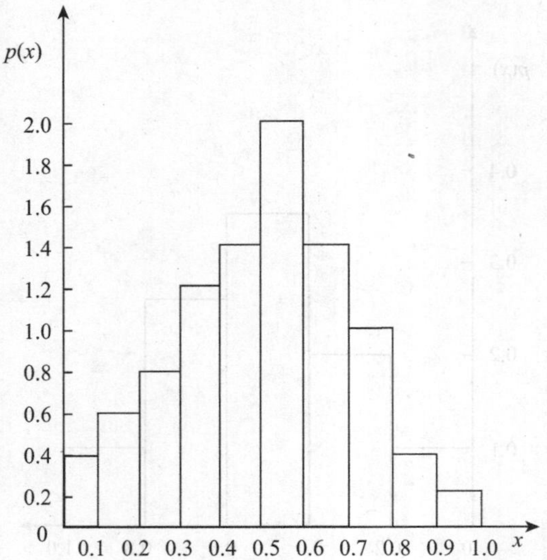

Figure 2.5 Graphical representation of data in Table 2.9.

Probability Density Function (pdf) $f(x)$
Consider the subinterval $[a, b]$. We require the probability that x lies in this interval. We get this by means of a pdf $f(x)$. Such a pdf is shown in Figure 2.6. This represents the area under the curve and between the ordinates $x = a$ and $x = b$ and gives $P(a \le x \le b)$, i.e.,

$$P(a \le x \le b) = \text{area above the interval } [a, b]$$

$$= \int_a^b f(x)\, dx$$

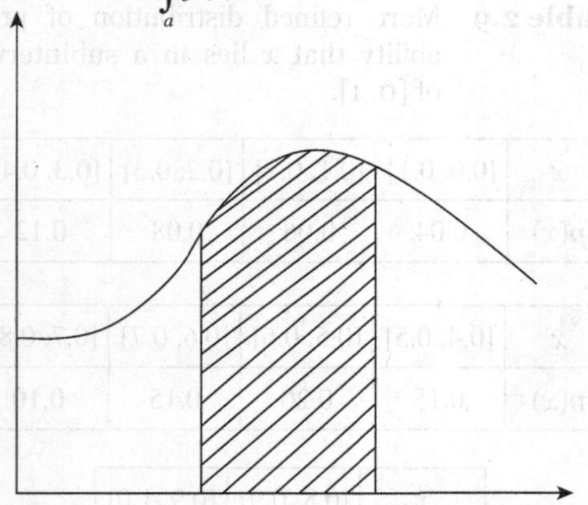

Figure 2.6 The shaded area represents $P(a \le x \le b)$. The total area under a pdf is always 1.

Example 2.12

Let x be a continuous random variable assuming any value on [1, 4]. Its pdf $f(x)$ is given by $f(x) = \frac{1}{2\sqrt{x}} \quad x \in [1,4]$

(a) Verify that $f(x)$ is a pdf.
(b) What is the probability that
 (i) $x \in [2, 3.5]$
 (ii) $x \ge 2$
 (iii) $x < 3^2$

Solution

(a) x can assume any value in [1, 4]. For $f(x)$ to be a pdf, the total area under it in [1, 4] should be exactly 1.
We have

$$\int_1^4 f(x)\, dx = \int_1^4 \frac{dx}{2\sqrt{x}} = [\sqrt{x}]_1^4 = \sqrt{4} - \sqrt{1} = 2 - 1 = 1$$

Hence, $f(x)$ is a suitable function for a pdf.

(b) (i) $P(2 \le x \le 3.5) = \int_2^{3.5} f(x)dx = [\sqrt{x}]_2^{3.5}$
$= \sqrt{3.5} - \sqrt{2} = 0.457$

(ii) $P(x \ge 2) = \int_2^4 f(x)dx = [\sqrt{x}]_2^4 = 2 - \sqrt{2}$
$= 0.586$

(iii) $P(x < 3) = \int_1^3 f(x)dx = [\sqrt{x}]_1^3 = \sqrt{3} - 1$
$= 0.732$

Example 2.13

Let x be a continuous random variable assuming any value x in $[0, \pi/2]$

(a) Verify if $f(x) = \cos x$ in $[0, \pi/2]$ is suitable for a pdf.
(b) Find the probability that
 (i) $x \in [0, \pi/4]$
 (ii) $x > \pi/4$
 (iii) $x < \pi/3$

Solution

(a) Let x be any value in $[0, \pi/2]$.
For $f(x) = \cos x$ to be a pdf, the total area under it in $[0, \pi/2]$ should be exactly 1. We have

$$\int_a^b f(x)dx = \int_0^{\pi/2} \cos x\, dx = [\sin x]_0^{\pi/2} = 1 - 0 = 1$$

Hence, $f(x) = \cos x$ is a suitable function for a pdf.

(b) (i) $P(0 \le x \le \pi/4) = \int_0^{\pi/4} \cos x\, dx$

$$= [\sin x]_0^{\pi/4} = \frac{1}{\sqrt{2}} - 0 = 0.707$$

(ii) $P(x > \pi/4) = \int_{\pi/4}^{\pi/2} \cos x\, dx = [\sin x]_{\pi/4}^{\pi/2} = 1 - \frac{1}{\sqrt{2}}$

$$= 1 - 0.707 = 0.293$$

(iii) $P(x < \pi/3) = \int_0^{\pi/3} \cos x\, dx = [\sin x]_0^{\pi/3}$

$$= \frac{\sqrt{3}}{2} - 0$$

$$= 0.866$$

Example 2.14

A random variable x has a p.d.f. $f(x)$ where $f(x) = e^{-x}$, $0 \le x < \infty$. Find the probability that

(a) $0 \le x \le 2$
(b) $x > 1$
(c) $x < 0.5$

Solution We have

$$\int_0^\infty f(x)\, dx = \int_0^\infty e^{-x}\, dx = (-e^{-x})_0^\infty = -(e^{-\infty} - e^0) = 1$$

Hence $f(x) = e^{-x}$ is a suitable function for a pdf.

(a) $P(0 \le x \le 2) = \int_0^2 e^{-x}\, dx = (-e^{-x})_0^2 = 1 - e^{-2}$
$= 0.865$

(b) $P(x > 1) = \int_1^\infty e^{-x}\, dx = (-e^{-x})_1^\infty = e^{-1} - 1$
$= 0.368$

(c) $P(x < 0.5) = \int_0^{0.5} e^{-x}\, dx = (-e^{-x})_0^{0.5} = 1 - e^{-0.5}$
$= 0.393$

Example 2.15

(a) Verify that $f(t) = \lambda e^{-\lambda t}$ $(t \ge 0)$ is suitable as a pdf.

(b) Find $P(t \ge 2)$, *if* $\lambda = 3$.

Solution

(a) $\int_0^\infty f(x)\, dx = \int_0^\infty \lambda e^{-\lambda t}\, dt = (-e^{-\lambda t})_0^\infty = 1 - 0 = 1$

Hence, $f(t) = \lambda e^{-\lambda t}$ $(t \ge 0)$ is suitable as a pdf.

(b) $P(t \ge 2) = \int_2^\infty 3e^{-3t}\, dt = (-e^{-3t})_2^\infty = -0 + e^{-6}$
$= 2.479 \times 10^{-3}$

2.7 CHEBYSHEV'S THEOREM

Theorem Let μ and σ denote the mean and standard deviation of a random variable X with probability density $f(X)$. Then the probability that X will assume a value within k standard deviations of the mean is at least $1 - \frac{1}{k^2}$ for any positive constant k.

Symbolically,

$$P(\mu - k\sigma < X < \mu + k\sigma) = P(|x - \mu| < k\sigma) \ge \left(1 - \frac{1}{k^2}\right) \quad (1)$$

Proof By definition,

$$\sigma^2 = \text{variance} = E[(X - \mu)^2] = \int_{-\infty}^{+\infty} (x - \mu)^2 f(x)dx$$

$$= \int_{-\infty}^{\mu-k\sigma} (x - \mu)^2 f(x)dx + \int_{\mu-k\sigma}^{\mu+k\sigma} (x - \mu)^2 f(x)dx + \int_{\mu+k\sigma}^{+\infty} (x - \mu)^2 f(x)dx$$

$$= I_1 + I_2 + I_3 \text{ (say)}, \quad (2)$$

dividing the variance into three: the first over $(-\infty, \mu - k\sigma)$, the second over $(\mu - k\sigma, \mu + k\sigma)$, and the third over $(\mu + k\sigma, +\infty)$. Now $I_2 \geq 0$, and for I_1, $x \leq \mu - k\sigma$ or $x - \mu \leq -k\sigma$ and for I_3, $x \geq \mu + k\sigma$ or $x - \mu \geq k\sigma$ so that, in either case, $|x - \mu| \geq k\sigma$ or $(x - \mu)^2 > k^2\sigma^2$

$$\therefore \sigma^2 \geq \int_{-\infty}^{\mu-k\sigma} k^2\sigma^2 f(x)\,dx + \int_{\mu+k\sigma}^{\infty} k^2\sigma^2 f(x)\,dx$$

Dividing throughout by non-negative quantity $k^2\sigma^2$, we get $P(|X - \mu| \geq k\sigma) = \int_{-\infty}^{\mu-k\sigma} f(x)dx + \int_{\mu+k\sigma}^{+\infty} f(x)dx < \frac{1}{k^2}$ (3)

By the complementation rule,

$$\int_{\mu-k\sigma}^{\mu+k\sigma} f(x)dx = 1 - [P(|X - \mu|) \geq k\sigma]$$

$$= 1 - \left[\int_{-\infty}^{\mu-k\sigma} f(x)dx + \int_{\mu+k\sigma}^{\infty} f(x)dx\right]$$

Hence

$$p(|X - \mu| < k\sigma) = P(\mu - k\sigma < X < \mu + k\sigma)$$

$$= \int_{\mu-k\sigma}^{\mu+k\sigma} f(x)dx \geq 1 - \frac{1}{k^2} \quad (4)$$

For $k = 2$, the theorem states that the random variable x has a probability of at least $1 - \frac{1}{2^2} = \frac{3}{4}$ of falling within two standard deviations of the mean. That is, three-fourths or more of the observation of any distribution lie in the interval $(\mu - 2\sigma, \mu + 2\sigma)$

Example 2.16

A random variable X has a mean $\mu = 8$ and a variance $\sigma^2 = 9$ and an unknown probability distribution. Find

(a) $P(-4 < x < 20)$

(b) $P(|x - 8| \geq 6)$

Solution

(a) Here $\mu - k\sigma = -4$

$$\Rightarrow k = \frac{\mu + 4}{\sigma} = \frac{8+4}{3} = 4 \because \mu = 8 \text{ and } \sigma = 3$$

$$\therefore P(-4 < X < 20) = P[8 - (4)(3) < X < 8 + (4)(3)] \geq 1 - \frac{1}{4^2} = \frac{15}{16}$$

(b) $P(|X - 8| \geq 6) = 1 - P(|X - 8| < 6)$

$= 1 - P(-6 < |X - 8| < 6)$ here $\mu = 8$ and $\sigma = 3$

$= 1 - P(2 < X < 14)$ $\mu - k\sigma = 2 \Rightarrow k\sigma = 6$

$\Rightarrow k = 2 \because \sigma = 3$

$= 1 - P[8 - (2)(3) < X < 8 + (2)(3)]$

$\Rightarrow k = 2$ and $\sigma = 3$

$\leq \frac{1}{2^2} = \frac{1}{4}$

Note Chebyshev's theorem holds for any distribution of observations and, for this reason, the results are usually weak. The value given by the theorem is lower bound only.

Example 2.17

The number of customers visiting a show room is a random variable with mean $\mu = 20$ and standard deviation $\sigma = 5$. Find the probability with which we can assert that there will be more than 10 but less than 30 customers.

Solution Let X be the number of customers. We are given that mean $\mu = 20$ and standard deviation $\sigma = 5$. We have

$$k = \frac{30 - 20}{5} = \frac{20 - 10}{5} = 2 \quad \text{Also, } P(|X - \mu| < k\sigma) \geq 1 - \frac{1}{k^2}$$

$$\therefore P(10 < X < 30) \geq 1 - \frac{1}{2^2} = \frac{3}{4}$$

Example 2.18

If X is the random variable of the number of heads obtained in tossing three coins, prove that $P(|X - 3/2| \geq 2) \leq 3/16$.

Solution When three coins are tossed simultaneously or a single coin is tossed three times, the sample space S consists of the following 8 elements:

S = {HHH, HHT, HTH, THH, HTT, THT, TTH, TTT}

Probability of getting no head {TTT} $= \frac{1}{8}$

Probability of getting 1 head {HTT, THT, TTH} $= \frac{3}{8}$

Probability of getting 2 heads {HHT, HTH, THH} $= \frac{3}{8}$

Probability of getting 3 heads {HHH} $= \frac{1}{8}$

Probability distribution is

x	0	1	2	3
$P(x)$	$\frac{1}{8}$	$\frac{3}{8}$	$\frac{3}{8}$	$\frac{1}{8}$

$$\mu = E(X) = 0 \times \frac{1}{8} + 1 \times \frac{3}{8} + 2 \times \frac{3}{8} + 3 \times \frac{1}{8} = \frac{12}{8} = \frac{3}{2}$$

$$E(X^2) = 0^2 \times \frac{1}{8} + 1^2 \times \frac{3}{8} + 2^2 \times \frac{3}{8} + 3^2 \times \frac{1}{8} = \frac{24}{8} = 3$$

$$\sigma^2 = E(X^2) - E^2(X) = 3 - \left(\frac{3}{2}\right)^2 = 3 - \frac{9}{4} = \frac{3}{4}$$

By Chebyshev's theorem, $P(|X - \mu| \geq k\sigma) \leq \frac{1}{k^2}$

Taking

$$k = \frac{c}{\sigma} \text{ or } k\sigma = c$$

we have $P(|X - \mu| \geq c) \leq \frac{\sigma^2}{c^2}$.

Here $\mu = 3/2$. Taking $c = 2$

we get $P(|X - 3/2| \geq 2) \leq \frac{3}{4} \cdot \frac{1}{2^2} = \frac{3}{16}$, which was to be proved.

Example 2.19

X is a random variable such that $E(X) = 3$ and $E(X^2) = 13$. Determine a lower bound for $P(-2 < X < 8)$ using Chebyshev's inequality.

Solution It is given that $\mu = E(X) = 3$ and $E(X^2) = 13$.

$$\sigma^2 = E(X^2) - E^2(X) = 13 - 9 = 4 \Rightarrow \sigma = 2.$$

By Chebyshev's inequality

$$P(|X - \mu| < k\sigma) \geq 1 - \frac{1}{k^2}$$

$$\Rightarrow P(|X - 3| < 2k) \geq 1 - \frac{1}{k^2}$$

$$\Rightarrow P(3 - 2k < X < 3 + 2k) \geq 1 - \frac{1}{k^2}$$

Taking $k = 2.5$ we get

$$P(-2 < X < 8) \geq \frac{21}{25}$$

$\therefore$ A lower bound for $P(-2 < X < 8)$ is $\frac{21}{25}$

Example 2.20

If X is the number scored in a draw of a fair die, show that $P(|X - 3.5| > 2.5) < 7/15$.

Solution We have

$$\mu = E(X) = \frac{1}{6}(1 + 2 + 3 + 4 + 5 + 6) = 3.5$$

$$E(X^2) = \frac{1}{6}(1^2 + 2^2 + 3^2 + 4^2 + 5^2 + 6^2) = \frac{91}{6}$$

Variance of X: $\sigma^2 = E(X^2) - \mu^2 = \frac{91}{6} - \left(\frac{7}{2}\right)^2$

$$= \frac{91}{6} - \frac{49}{4}$$

$$= \frac{182 - 147}{12} = \frac{35}{12}$$

By Chebyshev's theorem, we have for $k > 0$

$$P(|X - \mu| > k) < \frac{\sigma^2}{k^2}$$

Take $k = \frac{5}{2} = 2.5$, $\mu = 3.5$, $\sigma^2 = \frac{35}{12}$

$$P(|X - 3.5| > 2.5) < \frac{35}{12} \times \frac{4}{25} = \frac{7}{15}$$

EXERCISES

1. A die is tossed thrice. A success is "getting 1 or 6" on a toss. Find the mean and variance of the number of successes.

Ans: $\mu = 1$ and $\sigma^2 = \frac{2}{3}$

2. Find the standard deviation for the following discrete distribution:

x	8	12	16	20	24
$P(x)$	$\frac{1}{8}$	$\frac{1}{6}$	$\frac{3}{8}$	$\frac{1}{4}$	$\frac{1}{12}$

Ans: $\mu = 16$ and $\sigma^2 = 2\sqrt{5}$

3. The probability density $P(x)$ of a continuous random variable is given by $P(x) = y_0\, e^x \; -\infty < x < \infty$. Prove that $y_0 = \frac{1}{2}$. Find the mean and variance of the distribution.

Ans: $\mu = 0$ and $\sigma^2 = 2$

4. A random variable x has pdf $f(x)$ given by $f(x) = 5/4x^2 \mid 0 < x < 5$:
   (a) Compute the expected value of x.
   (b) Ten values of x are measured. They are 2.1, 1.9, 2.9, 2.8, 3.2, 3.4, 2.7, 2.3, 2.8 and 2.7. Calculate the mean of the observations.

Ans: (a) 2.012 and (b) 2.68

5. A pdf $h(x)$ is defined by

$$h(x) = \frac{3}{4}(1 - x^2), \; -1 \le x \le 1$$

Calculate the expected value of x.

Ans: 0

6. Calculate the expected value of the random variable x whose probability distribution is

x	2	2.5	3.0	3.5	4.0	4.5
$P(x)$	0.07	0.36	0.21	0.19	0.10	0.07

Ans: 3.05

7. A discrete random variable has probability distribution

x	1	2	3	4	5
$P(x)$	0.12	0.15	0.23	0.3	0.2

Calculate
(a) Expected value
(b) Standard deviation

Ans: (a) $\mu = 3.31$ and (b) $\sigma = 1.278$

8. A random variable t has pdf $H(t)$ given by
$H(t) = 3e^{-3t} \qquad t \ge 0$
Calculate
(a) Expected value of t
(b) Standard deviation of t

Ans: (a) $\frac{1}{3}$ and (b) $\frac{1}{3}$

9. A continuous random variable has pdf

$$f(x) = \begin{Bmatrix} x + 1, & -1 \le x \le 0 \\ -x + 1, & 0 < x \le 1 \end{Bmatrix}$$

Calculate
(a) Expected value of x
(b) Standard deviation of x

Ans: (a) 0 and (b) 0.4082

FILL IN THE BLANKS

1. A discrete random variable can take ______ number of values within its range.

Ans: Finite

2. A continuous random variable can take any value within its ______

Ans: Domain

3. If a discrete random variable has the following probability function

x	0	1	2	3	4
$P(x)$	k	$2k$	$3k$	$2k$	$2k$

then $k =$ ______

Ans: 0.1

4. If the pdf of a variable x is defined by $f(x) = cx(2 - x)$, $0\ x < 2$, then $c =$ ______

Ans: $\frac{3}{4}$

5. Given that $f(x) = \frac{k}{2^x}$ is a probability distribution for a random variable that can take on the values $x = 0, 1, 2, 3$ and 4, then $k =$ ______

Ans: 16/31

6. If ∞ for $x = 3, 4, 5$ and 6 defines a probability distribution then, $k =$ ______

Ans: 1

7. If $f(x) = ce^{-ax}$ is a pdf for $0 < x < \infty$, then $c =$ ______

Ans: a

8. If a random variable x has the probability distribution

x	3	2	1	0	−1	−2	3
$P(x)$	0.1	0.2	$3k$	k	$2k$	0	0.1

then $k =$ ______

Ans: 0.1

9. From Question 8, the mean of x is ______

Ans: 0.5

10. From Question 8, the variance of x is ______

Ans: 2.85

3

Special Distribution

3.1 Introduction

An experiment often consists of repeated trials, each with two possible outcomes—success or failure. For example, in a manufacturing industry, a test or trial may indicate a defective or non-defective item. We may choose to call either outcome as a success. This process is called a Bernoulli process. Each trial is called a Bernoulli trial.

3.2 Binomial (Bernoulli) Distribution

This process must possess the following properties:

1. The experiment consists of n repeated trials.
2. Each trial results in an outcome that may be classified as a success or a failure which are mutually exclusive.
3. The probability of success, denoted by p, remains constant from trial to trial.
4. The repeated trials are independent.

Assume that each Bernoulli trial can result in a success with probability p and a failure with probability $q = 1 - p$. Then the probability distribution of the binomial random variable X, the number of successes in n independent trials, is

$$b(x;\, n,\, p) = \binom{n}{x} p^x q^{n-x}, \quad x = 0, 1, 2, \ldots, n \qquad (3.1)$$

Example 3.1

The probability that a product will survive a shock test is $\frac{3}{4}$. Find the probability that exactly 2 of the next 4 products tested survive.

Solution We assume that the tests are independent.

Here $p = \frac{3}{4}$ for each of the 4 tests

We have

$$b\left(2;\, 4,\, \frac{3}{4}\right) = \binom{4}{2}\left(\frac{3}{4}\right)^2\left(\frac{1}{4}\right)^2 = \frac{4\times 3}{1\times 2}\times\frac{9}{16}\times\frac{1}{16}$$
$$= \frac{27}{128}$$

The binomial distribution derives its name from the fact that the $(n + 1)$ terms in the binomial expansion of $(q + p)^n$ correspond to the various values of $b(x; n, p)$ for $x = 0, 1, 2, \ldots, n$. That is,

$$(q+p)^n = \binom{n}{0} q^n + \binom{n}{1} pq^{n-1} + \binom{n}{2} p^2q^{n-2} + \cdots + \binom{n}{n} p^n$$
$$= b(0; n, p) + b(1; n, p) + b(2; n, p) + \ldots + b(n; n, p)$$

Since $q + p = 1$, we have

$$\sum_{x=0}^{n} b(x;\, n,\, p) = 1$$

which is the condition that must hold for any probability distribution.

The binomial distribution is characterized by the parameter p and the number of trials n.

3.2.1 Mean and Variance of Binomial Distribution

Mean $\mu = \text{Expectation} = \sum_{x=0}^{n} xp(x)$

$$= \sum_{x=0}^{n} xb(x;n,p) = \sum_{x=0}^{n} x \frac{n!}{x!(n-x)!} p^x q^{n-x}$$

$$= \sum_{x=1}^{n} \frac{n!}{(x-1)!(n-x)!} p^x q^{n-x}$$

$$= np \sum_{x=1}^{n} \frac{(n-1)!}{(x-1)!(n-x)!} p^{x-1} q^{n-x}$$

$$= np \sum_{y=0}^{n-1} \frac{(n-1)!}{y![(n-1)-y]!} p^y q^{n-1-y}$$

$\big[$Put $y = x - 1 \Rightarrow x = y + 1$ when $x = 1, y = 0$ or

$x = n, y = n - 1\big]$

$$= np(p+q)^{n-1} \quad \because p + q = 1$$

$$= np$$

$$\Rightarrow \mu = np$$

Variance $\sigma^2 = \sum_{x=0}^{n} (x-\mu)^2 p(x)$

$$= \sum_{x=0}^{n} (x^2 - 2\mu x + \mu^2)\, p(x)$$

$$= \sum_{x=0}^{n} x^2 p(x) - 2\mu \sum_{x=0}^{n} xp(x) + \mu^2 \sum_{x=0}^{n} p(x)$$

$$= \sum_{x=0}^{n} x^2 p(x) - 2\mu \times \mu + \mu^2 \times 1$$

$$\left(\because \sum xp(x) = \mu = np \text{ and } \sum p(x) = 1\right)$$

$$= \sum_{x=0}^{n} x^2 p(x) - \mu^2 = \sum_{x=0}^{n} x^2 p(x) - x^2 p^2 \quad (3.2)$$

Now

$$\sum_{x=0}^{n} x^2 p(x) = \sum_{x=0}^{n} \frac{x^2\, n!}{(n-x)!x!} p^x q^{n-x}$$

$$\because p(x) = \sum_{x=0}^{n} \frac{n!}{x!(n-x)!}$$

$$= \sum_{x=0}^{n} [x(x-1) + x] \frac{n!}{x!(n-x)!} p^x q^{n-x}$$

$$+ \sum_{x=0}^{n} x \frac{n!}{x!(n-x)!} p^x q^{n-x}$$

$$= \sum\nolimits_{x=0}^{n} \frac{n!}{(n-x)(x-2)!} p^x q^{n-x} + np$$

$$= n(n-1)\, p^2 \sum_{x=2}^{n} \frac{(n-2)!}{(n-x)!(x-2)!} p^{x-2} q^{n-x} + np$$

$$= n(n-1)p^2 \times 1 + np = n^2p^2 - np^2 + np$$

Substituting in (3.2), we get

$$\sigma = \sum_{x=0}^{n} x^2 p(x) - n^2 p^2 = -np^2 + np$$

$$= np\,(1-p) = npq \quad (3.3)$$

Example 3.2

A coin is tossed six times. Calculate the probability of obtaining four or more heads.

Solution Here $n = 6$ and $p = \frac{1}{2}$ and

$$q = 1 - p = \frac{1}{2}$$

Probability of getting 4 heads = $P(4)$

$$= \binom{6}{4}\left(\frac{1}{2}\right)^4\left(\frac{1}{2}\right)^2$$

$$\frac{6\times 5}{2} \times \frac{2}{26} = \frac{15}{64} = 0.234$$

Probability of getting 5 heads = $P(5)$

$$= \binom{6}{5}\left(\frac{1}{2}\right)^5\left(\frac{1}{2}\right)^1$$

$$= \frac{6}{64} = 0.095$$

Probability of getting 6 heads = $P(6) = \left(\frac{1}{2}\right)^6$

$$= \frac{1}{64} = 0.016$$

$\therefore$ Probability of getting 4 or more heads

$$= P(4) + P(5) + P(6) = 0.234 + 0.095 + 0.016$$

$$= 0.345$$

Example 3.3

If 1 out of every 10 bulbs is defective then find (a) mean and standard deviation for the distribution of defective bulbs in a total of 500 bulbs and (b) the coefficient of skewness γ_1 and the coefficient of kurtosis γ_2.

Solution

(a) $p = \frac{1}{10} = 0.1$, $q = 1 - p = 0.9$ and $n = 500$

Mean $= \mu = np = 50$

∴ We can expect that 50 bulbs to be defective out of 500.

Standard deviation $\sigma = \sqrt{npq}$

$$= \sqrt{500 \times 0.1 \times 0.9} = 6.71$$

(b) Coefficient of skewness

$$\gamma_1 = \sqrt{p_1} = \sqrt{\frac{(q-p)^2}{npq}} = \frac{q-p}{\sqrt{npq}} = \frac{0.9-0.1}{6.71}$$

$$= 0.119$$

Coefficient of kurtosis

$$\gamma_2 = 3 + \frac{1-6pq}{npq} = 3 + \frac{1-6\times(0.1)(0.9)}{4.5}$$

$$= 3 + \frac{1-0.54}{45} = 3.01$$

3·3 Poisson Distribution

The discrete distribution with infinitely many possible values and probability function

$$f(x, \lambda) = P(X = x) = \frac{\lambda^x e^{-\lambda}}{x!}, \quad x = 0, 1, 2, .. \quad (3.4)$$

is called the Poisson distribution in honour of S. D. Poisson[1]. Here $\lambda > 0$ is called the parameter of distribution. In the binomial distribution, the number of successes out of a definite total number of n trials is determined, while in Poisson distribution, the number of successes at a random point of time and space is determined. Poisson distribution is suitable for 'rare' events. In this case, the probability of occurrence p is very small and the number of trials n is very large. Further, binomial distribution can be approximated by Poisson distribution when $n \to \infty$ and $p \to 0$ such that $\lambda = np$ is constant.

Examples of Rare Events

1. Number of printing mistakes per page in a particular book.
2. Number of accidents on a high way during a month.
3. Number of defective components in a production unit.
4. Number of dishonoured cheques at a bank.

[1]Poisson, Simeon Denis (1781–1840) is a French mathematician and physicist professor in Paris from 1809. His work includes potential theory, partial differential equations and probability.

Results

1. $\sum_{x=0}^{\infty} f(x, \lambda) = \sum_{x=0}^{\infty} P(X = x) = \sum_{x=0}^{\infty} \frac{\lambda^x e^{-\lambda}}{x!} = e^{-\lambda} \sum_{x=0}^{\infty} \frac{\lambda^x}{x!}$

$$= e^{-\lambda} e^{\lambda} = 1$$

This shows that the function defined at Example 1 is a probability function.

2. Arithmetic mean of Poisson distribution

$$\overline{X} = E(X) = \sum_{x=0}^{\infty} xP(X = x) = \sum_{x=0}^{\infty} x \frac{e^{-\lambda}\lambda^x}{x!}$$

$$= \lambda e^{-\lambda} \sum_{x=1}^{\infty} \frac{\lambda^{x-1}}{(x-1)!} = \lambda e^{-\lambda} \sum_{m=0}^{\infty} \frac{\lambda^m}{m!}$$

$$= \lambda e^{-\lambda} \times e^{\lambda} = \lambda \quad (3.5)$$

This shows that the parameter λ is the arithmetic mean of the Poisson distribution.

3. Variance of Poisson distribution

$$= E\left[(x-\overline{X})^2\right] = \sum_{x=0}^{\infty} (x-\overline{X})^2 P(X = x)$$

$$= \sum_{x=0}^{\infty} (x^2 - 2x\overline{X} + \overline{X}^2)\, P(X = x)$$

$$= \sum x^2 P + \overline{X}^2 \sum P - 2\overline{X} \sum xP$$

$$= \sum x^2 P + \overline{X}^2 - 2\overline{X}^2 = \sum x^2 P + \lambda^2 - 2\lambda^2$$

$$= \sum x^2 P - \lambda^2 \quad (3.6)$$

But

$$\sum x^2 P = \sum_{x=0}^{\infty} x^2 \frac{\lambda^x e^{-\lambda}}{x!} = \sum_{x=0}^{\infty} [x(x-1) + x] \frac{\lambda^x e^{-\lambda}}{x!}$$

$$= \lambda^2 e^{-\lambda} \sum_{x=2}^{\infty} \frac{\lambda^{x-2}}{(x-2)!} + \lambda e^{-\lambda} \sum_{x=0}^{\infty} \frac{\lambda^{x-1}}{(x-1)!}$$

$$= \lambda^2 e^{-\lambda} e^{\lambda} + \lambda e^{-\lambda} e^{\lambda} = \lambda^2 + \lambda \quad (3.7)$$

Hence

$$\text{Variance} = \sum x^2 P - \lambda^2 = (\lambda^2 + \lambda) - \lambda^2$$

$$= \lambda \quad (3.8)$$

Therefore, variance of Poisson distribution = mean of Poisson distribution.
Also, standard deviation of the Poisson distribution = $\sqrt{\lambda}$.

4. Recurrence relation satisfied by $P(x)$:

We have $\dfrac{P(x+1)}{P(x)} = \dfrac{\lambda^{x+1} e^{-\lambda}}{(x+1)!} \times \dfrac{x!}{\lambda^x e^{-\lambda}}$

$$= \frac{\lambda}{x+1} \quad (3.9)$$

so that $P(x+1) = \left(\dfrac{\lambda}{x+1}\right) P(x) \quad (3.10)$

5. Cumulative Poisson distribution function $F(x; \lambda)$ defined by

$$F(x;\lambda) = \sum_{k=0}^{x} \frac{\lambda^k e^{-\lambda}}{k!} \quad (3.11)$$

is related to

$$f(x;\lambda) = P(X = x) = \frac{\lambda^x e^{-\lambda}}{x!}$$

for $x = 0, 1, 2, \ldots$ (3.12)
by the formula

$$f(x;\lambda) = F(x;\lambda) - F(x-1;\lambda) \quad (3.13)$$

3.3.1 Poisson Approximation to the Binomial Distribution

Theorem Prove that the Poisson distribution is the limiting case of the binomial distribution for very large trials has very small probability. That is, as $n \to \infty$ and $p \to 0$ such that $\lambda = np$ = constant:

$$f(x; \lambda) = \lim_{\substack{n \to \infty \\ p \to 0}} b(x; n, p)$$

Proof We know that

$$b(x; n, p) = \frac{n!}{x!(n-x)!} p^x q^{n-x} \quad (3.14)$$

Put $p = \dfrac{\lambda}{n}$ in Eq. (3.1), we get

$$b(x; n, p) = \frac{n!}{x!(n-x)!} \left(\frac{\lambda}{n}\right)^x \left(1 - \frac{\lambda}{n}\right)^{n-x}$$

$$[\because q = 1 - p]$$

$$= \frac{(n(n-1)(n-2)\cdots\cdots(n-\overline{x-1})}{x!} \times \frac{\lambda^x}{n^x} \times \left(1 - \frac{\lambda}{n}\right)^{n-x}$$

$$= \frac{1\left(1-\frac{1}{n}\right)\left(1-\frac{2}{n}\right)\cdots\cdots 1-\left(\frac{x-1}{n}\right)}{x!} \lambda^x,$$

$$\left(1-\frac{\lambda}{n}\right)^n \times \left(1-\frac{\lambda}{n}\right)^{-x}$$

Letting $n \to \infty$, we get

$$\lim_{n\to\infty} b(x; n, p) = \frac{1}{x!} \lambda^x e^{-\lambda} \quad (3.15)$$

since

$$\lim_{n\to\infty} \left(1 - \frac{x}{r}\right) = e^{-x}$$

Example 3.4

Suppose that on an average 1 out of 1000 houses catch fire in a year in a district. If there are 2000 houses in the district, find the probability that exactly 5 houses will catch fire during that year.

Solution It is a Poisson distribution with $n = 2000$ and $p = \dfrac{1}{1000}$.

$\therefore$ Mean $m = np = 2$

$$P(r) = \frac{e^{-m} m^r}{r!} (e = 2.7183)$$

$$P(5) = (2.7183)^{-2} \frac{2^5}{5!} = \frac{32}{120(2.7183)^2}$$

$$= \frac{32}{120 \times (7.389)} = 0.036$$

Example 3.5

In a factory manufacturing razor blades, there is a small chance of $\dfrac{1}{50}$ for any blade to be defective. The blades are placed in packets of 10 blades. Using Poisson distribution, calculate the approximate number of pockets containing not more than 2 defective blades in a consignment of 10,000 packets.

Solution Here $N = 10{,}000$, $p = \frac{1}{50}$ and $n = 10$

$\therefore$ Mean $m = np = 0.2$

Since $P(r) = e^{-m}\frac{m^r}{r!} = e^{-0.2}\frac{(0.2)^r}{r!}$

$P(r = 0) = e^{-0.2} = 0.8187$

$N \times P(r = 0) = 10{,}000 \times 0.8187 = 8187$

Number of packets having 1 defective blade is

$$N \times P(r = 1) = Ne^{-0.2}\frac{0.2}{1!}$$

$$= 10{,}000 \times 0.2 \times e^{-0.2}$$
$$= 2000 \times 0.8187 = 1637.4$$

Number of packets having 2 defective blades is

$$N \times P(r = 2) = Ne^{-0.2}\frac{(0.2)^2}{2!}$$

$$= \frac{e^{-0.2} \times 0.04}{2} \times 10{,}000 = 163.74$$

The approximate number of packets containing not more than 2 defective blades in a consignment of 10,000 packets is

$$10{,}000 - (8187 + 1637.4 + 163.74) = 11.86$$
$$= 12 \text{ packets}$$

Example 3.6

Fit a Poisson distribution to the set of observation

x	0	1	2	3	4
f	122	60	15	2	1

Solution Mean $= \frac{\Sigma f_i x_i}{\Sigma f_i} = \frac{60 + 30 + 6 + 4}{200} = 0.5$

Mean of the Poisson distribution $m = 0.5$

Hence the theoretical frequency for r successes is

$$\frac{Ne^{-m}(m)^r}{r!} = \frac{200e^{-0.5}(0.5)^r}{r!},\ r = 0, 1, 2, 3, 4$$

and $e^{-5} = 0.61$

The theoretical frequencies are

x	0	1	2	3	4
f	121	61	15	2	0

Note In Examples 3.4–3.6, m has been used in place of λ.

3.3.2 Poisson Processes

In general, a random process is a physical process that is wholly or partly controlled by some sort of chance mechanism. It may be a sequence of repeated tossing of a coin, the vibrations of the wings of an aeroplane, etc. These processes depend upon time. Certain events do or do not take place at regular intervals of time or throughout continuous intervals of time.

Here we shall be concerned with processes occurring over continuous intervals of time or space—e.g., the arrival of telephone calls at a switchboard or the passing by cars over an electronic counting device. We will now show that the mathematical model which can be used to describe situations like these is that of Poisson distribution. Let us find the probability of x successes during a time interval of length T. We divide the interval into n equal parts each of length Δt. So, $T = n \times \Delta t$. We assume that

1. The probability of a success during a very small interval of time Δt is given by $\alpha \times \Delta t$.
2. The probability of more than one success during such a small time interval Δt is negligible.
3. The probability of a success during such a time interval does not depend on what happened prior to that time.

Hence the assumptions relating to the binomial distribution are satisfied. So, the probability of x successes in the time interval T is given by the binomial probability

$$b(x; n, p) \text{ with } n = \frac{T}{\Delta t} \text{ and } p = \alpha \times \Delta t$$

Letting $n \to \infty$, we find that the probability of x successes during the time interval T is given by the corresponding Poisson probability with the parameter $\lambda = np = \frac{T}{\Delta t}(\alpha \times \Delta t) = \alpha T$ since λ is the mean of this Poisson distribution and α is the average (mean) number of successes per unit time.

3.4 Uniform Distribution

3.4.1 Discrete Uniform Distribution

It is the simplest of all discrete probability distributions. In this case, the discrete random variable assumes each of its values with the same probability. In this case, each sample point is assigned equal probabilities such that their sum is unity.

Suppose the sample space S contains points each with probability $\frac{1}{m}$. Thus, in the discrete uniform distribution, the discrete random variable X assigns equal probabilities to all possible values of x. Therefore, the probability mass function $f(x)$ has the form

$$f(x) = \frac{1}{m}, \text{ for } x = x_1, x_2, \ldots, x_m \quad (3.16)$$

That is,

x	x_1	x_2	x_3	...	x_m
f	$\frac{1}{m}$	$\frac{1}{m}$	$\frac{1}{m}$	...	$\frac{1}{m}$

The mean and variance of discrete uniform distribution are given by

$$\mu = E(X) = \sum_{i=1}^{m} x_i f(x_i) = \frac{1}{m}\sum_{i=1}^{m} x_i \quad (3.17)$$

and

$$\sigma^2 = E[(X-\mu)^2] = \sum_{i=1}^{m}(x_i - \mu)^2 f(x_i)$$

$$= \frac{1}{m}\sum_{i=1}^{m}(x_i - \mu)^2 \quad (3.18)$$

3.4.2 Continuous Uniform Distribution

Suppose the probability of an event occurring remains constant across a given time interval.

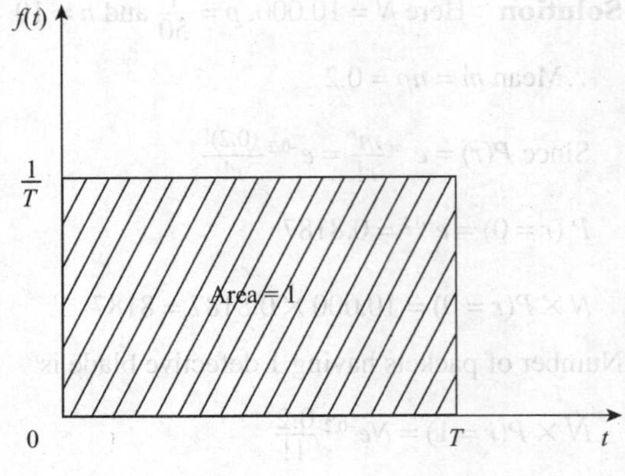

Figure 3.1 Probability density function for continuous uniform distribution.

The probability density function (pdf) $f(t)$ of such a distribution takes the form shown in Figure 3.1.

The area under $f(t)$ must be equal to 1 and so, if the interval is of length T, the height of the rectangle is $1/T$, then pdf for the continuous uniform distribution is given by

$$f(t) = \begin{cases} \frac{1}{T}, & 0 < t < T \\ 0, & \text{otherwise} \end{cases}$$

The probability that an event occurs in an interval $[a, b]$ is $\int_a^b f(t)\,dt$. If $0 \le a \le b \le T$, this probability is simply $\frac{b-a}{T}$.

Example 3.7

A random variable x has a uniform pdf with $T = 10$. Find the probability that

(a) $1 \le x \le 3$ (b) $1.6 \le x \le 9.3$

(c) $x \ge 2.9$ (d) $x < 7.2$

(e) $-1 < x < 2$ (f) $9.1 < x < 12.3$

Solution

$$P(a \le x \le b) = \int_a^b f(t)\,dt = \frac{b-a}{T}$$

(a) $P(1 \le x \le 3) = \frac{3-1}{10} = 0.2$

(b) $P(1.6 \le x \le 9.3) = \frac{9.3-1.6}{10} = 0.77$

(c) $P(x \ge 2.9) = 1 - P(0 \le x \le 2.9)$

$= 1 - \frac{2.9-0}{10} = 0.71$

(d) $P(x < 7.2) = \frac{7.2-0}{10} = 0.72$

(e) $P(-1 < x < 2) = \frac{2-0}{10} = 0.2$

(f) $P(9.1 < x < 12.3) = \frac{12.3-9.1}{10} = 0.32$

3.5 Exponential Distribution

Let a random variable have a Poisson distribution. For example, the random variable could be the number of customers arriving at a service point, the number of telephone calls received at a switchboard or the number of machines breaking down in a week. Then the time between events happening is a random variable which follows an exponential distribution. Note that whereas the number of events is a discrete variable, the time between events is a continuous variable.

Let t be the time between events happening. The exponential pdf $f(t)$ is given by

$$f(t) = \begin{cases} \alpha e^{-\alpha t}, & t \ge 0 \\ 0, & \text{otherwise} \end{cases} \quad (3.19)$$

where $\alpha > 0$.

Figure 3.2 Exponential probability density function for various values of α.

The probability of an event occurring in a time interval T is given by $\int_0^T f(t)\,dt$. Figure 3.2 shows $f(t)$ for different values of α. The expected value of the distribution is given by

$$\text{Expected value} = \mu = \int_0^\infty \alpha . t e^{-\alpha t}\,dt = \frac{1}{\alpha} \quad (3.20)$$

For example, if $f(t) = 3e^{-3t}$; t in secs. $t \ge 0$, then the mean time between events is $\frac{1}{3}$ sec.

That is, on average there are three events happening per second.

Example 3.8

The time between breakdowns of a machine follows an exponential distribution, with a mean of 17 days. Find the probability that a machine breaks down in a 15 day period.

Solution The mean time between breakdowns

$$= 17$$

$$= \frac{1}{\alpha}$$

$$\Rightarrow \alpha = \frac{1}{17}$$

Thus, the pdf $f(t)$ is given by

$$f(t) = \alpha e^{-\alpha t} = \frac{1}{17} e^{-t/17}, \quad t \ge 0$$

We require the probability that the machine breaks down in a 15 day period

$$P(0 \le t \le 15) = \int_0^{15} f(t)\,dt$$

$$= \int_0^{15} \frac{1}{17} e^{-t/17}\,dt$$

$$= \left[-e^{-t/17}\right]_0^{15} = 1 - e^{-15/17} = 0.5862$$

There is a 58.62% chance that the machine will break down in a 15 day period.

3.6 Normal Distribution

The normal probability distribution or simply the normal distribution is one of the most important and widely used. It is used to calculate the probable values of continuous random variables, e.g., weight, length, density and error measurement. It is defined by

$$N(\overline{X}, \sigma) = Y(X) = \frac{1}{\sigma\sqrt{2\pi}} \exp\left[-\frac{1}{2}\left(\frac{X-\overline{X}}{\sigma}\right)^2\right]$$

Here $\overline{X}$ = arithmetic mean and σ = standard deviation are the two parameters of the continuous distribution. Normal distribution is also known as Gaussian distribution. Discrete probability distributions such as binomial, Poisson, hypergeometric distributions can be approximated by the normal distribution. Sampling distributions t, F and χ^2 tend to be normal for large samples. Further, they are applicable in statistical quality control in industry.

The normal distribution is obtained if we let the number of trials n of a binomial distribution tend to infinity keeping p and q very small. The limit is approached more rapidly if p and q are nearly equal and are close to 0.5. In fact, using Stirling's formula, the following theorem can be proved.

Theorem A binomial pdf

$$P(x) = \binom{n}{x} p^x q^{n-x} \tag{3.22}$$

in which n becomes infinitely large, tends in the limit, to the normal density function

$$f(x) = \frac{1}{\sqrt{2\pi npq}} \exp\left[\frac{-(x-np)^2}{2npq}\right]$$

For a binomial distribution, the mean and standard deviation are given by

$$\mu = np \tag{3.23}$$

and

$$\sigma = \sqrt{npq} \tag{3.24}$$

and hence the normal frequency function becomes

$$f(x) = \frac{1}{\sigma\sqrt{2\pi}} \exp\left[-\frac{1}{2}\left(\frac{x-\mu}{\sigma}\right)^2\right] \tag{3.25}$$

where the variable x can assume all values from $-\infty$ to ∞.

The graph of the normal frequency function is called the normal curve (Figure 3.3). The normal curve is bell-shaped and is symmetrical about the mean μ. This curve is unimodal and its mode coincides with its mean μ. The two ends of this open curve extend to $+\infty$ *and* $-\infty$ along the positive and the negative parts of the x-axis, which is an asymptote to the curve. Since the $f(x)$ curve is symmetrical about $x = \mu$, its mean, median and mode are the same.

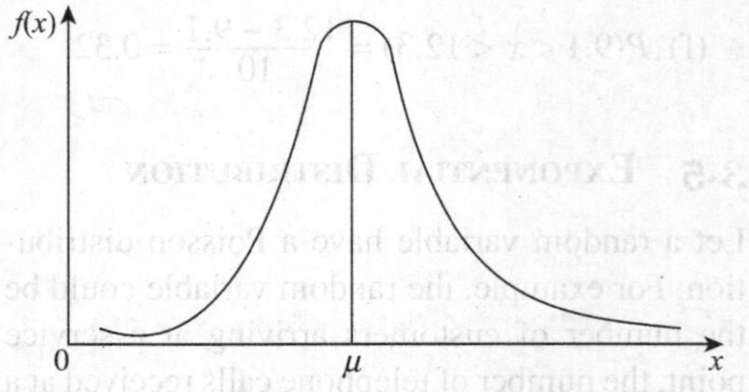

Figure 3·3 Normal curve.

Its points of inflexion are found to be $x = \mu \pm \sigma$, which are equidistant from the mean and are on either side of it. The total area under the normal curve and above the x-axis is unity.

The parameters μ and σ determine the position and the relative proportions of the normal curve.

If two populations defined by normal frequency functions have different means μ_1 and μ_2 but equal standard deviations $\sigma_1 = \sigma_2$ then their graphs appear as shown in Figure 3.4.

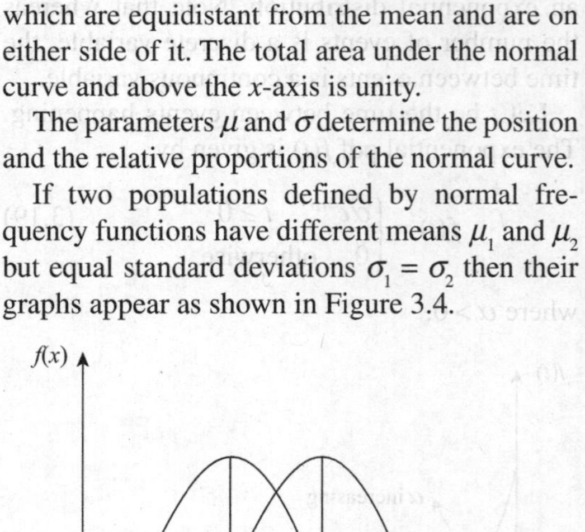

Figure 3·4 Normal frequency curves with unequal means but equal standard deviations.

On the other hand, if the two populations have equal means but unequal standard deviations σ_1 and σ_2 then their graphs would appear as shown in Figure 3.5.

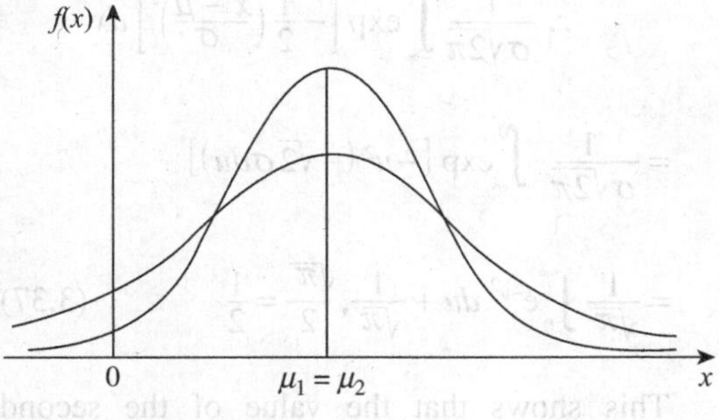

Figure 3.5 Normal frequency curves with equal means but with unequal standard deviations.

3.6.1 Characteristics of the Normal Distribution

Normal distribution is continuous The pdf of the normal distribution is

$$f(x) = \frac{1}{\sigma\sqrt{2\pi}} \exp\left[-\frac{1}{2}\left(\frac{x-\mu}{\sigma}\right)^2\right]$$

Hence the area under the curve above the x-axis is

$$\int_{-\infty}^{\infty} f(x)\,dx = \frac{1}{\sigma\sqrt{2\pi}} \int_{-\infty}^{\infty} \exp\left[-\frac{1}{2}\left(\frac{x-\mu}{\sigma}\right)^2\right] dx \tag{3.26}$$

Putting $\frac{x-\mu}{\sigma\sqrt{2}} = u$, we have $dx = \sigma\sqrt{2}\,du$ and the limits are the same so that we obtain

$$\int_{-\infty}^{\infty} f(x)\,dx = \frac{1}{\sigma\sqrt{2\pi}} \int_{-\infty}^{\infty} e^{-u^2}\,\sigma\sqrt{2}du$$

$$= \frac{1}{\sqrt{\pi}} \int_{-\infty}^{\infty} e^{-u^2}\,du = \frac{1}{\sqrt{\pi}}.\sqrt{\pi} = 1 \tag{3.27}$$

Therefore, $f(x) \geq 0$ and $\int_{-\infty}^{\infty} f(x)dx = 1$

Hence the normal distribution is continuous.

$$\int_{-\infty}^{\infty} e^{-u^2}\,du = 2\int_{0}^{\infty} e^{-u^2}\,du \because e^{-u^2} \text{ even in } (-\infty, \infty)$$

$$= \int_{0}^{\infty} t^{-\frac{1}{2}} e^{-t}\,dt \quad u^2 = t,\ 2udu = dt \text{ or } 2du = \frac{dt}{\sqrt{t}}$$

$$= \Gamma\left(\frac{1}{2}\right) = \sqrt{\pi}\ \left[\textit{Gamma Function}\right.$$

$$\left.\Gamma(p) = \int_{0}^{\infty} e^{-t}\,t^{p-1}\,dt\ (p > 0)\right]$$

3.6.2 Mean, Mode and Median of the Normal Distribution

Mean By definition, the arithmetic mean of a continuous distribution $f(x)$ is given by

$$\overline{X} = \frac{\int_{-\infty}^{\infty} xf(x)\,dx}{\int_{-\infty}^{\infty} f(x)\,dx} \tag{3.28}$$

The normal distribution with b and c as parameters is

$$N(b, c) = f(x)\,dx = \frac{1}{c\sqrt{2\pi}} \exp\left[-\frac{1}{2}\left(\frac{x-b}{c}\right)^2\right] \tag{3.29}$$

Then, since $\int_{-\infty}^{\infty} f(x)\,dx$ = area under the normal curve = 1,

$$\overline{X} = \int_{-\infty}^{\infty} x \frac{1}{c\sqrt{2\pi}} \exp\left[-\frac{1}{2}\left(\frac{x-b}{c}\right)^2\right] dx \tag{3.30}$$

[Put $(x - b)/c = z$ so that $x = b + cz$ and $dx = cdz$]

$$= \int_{-\infty}^{\infty} (b + cz) \frac{1}{c\sqrt{2\pi}} e^{-\frac{1}{2}z^2}\,cdz$$

$$= b\int_{-\infty}^{\infty} \frac{1}{c\sqrt{2\pi}} e^{-\frac{1}{2}z^2}\,dz + \frac{c}{\sqrt{2\pi}} \int_{-\infty}^{\infty} ze^{-\frac{b}{2}z^2}\,dz$$

$$= b \times 1 + c \times 0$$

$$\text{since} \int_{-\infty}^{\infty} f(x)dx = \int_{-\infty}^{\infty} \frac{1}{\sqrt{2\pi}} e^{-\frac{1}{2}z^2}\,dz = 1$$

$$\Rightarrow \overline{X} = b \tag{3.31}$$

and

$$\int_{-\infty}^{\infty} ze^{-\frac{1}{2}z^2}\,dz = 0$$

the integrand being an odd function.

Mode It is the value of x for which f is maximum. Thus, it is the solution of $f'(x) = 0$ and $f''(x) < 0$. For normal distribution, we have

$$f(x) = -\frac{1}{\sigma\sqrt{2\pi}} \exp\left[-1/2\left(\frac{x-\mu}{\sigma}\right)^2\right]$$

Taking logs on both sides

$$\log f(x) = \log \frac{1}{\sigma\sqrt{2\pi}} - \frac{1}{2\sigma^2}(x-\mu)^2 \quad (3.32)$$

Differentiating w.r.t. x, we get

$$\left(\frac{f'(x)}{f(x)}\right) = \frac{-1}{\sigma^2}(x-\mu) \Rightarrow f'(x) = -\frac{1}{\sigma^2}(x-\mu) f(x)$$

Differentiating w.r.t. x, we get

$$f''(x) = -\frac{1}{\sigma^2}[f(x) + (x-\mu) f'(x)]$$

$$= \frac{-f(x)}{\sigma^2}\left[1 - \frac{(x-\mu)^2}{\sigma^2}\right] \quad (3.33)$$

Now $f'(x) = 0 \Rightarrow x = \mu$

For this value of x, we get

$$f''(\mu) = \frac{-1}{\sigma^2} \times \frac{1}{\sigma\sqrt{2\pi}} < 0$$

Hence $x = \mu$ is the mode of the normal distribution.

Median We know that

$$\int_{-\infty}^{\infty} f(x)\, dx = 1 \quad (3.34)$$

Hence, if M denotes the median of the normal distribution, we must have

$$\int_{-\infty}^{M} f(x)\, dx = \frac{1}{2}$$

$$\Rightarrow \int_{-\infty}^{M} \frac{1}{\sigma\sqrt{2\pi}} \exp\left[-\frac{1}{2}\left(\frac{x-\mu}{\sigma}\right)^2\right] dx$$

$$= \frac{1}{2} \quad (3.35)$$

$$\Rightarrow \frac{1}{\sigma\sqrt{2\pi}} \int_{-\infty}^{M} \exp\left[-\frac{1}{2}\left(\frac{x-\mu}{\sigma}\right)^2\right] dx$$

$$+ \frac{1}{\sigma\sqrt{2\pi}} \int_{\mu}^{M} \exp\left[-\frac{1}{2}\left(\frac{x-\mu}{\sigma}\right)^2\right] dx$$

$$= \frac{1}{2} \quad (3.36)$$

To evaluate the first integral, put $\frac{x-\mu}{\sqrt{2}\sigma} = -u$ and $dx = -\sqrt{2}\sigma\, du$, and the limits are ∞, 0.

$$\therefore \frac{1}{\sigma\sqrt{2\pi}} \int_{-\infty}^{\mu} \exp\left[-\frac{1}{2}\left(\frac{x-\mu}{\sigma}\right)^2\right] dx$$

$$= \frac{1}{\sigma\sqrt{2\pi}} \int_{\infty}^{0} \exp[-u^2(-\sqrt{2}\sigma du)]$$

$$= \frac{1}{\sqrt{\pi}} \int_{0}^{\infty} e^{-u^2} du + \frac{1}{\sqrt{\pi}}, \frac{\sqrt{\pi}}{2} = \frac{1}{2} \quad (3.37)$$

This shows that the value of the second integral = 0.

This implies that $M = \mu$.

Note The mean, mode and median coincide in the case of the normal distribution. Hence the normal curve is symmetrical.

3.6.3 Variance of the Normal Distribution

$$\sigma^2 = 2\int_{-\infty}^{\infty} (x-\mu)^2 f(x)\, dx \quad (3.38)$$

$$\int_{-\infty}^{\infty} (x-\mu)^2 \frac{1}{\sigma\sqrt{2\pi}} \exp\left[-\frac{1}{2}\left(\frac{x-\mu}{\sigma}\right)^2\right] dx$$

$$= \frac{2}{\sigma\sqrt{2\pi}} \int_{0}^{\infty} (x-\mu)^2 \exp\left[-\frac{1}{2}\left(\frac{x-\mu}{\sigma}\right)^2\right] dx$$

since the integral is even, put $\frac{x-\mu}{\sqrt{2}\sigma} = u$ and $dx = -\sqrt{2}\sigma du$, and the limits are ∞, 0.

$$= \frac{2}{\sigma\sqrt{2\pi}} \int_{0}^{\infty} 2\sigma^2 u^2 e^{-u^2} \sqrt{2}\, \sigma du$$

$$= \frac{4\sigma^2}{\sqrt{\pi}} \int_{0}^{\infty} u^2 e^{-u^2} du$$

$$\left[\text{Put } u = \sqrt{t},\ du = \frac{1}{2\sqrt{t}} dt; \text{ the limits are the same}\right]$$

$$= \frac{4\sigma^2}{\sqrt{\pi}} \int_0^\infty t\, e^{-t} \frac{1}{2\sqrt{t}}\, dt = \frac{2\sigma^2}{\sqrt{\pi}} \int_0^\infty t^{3/2-1} e^{-t}\, dt$$

$$= \frac{2\sigma^2}{\sqrt{\pi}} \Gamma(3/2) = \frac{2\sigma^2}{\sqrt{\pi}} \frac{1}{2} \sqrt{\pi} = \sigma^2 \quad (3.39)$$

$$\because \Gamma(p+1) = p\,\Gamma(p), \Gamma\left(\frac{1}{2}\right) = \sqrt{\pi}$$

$$\therefore \text{Standard deviation} = \sqrt{\text{Variance}} = \sigma \quad (3.40)$$

3.6.4 Points of Inflexion of the Normal Curve

At a point of inflexion, we should have $f''(x) = 0$ and $f'''(x) \neq 0$.

As shown above,

$$f(x) = \frac{1}{\sigma\sqrt{2\pi}} \exp\left[-\frac{1}{2}\left(\frac{x-\mu}{\sigma}\right)^2\right] \quad (3.41)$$

$$f'(x) = -\frac{1}{\sigma^2}(x-\mu) f(x) \quad (3.42)$$

$$\Rightarrow \frac{f'(x)}{f(x)} = -\frac{1}{\sigma^2}(x-\mu) \quad (3.43)$$

$$f''(x) = -\frac{1}{\sigma^2}[f(x) + (x - \tau\, \mu) f'(x)]$$

$$= \frac{1}{\sigma^2} f(x) \left[1 - \frac{(x-\mu^2)}{\sigma^2}\right] \quad (3.44)$$

$$f'''(x) = -\frac{1}{\sigma^2} f(x) \left[0 - 2\frac{(x-\mu)}{\sigma^2}\right] - \frac{1}{\sigma^2} f'(x)$$

$$\left[1 - \frac{(x-\mu)^2}{\sigma^2}\right]$$

$$= \frac{1}{\sigma^2} f(x) \left\{2\frac{(x-\mu)}{\sigma^2} - \frac{f'(x)}{fx}\left[1 - \frac{(x-\mu)^2}{\sigma^2}\right]\right\}$$

$$= \frac{1}{\sigma^2} f(x) \left\{\frac{2(x-\mu)}{\sigma^2} + \frac{(x-\mu)}{\sigma^2}\left[1 - \frac{(x-\mu)^2}{\sigma^2}\right]\right\}$$

$$= \frac{(x-\mu) f(x)}{\sigma^4}\left[3 - \frac{(x-\mu)^2}{\sigma^2}\right] \quad (3.45)$$

Now

$$f'''(\mu \pm \sigma) = \frac{\sigma}{\sigma^4} f(\mu \pm \sigma)\,(3-1)$$

$$= \frac{2}{\sigma^2} \frac{1}{\sigma\sqrt{2\pi}} e^{-\frac{1}{2}} \neq 0 \quad (3.46)$$

since

$$f(x)\Big|_{x=\mu\pm\sigma} = \frac{1}{\sigma\sqrt{2\pi}} \exp\left[-\frac{1}{2}\left(\frac{x-\mu}{\sigma^2}\right)^2\right]$$

$$x = \mu \pm \sigma = \frac{1}{\sigma\sqrt{2\pi}} e^{-\frac{1}{2}}$$

3.6.5 Mean Deviation about the Mean

This is given by

$$= \int_{-\infty}^{\infty} |x-\mu| f(x)\, dx = \frac{1}{\sigma\sqrt{2\pi}} \int_{-\infty}^{\infty} |x-\mu|$$

$$\exp\left[-\frac{1}{2}\left(\frac{x-\mu}{\sigma}\right)^2\right] dx$$

$$= \frac{2}{\sigma\sqrt{2\pi}} \int_0^\infty |x-\mu| \exp\left[-\frac{1}{2}\left(\frac{x-\mu}{\sigma}\right)^2\right] dx$$

Put $\frac{x-\mu}{\sigma} = \mu$ and $dx = \sigma du$

$$= \frac{2\sigma^2}{\sigma\sqrt{2\pi}} \int_0^\infty |u|\, e^{-\frac{1}{2}\mu^2}\, du$$

Put $u = \sqrt{2t}$ and $du = \frac{1}{\sqrt{2t}}\, dt$

$$= \sigma\sqrt{\frac{2}{\pi}} \int_0^\infty e^{-t}\, dt = \sigma\sqrt{\frac{2}{\pi}}\,(-e^{-t})_0^\infty$$

$$= \sigma\sqrt{\frac{2}{\pi}} = \frac{4}{5}\sigma \cong 0.8\sigma \quad (3.47)$$

3.6.6 Normal Probability Integral

If X is a normal random variable with mean μ and variance σ^2, then the probability that random value of X will be between $X = \mu$ and $X = x_1$ is given by

$$P(\mu < X < x_1) = \int_\mu^{x_1} f(x)\, dx$$

$$= \frac{1}{\sigma\sqrt{2\pi}} \int_\mu^{x_1} \exp\left[\frac{1}{2}\left(\frac{x-\mu}{\sigma}\right)^2\right] dx$$

To evaluate the integral, put $\frac{x-\mu}{\sigma} = z$ or $x - \mu = \sigma z$ and $dx = \sigma dz$.

The limits are $x = \mu \Rightarrow z = 0$ and $x = x_1$

$$\Rightarrow z = z_1 = \frac{x_1 - \mu}{\sigma}$$

Now

$$P(\mu < X < x_1) = P(0 < z < z_1)$$

$$= \frac{1}{\sqrt{2\pi}} \int_0^{z_1} e^{-\frac{1}{2}z^2} dz$$

$$= \int_0^{z_1} \Phi(z)dz \qquad (3.48)$$

where $\Phi(z) = \frac{1}{\sqrt{2\pi}} e^{-\frac{1}{2}z^2}$ is the probability function of the standard normal variate $z = \frac{X-\mu}{\sigma}$. The definite integral $\int_0^{z_1} \Phi(z)dz$ is called the normal probability integral which gives the area under the standard normal curve between the ordinates at $z = 0$ and $z = z_1$ (Figure 3.6).

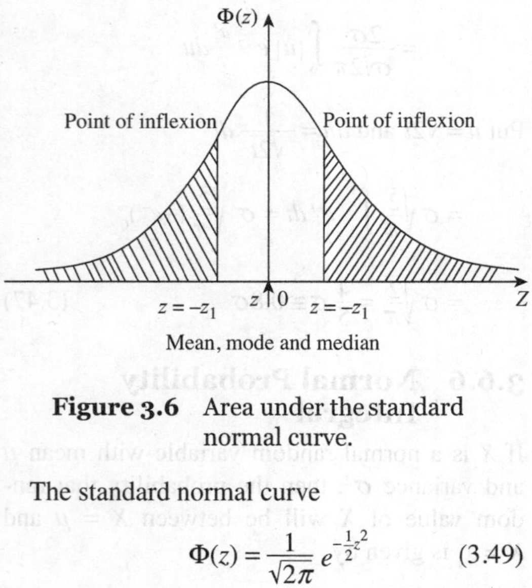

Figure 3.6 Area under the standard normal curve.

The standard normal curve

$$\Phi(z) = \frac{1}{\sqrt{2\pi}} e^{-\frac{1}{2}z^2} \qquad (3.49)$$

is symmetrical w.r.t. $\Phi(z)$ axis since $\Phi(z)$ remains unchanged when we replace z by $-z$. Thus, the arithmetic mean and the median of a normal frequency distribution coincide at its central point: The exponent of e in $\Phi(z)$ is, $-\frac{1}{2}z^2$ which is negative. Hence $\Phi(z)$ is maximum when $z = 0$. For all other values of z, $\Phi(z)$ is smaller since $e^{-\frac{1}{2}z^2} = \frac{1}{e^{\frac{1}{2}z^2}}$. Therefore, the maximum value of $\Phi(z)$ is

$$\Phi(0) = \frac{1}{\sqrt{2\pi}} = 0.3989$$

As z increases numerically, $e^{-\frac{1}{2}z^2}$ decreases and approaches zero when z becomes infinite. Thus, the z-axis is an asymptote to the standard normal curve.

Now, differentiating $\Phi(z)$ w.r.t. z,we get $\Phi(z) = z\Phi(z)$ and $\Phi''(z) = -\Phi(z) - z\Phi'(z) = (z^2 - 1)\Phi(z)$. Therefore $\Phi''(z) = 0 \Rightarrow z = \pm 1$. These are the points of inflexion. That is, the points at which the curve changes from concave downward to concave upward; and these points are situated at a unit distance from the $\Phi(z)$ axis, on either side of it.

3.6.7 Area under the Standard Normal Curve

The equation of the standard normal curve is

$$\Phi(z) = \frac{1}{\sqrt{2\pi}} e^{-\frac{1}{2}z^2} \qquad (3.50)$$

The area under this curve is given by

$$\int_{-\infty}^{\infty} \Phi(z)dz = \frac{1}{\sqrt{2\pi}} \int_{-\infty}^{\infty} e^{-\frac{1}{2}z^2} dz$$

$$= \sqrt{\frac{2}{\pi}} \int_0^{\infty} e^{-\frac{1}{2}z^2} dz, \qquad (3.51)$$

since the integrand is even in $(-\infty, \infty)$

$$= \frac{1}{\sqrt{\pi}} \int_0^{\infty} t^{-\frac{1}{2}} e^{-t} dt$$

where $\frac{1}{2}z^2 = t \Rightarrow z = \sqrt{2t}$ and $dz = \frac{1}{\sqrt{2t}} dt$

$$= \frac{1}{\sqrt{\pi}} \Gamma\left(\frac{1}{2}\right) = \frac{1}{\sqrt{\pi}} \sqrt{\pi} = 1$$

It follows, therefore, that the area under the standard normal curve $\Phi(z)$ from $z = z_1$ to $z = z_2$ is always less than 1 where z_1 and z_2 are finite.

Consequently, $\int_{z_1}^{z_2} \Phi(z)\,dz$ is always less than unity.

Because of the symmetry of the curve w.r.t. the $\Phi(z)$ axis, the area from any point $z = z_1$ to $+\infty$ is equal to the area from $-\infty$ to $-z_1$. Therefore,

$$\int_{z_1}^{\infty} \Phi(z)\,dz = \int_{-\infty}^{-z} \Phi(z)dz$$

Also, the area under the curve from $z = 1$ to $z = \infty$ is 0.1587. So, we have

$$\int_{-1}^{1} \Phi(z)\,dz = \int_{-\infty}^{\infty} \Phi(z)\,dz - \int_{-\infty}^{-1} \Phi(z)\,dz - \int_{1}^{\infty} \Phi(z)\,dz$$

$$= \int_{-\infty}^{\infty} \Phi(z)\,dz - 2\int_{1}^{\infty} \Phi(z)\,dz$$

$$= 1 - 2\,(0.1587) = 0.6826 \qquad (3.52)$$

In terms of statistics, this means that 68% of the normal variates deviate from their mean by less than one standard deviation. Similarly,

$$\int_{-2}^{2} \Phi(z)\,dz = 0.9544 \qquad (3.53)$$

$$\int_{-3}^{3} \Phi(z)\,dz = 0.9974 \qquad (3.54)$$

Thus, over 95% of the area is included between the limits −2 and 2 and over 99% of the area is included between −3 and 3 (Figure 3.7)

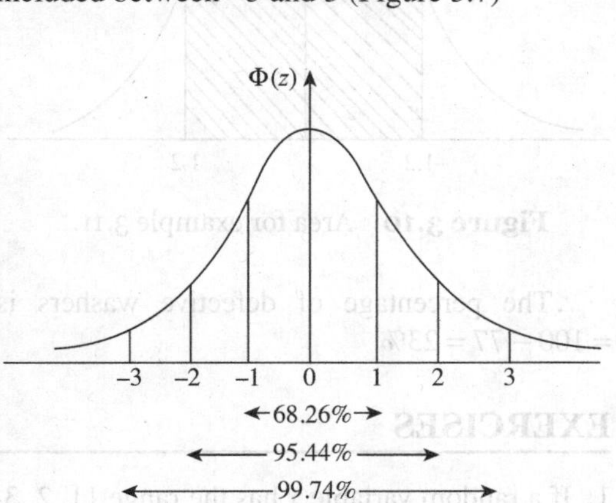

Figure 3.7 Percentages of area under the normal curve.

3.6.8 Fitting of Normal Distribution to Given Data

The equation of the normal curve fitting to given data is

$$y = f(x) = \frac{1}{\sigma\sqrt{2\pi}} \exp\left[-\frac{1}{2}\left(\frac{x-\mu}{\sigma}\right)^2\right], -\infty < x < \infty \qquad (3.55)$$

Therefore, first calculate the mean μ and the standard deviation σ. Then find the standard normal variate $Z = \dfrac{X-\mu}{\sigma}$ corresponding to the lower limits of each of the class interval. That is, determine, $z_1 = \dfrac{x_i' - \mu}{\sigma}$, where x_i' is the lower limit of the ith class. The final step is to calculate the area under the normal curve to the left of the ordinate $z = z_1$ (SRY), $\Phi(z_i)$ from the tables. Then areas for the successive class intervals are obtained by subtraction:

viz. $\Phi(z_i + 1) - \Phi(z_i)\ i = 1, 2, 3\ldots$

Then, expected frequency is

$$N = \left[\Phi(z_i + 1) - \Phi(z_i)\right] \qquad (3.56)$$

3.6.9 Application of Normal Distribution

Example 3.9

A certain type of storage battery lasts on an average for 3.0 years, with a standard deviation of 0.5 year. Assuming that the battery lives are normally distributed, find the probability that a given battery will last less than 2.3 years.

Solution First let us construct a diagram as in Figure 3.8 showing the given distribution of battery lives and the desired area.

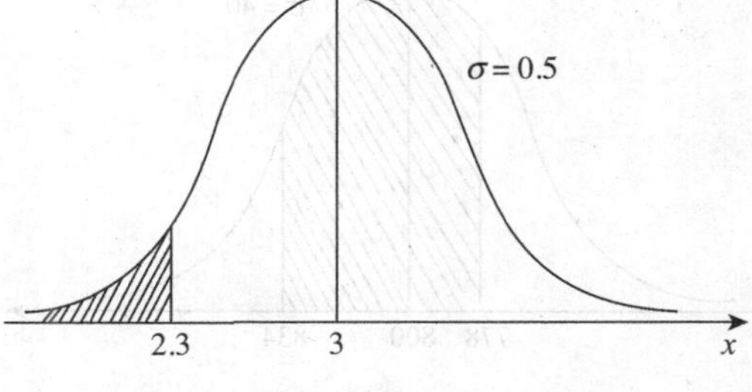

Figure 3.8 Area for example 3.9.

To find $P(X < 2.3)$, we have to evaluate the area under the normal curve to the left of 2.3. This is obtained by finding the area to the left of the corresponding z value. It is found from

$$Z = \frac{X-\mu}{\sigma} = \frac{2.3 - 3.0}{0.5} = 1.4$$

Using the relevant table, we have

$$P\,(X < 2.3) = P\,(Z < -1.4) = 0.0808$$

Example 3.10

An electrical firm manufactures light bulbs that have a life, before burnout, which is normally distributed with mean equal to 800 h and a standard deviation of 40 h. Find the probability that a bulb burns between 778 and 834 h.

Solution The distribution of light bulbs is shown in Figure 3.9.

Here $\mu = 800$ and $\sigma = 40$.

The z values corresponding to $x_1 = 778$ and $x_2 = 834$ are

$$z_1 = \frac{x_1 - \mu}{\sigma} = \frac{778 - 800}{40} = 0.55;$$

$$z_2 = \frac{x_2 - \mu}{\sigma} = \frac{834 - 800}{40} = 0.85$$

Hence,

$$P\,(778 < x < 834) = P\,(-\,0.55 < z < 0.85)$$

$$= P\,(z < 0.85) - P\,(z < -\,0.55)$$

$$= 0.8023 - 0.2912 = 0.5111$$

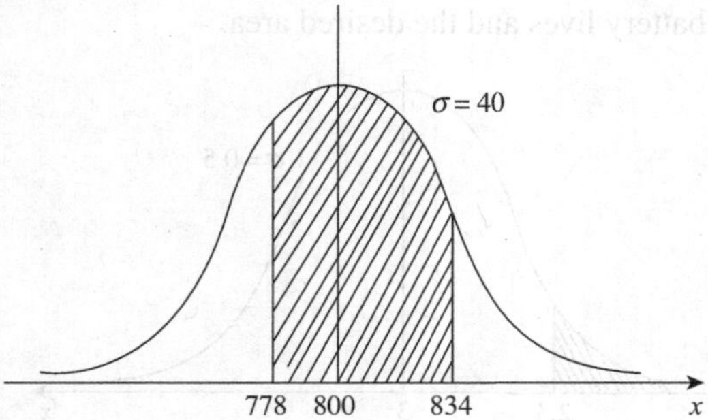

Figure 3.9 Area for example 3.10.

Example 3.11

The mean inside diameter of a sample of 200 washers produced by a machine is 0.502 in. and the standard deviation is 0.005 in. The purpose for which these washers are intended allows a maximum tolerance in the diameter of 0.496 – 0.508 in., otherwise, the washers are considered defective. Determine the percentage of defective washers produced by the machine, assuming the diameters are normally distributed.

Solution The distribution of washers is shown in Figure 3.10.

Here $n = 200$, $\sigma = 0.005$, and $\mu = 0.502$.

$$0.496 \text{ in. standard units} = \frac{0.496 - 0.502}{0.005}$$

$$= -1.2$$

$$0.508 \text{ in. standard units} = f(x) = kx^2\ (-1 \le x \le 1)$$

Proportion of non-defective washers = area under normal curve between $z = -1.2$ and $z = 1.2$

= 2 (area between $z = 0$ and $z = 1, 2$)

= 2 (0.3849) = 0.7698 or 77%

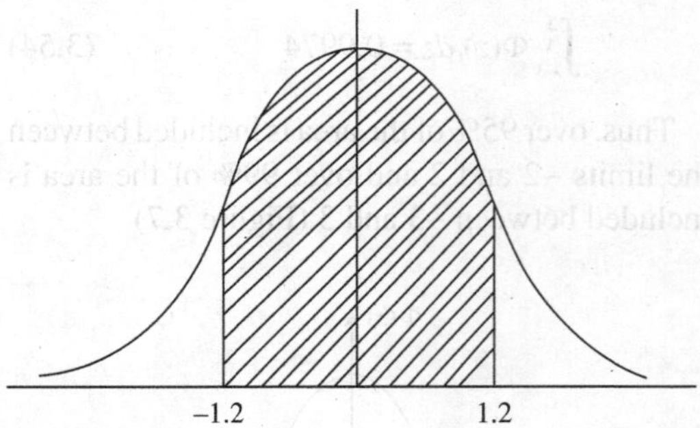

Figure 3.10 Area for example 3.11.

∴The percentage of defective washers is $= 100 - 77 = 23\%$

EXERCISES

1. If a random variable x has the range {1, 2, 3, ...} of $P\,(x = k)\frac{c^k}{k!}$ for $k = 1, 2, 3....$, then find c and $P(0 < x < 3)$.

Ans: $c = \log_e 2$ and
$P(0 < x, 3) = \dfrac{(\log e\, 2)^2 + 2\log_e 2}{2}$

2. If a random variable x has the range $\{0, 1, 2, 3, \ldots\}$ of $P(x = k) = \dfrac{c\,(k+1)}{2^k}$ for $k = 0, 1, 2, 3, \ldots$, then find c.

Ans: $c = \dfrac{1}{4}$

3. The probability distribution of a random variable x is given as follows:

$X = x$	0	1	2	3	4
$P(X = x)$	0.4	0.3	0.1	0.1	0.1

Find the mean and variance of x.

Ans: $\mu = 1.2$ and $\sigma^2 = 1.76$

4. Let x be a random variable such that $P(x = -2) = P(x = -1) = P(x = 2) = P(x = 1) = \dfrac{1}{6}$ and $P(x = 0) = \dfrac{1}{3}$. Find the mean and the variance of x.

Ans: $\mu = 0$ and $\sigma^2 = 5/3$

5. The probability distribution of a random variable x is given as follows:

$X = x$	1	2	3	4	5
$P(X = x)$	k	$2k$	$3k$	$4k$	$5k$

Find the value of k, the mean and the variance of x.

Ans: $k = \dfrac{1}{15}$, $\mu = \dfrac{11}{3}$ and $\sigma^2 = \dfrac{14}{9}$

6. A random variable x has the following probability distribution:

$X = x$	−2	−1	0	1	2	3
$P(X = x)$	0.1	k	0.2	$2k$	0.3	k

Find k, the mean and the variance of x.

Ans: $k = 0.1$, $\mu = 0.8$ and $\sigma^2 = 2.16$

7. A range of a random variable x is $\{0, 1, 2\}$. Given that $P(x = 0) = 3c^3$, $P(x = 1) = 4c - 10c^2$ and $P(x = 2) = 5c - 1$, where c is a constant, find the following:

(a) Value of c

(b) $P(x < 1)$

(c) $P(1 < x \le 2)$

(d) $P(0 < x \le 3)$.

Ans: (a) $\dfrac{1}{3}$, (b) $\dfrac{1}{9}$, (c) $\dfrac{2}{3}$ and (d) $\dfrac{8}{9}$

8. A pdf is given by $f(x) = 2e^{-2z}$, $0 \le z < \infty$

(a) If 200 measurement of z are made, how many on an average will be > 1?

(b) If 50% of measurements are less than k, find k.

Ans: (a) 27 and (b) 0.3466

9. A pdf is defined by

$$h(x) = \frac{3}{4}(1 - x^2) \quad -1 \le x \le 1$$

Find (a) $P(0 \le x \le 0.5)$

(b) $P(-0.3 \le x \le 0.7)$

(c) $P(|x| < 0.5)$

(d) $P(x > 0.5)$

(e) $P(x < 0.7)$

Ans: (a) 11/32, (b) 0.13775, (c) 11/16, (d) 21/32, (e) 0.21825

10. Let, $f(x) = kx^2$ $(-1 \le x \le 1)$, where k is a constant.

(a) If $f(x)$ is a pdf, find k.

(b) Calculate the probability that $x > 0.5$.

(c) If $P(x > c) = 0.6$, find c.

Ans: (a) 1.5, (b) 0.4375 and (c) − 0.5848

11. In a binomial distribution $n = 5$, the sum of the mean and the variance is 1.8. Find the distribution.

Ans: $\left(\dfrac{4}{5} + \dfrac{1}{5}\right)^5$

12. In a binomial distribution the sum and the difference of the mean and the variance

are 1.8 and 0.2, respectively. Find the parameters.

Ans: $n = 5$ and $p = 1/5$

13. If x is a binomial variate with the mean 10 and the variance 5 such that $P(x = 0) = P(x = 1)$, find the parameter x.

Ans: $2^{\frac{1}{20}} + 2^{\frac{20}{20}}$

14. If x is a Poisson variate such that $P(x = 0) = P(x = 1)$, find the parameter x.

Ans: $\lambda = 1$

15. If a random variable x has Poisson distribution with parameter 2, find $P(x > 3)$.

Ans: $1 - \frac{19}{3e^2}$

16. In a box containing 15 identical bulbs, 5 are defective. If 5 bulbs are drawn at random from the box with replacement, find the probability that

(a) None are defective

(b) Only one of them is defective

(c) At least one of them is defective

Ans: (a) $\frac{32}{243}$, (b) $\frac{80}{243}$ and (c) $\frac{211}{243}$

17. If x is a Poisson variate such that $2P(x = 0) = 3P(x = 1)$, find λ.

Ans: $\lambda = \frac{2}{3}$

18. In a Poisson distribution, $P(x=2) = P(x=3)$. Find the variance of x and $P(x = 4)$.

Ans: $\lambda = 3$ and $P(x = 4) = \frac{27}{8e^2}$

19. In a factory 2% of items are defective. By using Poisson distribution, find the probability of having more than 2 defective items in a sample of 100 items.

Ans: $1 - 5/e^2$

20. Given that $P(x = 2) = 45P(x = 6) - 3P(x = 4)$ for a Poisson variate x, find the probability that (a) $x \geq 1$ and (b) $x < 2$.

Ans: (a) 0.864 and (b) 0.408

21. If a bank receives on the average $a = 2$ bad cheques per day, what are the probabilities that it will receive

(a) 2 bad cheques on any given day

(b) 3 bad cheques over any two consecutive days

Ans: (a) 0.272 and (b) 0.192

22. Fit a Poisson distribution to the following data:

x_i	0	1	2	3	4
f_i	122	60	15	2	1
Ans:	121	61	15	2	0

$\left[\text{Hint: } \lambda = \Sigma \frac{f_i x_i}{N} = \frac{60 + 36 + 6 + 1}{200} = 0.5; e^{-0.5} = 0.61\right]$

23. Find the number of pages expected with 0–4 errors in 1000 pages of a book if, on the average, 2 errors are found for 5 pages.

x_i	0	1	2	3	4
f_i	109	65	22	3	1
Ans:	108.7	66.3	20.2	4.1	0.7

$\left[\text{Hint: } \lambda = \frac{65 + 44 + 9 + 4}{200} = 0.61\right]$

24. Find the probability that a random variable having the standard normal distribution will take on a value between 0.87 and 1.28.

Ans: $P(0.87 \leq z \leq 1.28) = 0.0919$

25. The mean of the height of students in a class is 158 cm with the standard deviation 20 cm. Find how many students' heights are between 150 and 170 cm, if there are 100 students in the class.

Ans: 38

26. Let x be a random variable with the standard normal distribution. Find the value of z_1 in each of the following cases:

(a) $P(0 \leq z \leq z_1) = 0.4938$

(b) $P(x \leq z_1) = 0.834$

(c) $P(z_1 \leq z \leq 2) = 0.2857$

Ans: (a) $z_1 = 2.5$, and (b) $z_1 = 0.97$ and (c) $z_1 = 0.5$

27. Let x be a normal variate with mean 30 and standard deviation 5. Then find

    (a) $P(26 \le x \le 40)$

    (b) $P(x \ge 45)$

Ans: (a) 0.7653 and (b) 0.0014

28. Ten cards are drawn from a deck of 2 cards. Find the probability of getting 2–5 diamonds using normal distribution.

Ans: 0.753

29. Find the probability of getting an even number 3–5 times when 10 dice are thrown simultaneously, using (a) binomial distribution and (b) normal distribution.

Ans: (a) 0.5683 and (b) 0.5684

30. The average grade for an examination is 74 and the standard deviation is 7. If 12% of the class are given A's and the grades follow a normal distribution, what is the lowest possible A and the highest possible B?

Ans: Lowest A is 83 and highest B is 82

31. A certain machine makes electrical resistors having a mean resistance of 40 ohms and a standard deviation of 2 ohms. Assuming that the resistance follows normal distribution and can be measured to any degree of accuracy, what percentage of resistors will have a resistance exceeding 43 ohms.

Ans: 6.68%

32. Find k such that $P(k < T < -1.761) = 0.045$ for a random sample of size 15 selected from a normal distribution and $T = \dfrac{\bar{x}-\mu}{s/\sqrt{\pi}}$

Ans: $k = -t_{0.005} = -2.977$ and $P(-2.977 < T < -1.761) = 0.045$

33. For an f distribution, find

    (a) $f_{0.05}$ with $\upsilon_1 = 7$ and $\upsilon_2 = 15$

    (b) $f_{0.09}$ with $\upsilon_1 = 19$ and $\upsilon_2 = 24$

Ans: (a) 2.71 and (b) 0.47

Continuous Uniform Distribution

34. A random variable t has a uniform pdf with $T = 1.5$. Find the probability that

    (a) $0.7 \le t \le 1.3$

    (b) $1 \le t \le 2$

    (c) $|t| < 0.5$

    (d) $|t| < 1$

Ans: (a) 0.4, (b) 0.3333, (c) 0.3333 and (d) 0.3333

Exponential Distribution

35. The mean time between breakdowns for a machine is 400 h. Find the probability that the time between breakdowns for the machine is

    (a) Greater than 450 h

    (b) Less than 350 h

Ans: (a) 0.3247 s and (b) 0.5831 s

MULTIPLE CHOICE QUESTIONS

1. The mean of the binomial distribution with n observations and the probability of success P is

   (a) pq (b) np (c) $\sqrt{np}$ (d) $\sqrt{pq}$

Ans: (b)

2. If the mean of a Poisson distribution is M then the standard deviation of this distribution is

   (a) m^2 (b) $\sqrt{m}$ (c) m (d) $2m$

Ans: (b)

3. The standard deviation of the binomial distribution is

(a) $\sqrt{npq}$ (b) $\sqrt{np}$ (c) npq (d) pq

Ans: (a)

4. In a Poisson distribution, if $2P(x = 1) = P(x = 2)$ then the variance is

(a) 0 (b) −1 (c) 4 (d) 2

Ans: (c)

5. If the probability distribution of a random variable x is

x	0	1	2	3	4
$P(x)$	0.5	k	−1	$3k$	k

Then the value of k is

(a) 0.3 (b) 0.2 (c) 0.1 (d) 0

[Hint: $0.5 + k - 1 + 3k + k = 1 \Rightarrow 5k = 0.5 \Rightarrow k = 0.1$]

Ans: (c)

6. A random variable x has the following probability distribution:

x	1	2	3	4
$P(x)$	c	$2c$	$3c$	$4c$

Then c is

(a) 0.2 (b) 0.3 (c) 0.4 (d) 0.1

$\left[\text{Hint: } 10c = 1 \Rightarrow c = \frac{1}{10} = 0.1\right]$

Ans: (d)

7. The probability distribution of a random variable x is as follows:

$X = x_i$	1	2	3	4
$P(X = x_i)$	$2k$	$4k$	$3k$	k

Then k is

(a) 0.4 (b) 0.1 (c) 0.2 (d) 0.3

[Hint: $2k + 4k + 3k + k = 10k = 1 \Rightarrow k = 0.1$]

Ans: (b)

8. A random variable x has the range {1, 2, 3, …}. If $P(x = r) = \frac{c^r}{r!}$ for $r = 1, 2, 3,..$ then c is

(a) $e2$ (b) $2e$ (c) $\log e2$ (d) 0.5

$\left[\text{Hint: } \sum_{n=1}^{\infty} \frac{c^r}{r!}\ \sum_{n=0}^{\infty} \frac{c^r}{r!} - 1 = 1 \Rightarrow e^c = 2 \Rightarrow c = \log e^2\right]$

Ans: (c)

9. The probability distribution of a random variable x is as follows:

$X = x_i$	1	2	3
$P(X = x_i)$	1/4	1/8	5/8

Then its mean is

(a) 19/8 (b) 5/4 (c) 1 (d) 4/5

Ans: (a)

10. A random variable x has the following probability distribution:

x	1	2	3	4
$P(x)$	0.15	0.23	0.12	0.10

x	5	6	7	8
$P(x)$	0.20	0.08	0.07	0.05

For the events E = {x is a prime} and $F = \{x < 4\}$, the probability $P(E \cup F)$ is

(a) 0.87 (b) 0.50 (c) 0.35 (d) 0.77

$\left[\text{Hint: } P(E \cup F) = P(E) + P(F) - P(E \cap F) = 0.62 + 0.50 - 0.35 = 0.77\right]$

Ans: (d)

11. If x is a random variable with the following distribution:

$X = x$	0	1	2	3
$P(X = x)$	k	$3k$	$3k$	k

The values of k and variance are

(a) $\frac{1}{8}, \frac{1}{4}$ (b) $\frac{3}{8}, \frac{1}{4}$ (c) $\frac{1}{8}, \frac{3}{4}$ (d) 1

$\left[\text{Hint: } k + 3k + 3k + k = 8k = 1 \Rightarrow k = \frac{1}{8}\right.$

$\mu = 0 \times \frac{1}{8} + 1 \times \frac{3}{8} + 2 \times \frac{3}{8} + 3 \times \frac{1}{8} = \frac{3}{2}$

$$\sigma^2 = 0^2 \times \frac{1}{8} + 1^2 \times \frac{3}{8} + 2^2 \times \frac{3}{8} + 3^2 \times \frac{1}{8}$$

$$= \frac{3+12+9}{8} = \frac{24}{8} = 3 \Rightarrow 3 - \frac{9}{4} = \frac{3}{4}\Big]$$

Ans: (c)

12. A random variable x has the following distribution:

$X = x_i$	0	1	2	3
$P(X = x_i)$	2/6	3/6	0/6	1/6

The mean and variance are

(a) 0, 1 (b) $\frac{3}{6}, \frac{2}{6}$ (c) 1, 1 (d) $\frac{3}{6}, 0$

[Hint:

$$\mu = 0 \times \frac{2}{6} + 1 \times \frac{3}{6} + 2 \times \frac{0}{6} + 3 \times \frac{1}{6}$$

$$= \frac{0+3+0+3}{6} = 1$$

$$\sigma^2 = 0^2 \times \frac{2}{6} + 1^2 \times \frac{3}{6} + 2^2 \times \frac{0}{6} + 3^2 \times \frac{1}{6}$$

$$= \frac{0+3+0+9}{6} = 2 - 1 = 1\Big]$$

Ans: (c)

13. A random variable x takes on the values 0, 1 and 2. If $P(x = 1) = P(x = 2)$ and $P(x = 0) = 0.4$, then the mean of the random variable x is

(a) 0.2 (b) 0.5 (c) 0.7 (d) 0.9

[Hint: $0.4 + p + p = 1$, $\therefore p = 0.3 \Rightarrow \mu = 0 \times 0.4 + 1 \times 0.3 + 2 \times 0.3 = 0.9$ s]

Ans: (d)

14. In a binomial distribution $n = 20$ and $q = 0.75$, then the mean is

(a) 5 (b) 15 (c) 10 (d) 7.5

[Hint: $p = 1 - q = 1 - 0.75 = 0.25 \Rightarrow \mu = np = 20 \times 0.25 = 5$]

Ans: (a)

15. For a binomial distribution, the probability of getting a success is ¼ and the variance is 3. Then its mean is

(a) 12 (b) 10 (c) 4 (d) 6

[Hint: $p = \frac{1}{4}$ and $q = \frac{3}{4}$

$\sigma^2 = npq = 3$, $\therefore n \times \frac{1}{4} \times \frac{3}{4} = 3 \Rightarrow n = 16$

$\mu = np = 16 \times \frac{1}{4} = 4$]

Ans: (c)

16. For a binomial distribution $n = 10$ and $q = 0.4$, then mean is

(a) 1 (b) 4 (c) 6 (d) 10

[Hint: $\mu = np = 10\,(1 - 0.4) = 6$]

Ans: (c)

17. The mean for a binomial distribution is 6 and standard deviation is $\sqrt{2}$ then

(a) ${}^9C_\lambda \left(\frac{2}{3}\right)^\lambda \left(\frac{1}{3}\right)^{9-\lambda}$ (b) ${}^9C_\lambda \left(\frac{1}{3}\right)^\lambda \left(\frac{2}{3}\right)^{9-\lambda}$

(c) ${}^{12}C_\lambda \left(\frac{2}{3}\right)^\lambda \left(\frac{1}{3}\right)^{12-\lambda}$ (d) ${}^{12}C_\lambda \left(\frac{1}{3}\right)^\lambda \left(\frac{2}{3}\right)^{9-\lambda}$

$P(x = \lambda)$ is

(a) ${}^9C_\lambda \left(\frac{2}{3}\right)^\lambda \left(\frac{1}{3}\right)^{9-\lambda}$ (b) ${}^9C_\lambda \left(\frac{1}{3}\right)^\lambda \left(\frac{2}{3}\right)^{9-\lambda}$

(c) ${}^{12}C_\lambda \left(\frac{2}{3}\right)^\lambda \left(\frac{1}{3}\right)^{12-\lambda}$ (d) ${}^{12}C_\lambda \left(\frac{1}{3}\right)^\lambda \left(\frac{2}{3}\right)^{9-\lambda}$

Ans: (a)

18. The probability of a man hitting a target is $\frac{1}{4}$. If he fixes 7 times, the probability of hitting the target at least twice is

(a) $1 - \frac{5}{4}\left(\frac{3}{4}\right)^6$ (b) $1 - \frac{15}{2}\left(\frac{3}{4}\right)^6$

(c) $1 - \frac{5 \times 3^5}{6}$ (d) $1 - \left(\frac{3}{9}\right)^6$

Ans: (a)

19. If the mean of Poisson distribution is $\frac{1}{2}$, the ratio of $P(x = 3)$ to $P(x = 2)$ is

(a) 1:2 (b) 1:4 (c) 1:6 (d) 1:8

[Hint: $P(x = 3) : P(x = 2) = \frac{e^{-\lambda}\lambda^3}{3!} : \frac{e^{-\lambda}\lambda^2}{2!} \Rightarrow \lambda = 3$]

Ans: (c)

20. In a Poisson distribution, $P(x=2) = P(x=3)$. Then the variance of x such that $P(x = 4) = -e^{-3}$ is

(a) $\frac{9}{16}$ (b) $\frac{27}{16}$ (c) $\frac{9}{8}$ (d) $\frac{27}{8}$

[Hint: $P(x = r) = \frac{e^{-\lambda}\lambda^r}{r!}$ for $r = 0, 1, 2, ...$

$P(x = 2) = P(x = 3) \Rightarrow \frac{e^{-\lambda}\lambda^r}{2!} = \frac{e^{-\lambda}\lambda^r}{3!}$

$\Rightarrow \lambda = 3$

$P(x = 4) = \frac{e^{-3}3^4}{4!} = \frac{27}{8}e^{-3}$]

Ans: (d)

21. A random variable x has Poisson distribution with mean 2. Then $P(x > 1.5)$ is

(a) $2e^{-2}$ (b) 0 (c) $1-2e^{-2}$ (d) $3e^{-2}$

[Hint: $\lambda = 2$ and $P(x = r) = \frac{e^{-\lambda}\lambda^r}{r!} = \frac{e^{-2}2^r}{r!} =$

$P(x = 1) = \frac{e^{-2}2^1}{1!} = 2e^{-2}$

$P(x > 1.5) = 1 - P(x = 1) = 1 - 2e^{-2}$]

Ans: (c)

22. In a Poisson distribution, $P(x = 0)$ is twice $P(x = 1)$. Then the mean is

(a) 1 (b) 2 (c) 0.5 (d) 1.5

[Hint: $P(x = r) = \frac{e^{-\lambda}\lambda^r}{r!}$ for $r = 0, 1, 2, ...$

$P(x = 0) = \frac{e^{-\lambda}\lambda^0}{0!} = e^{-\lambda}$

$P(x = 1) = \frac{e^{-\lambda}\lambda^1}{0!} = \lambda e^{-\lambda}$

$e^{-\lambda} = 2\lambda e^{-\lambda} \Rightarrow \lambda = \frac{1}{2} = 0.5$]

23. The number of vehicles passing a certain point on a road per minute is Poisson distributed with mean 4. Find the probability that 5 vehicles pass in a minute.

(a) $\frac{e^{-4}5^4}{5!}$ (b) $\frac{e^{-4}5^5}{5!}$

(c) $\frac{e^{-4}5^4}{4!}$ (d) $\frac{e^{-4}4^5}{4!}$

[Hint: $\lambda = 4$ and $r = 5 \Rightarrow P(x = r) = \frac{e^{-\lambda}\lambda^r}{r!}$

$\Rightarrow P(x = 5) = \frac{e^{-4}4^5}{5!}$]

Ans: (b)

24. If

$$f(x) = \begin{cases} cxe^{-4}, & 0 < x < \infty \\ 0, & \text{otherwise} \end{cases}$$

is a pdf of a continuous random variable, then c is

(a) 2 (b) 0 (c) 1 (d) 4

Hint: $\int_0^\infty cxe^{-x}dx = c(-xe^{-x} - e^{-x})_0^\infty = c = 1$

Ans: (c)

25. A binomial distribution has mean 4 and variance 3, then $P(x \geq 1)$ is

[Hint: $np = 4, npq = 3 \Rightarrow q = \frac{3}{4}, p = 1 - \frac{3}{4} = \frac{1}{4}$ and $\therefore n = 16$]

$P(x \geq 1) = 1 - P(x = 0) = 1 - q^n = 1 - \left(\frac{3}{4}\right)^{16}$

(a) $1 - \left(\frac{3}{4}\right)^{16}$ (b) $1 - \left(\frac{3}{4}\right)^6$

(c) $1 - \left(\frac{1}{4}\right)^{16}$ (d) $1 - \left(\frac{1}{4}\right)^6$

Ans: (a)

26. If x is a standard normal variate then $P(0 \leq x \leq 1.2)$ is

(a) 0.5 (b) 0.3849 (c) 0.1151 (d) 1

Ans: (b)

FILL IN THE BLANKS

1. If the probability distribution of the random variable x is as follows:

$X = x_i$	0	1	2	3	4	5
$P(X = x_i)$	$\frac{1}{25}$	$\frac{2}{25}$	$\frac{3}{25}$	$\frac{1}{5}$	$\frac{1}{5}$	$\frac{9}{25}$

Then $P(1 < x < 4) =$ _____.

Ans: $\frac{8}{25}$

2. In Question 1, $P(2 \le x \le 4) =$ _______.

Ans: $\frac{13}{25}$

3. In Question 1, the mean of the probability distribution is _____.

Ans: $\frac{88}{25}$

4. In Question 1, the variance of the probability distribution is _____.

Ans: 27

5. The probability distribution of a random variable x is as follows:

$X = x$	0	1	2	3	4
$P(X = x)$	0.4	0.3	0.1	0.1	0.1

The mean of the probability distribution is _____.

Ans: 1.2

6. In Question 5, the variance of the distribution is _____.

Ans: 1.76

7. Let x be a random variable such that $P(x = -2) = P(x = -1) = P(x = 2) = P(x = 1) = \frac{1}{6}$ and $P(x = 0) = \frac{1}{3}$. Then the mean of the probability distribution is _____.

Ans: 0

8. In Question 8, the variance is _____.

Ans: 5/3

9. If x follows Poisson distribution such that $P(x = 1) = 3P(x = 2)$, then the variance is _____.

Ans: 2/3

10. If a random variable x follows Poisson distribution such that $eP(x = 2) = 2P(x = 1)$, then $P(x = 0) =$ _____.

Ans: $e^{-\frac{4}{3}}$

11. A binomial distribution has mean 20 and variance 15. Then $p =$ _____.

Ans: ¼

[Hint: Mean $= np = 20$ and variance $= npq = 15 \Rightarrow q = \frac{15}{20} = \frac{3}{4}$ and $p = 1 - \frac{3}{4} = \frac{1}{4}$]

12. If x is a Poisson variate such that $P(x = 0) = P(x = 1)$, then the parameter is _______.

Ans: 1

[Hint: $P(x = r) = \frac{e^{-\lambda}\lambda^r}{r!}$ for $r = 0, 1, 2, \ldots$

$P(x = 0) = P(x = 1) \Rightarrow \frac{e^{-\lambda}\lambda^0}{0!} = \frac{e^{-\lambda}\lambda^1}{1!} \Rightarrow \lambda = 1$]

13. If the range of a random variable x is $\{0, 1, 2, 3, \ldots\}$ with $P(x = k) = \frac{(k+1)a}{3^k}$ for $k \ge 0$ then a is _____.

Ans: $\frac{4}{9}$

[Hint: $\Sigma r^{k+1} = \frac{1}{1-r} \Rightarrow s = \frac{d}{dr}\Sigma r^{k+1} = \Sigma(k+1)r^k$

$= \frac{1}{(1-r)^2} = \frac{1}{\left(1-\frac{1}{2}\right)^2} = \frac{9}{4}$ where $r = \frac{1}{3}$ as $= \frac{9}{4}a$

$= 1 \Rightarrow a = \frac{4}{9}$]

14. The mean and variance of a binomial distribution are 4 and 2, respectively. Then the probability of 2 successes is _____.

Ans: $\frac{7}{64}$

[Hint: Mean = $np = 4$ and variance = $npq = 2 \Rightarrow q = \frac{2}{4} = \frac{1}{2}$, $p = \frac{1}{2}$ and $n = 4 \times 2 = 8$

$P(2 \text{ successes}) = \frac{8}{2}\left(\frac{1}{2}\right)^2\left(\frac{1}{2}\right)^{8-2} = \frac{7}{64}$]

15. If x is a random Poisson variate such that $\alpha = P(x = 1) = P(x = 2)$ then $P(x = 4)$ is ______.

Ans: $\frac{1}{3}\alpha$

16. If 3% of electric bulbs manufactured by a company are defective, then the probability that a sample of 100 bulbs are not defective is ______.

Ans: e^{-3}

17. The mean and variance of a binomial distribution are 4 and, 3 respectively. Then $P(x \geq 1)$ is ______.

Ans: $1 - \left(\frac{3}{4}\right)^{16}$

[Hint: $np = 4, npq = 3 \Rightarrow q = \frac{3}{4}, p = 1 - \frac{3}{4} = \frac{1}{4}$ and $\therefore n = 16$]

$P(x \geq 1) = 1 - P(x = 0) = 1 - q^n = 1 - \left(\frac{3}{4}\right)^{16}$

18. If $\mu = 1$, the probability that $P(x = 1)$ if x is a Poisson variate is ______.

Ans: $\frac{1}{e}$

19. If x is a Poisson variate which is binomial at $x = 3$ and at $x = 4$, then $\mu =$ ______.

Ans: 6

20. If the variance of a Poisson distribution is 2, then $P(x = 0)$ is ______.

Ans: 0.135

Sampling Distributions

4

4.1 Introduction

In statistics, we deal with the method of collection, classification and analysis of quantitative data for drawing inferences and making decisions. Instead of examining the entire collection called the *population*, which may be difficult or impossible, we may examine a small part of it called a *sample*. We deal with a particular kind of sample called random sample. We do this with the aim of drawing certain inferences about the population from results found in the sample, a process known as *statistical inference*. The process of obtaining samples is called *sampling*. The distribution of a statistic calculated on the basis of random sample is called *sampling distribution* and it is basic to all of statistical inference.

4.2 Population and Sample

A *variable* is a characteristic that varies or changes over time and/or for different individuals or objects under consideration—e.g., body temperature of an individual at different times and body temperature of different individuals at a particular time.

An *experimental unit* is the individual or object on which a variable is measured. A simple measurement or data value results when a variable is actually measured on an experimental unit.

If a measurement is generated in every experiment unit in the entire collection, the resulting data set constitutes the *population* of interest. Any smaller subset of measurements is a *sample*.

A *population* is the set of all measurements of interest to the investigator.

A *sample* is a subset of measurements selected from the population of interest.

The size of the population N is the number of objects or observations in the population. The size of a sample is denoted by n. If $n \geq 30$, the sampling is said to be large and if $n < 30$, the sampling is said to be small or exact.

Statistical measures or constants obtained from the population such as mean and variance are known as population parameters or simply parameters; statistical quantities calculated from sample observations are known as sample statistics or briefly statistics.

Population mean, population standard deviation (SD) and population proportion are denoted by μ, s and p respectively; while $\overline{X}$, s and p stand for sample mean, sample SD and sample proportion.

Population $f(x)$ means a population whose probability distribution is $f(x)$.

4.2.1 Sampling with or without Replacement

In sampling with replacement from a population of size N, the probability of drawing a unit at each draw remains $\frac{1}{N}$. Thus, it can be considered theoretically as sampling from infinite population. In this case, we can draw N^n samples.

In sampling without replacement, a unit cannot be drawn more than once so that the number of samples that could be drawn is $\binom{N}{n}$.

Hence, the probability of drawing a unit from a population of size N at rth draw is $\frac{1}{N-r+1}$.

4.2.2 Sample Mean and Sample Variance

Let $x_1, x_2, x_3, \ldots, x_n$ be a random sample of size n from a population. Then

$$\text{Sample mean} = \overline{X} = \frac{1}{n}\sum_{i=1}^{n} x_i \tag{4.1}$$

$$\text{Sample variance} = s^2 = \frac{1}{n-1}\sum_{i=1}^{n}(x_i - \overline{X})^2 \tag{4.2}$$

Sample SD is the positive square root of the sample variance.

4.3 SAMPLING DISTRIBUTION

Suppose all possible samples of size n are drawn from a given finite population of size N. Then the total number of all possible samples each of size n that could be drawn from the population is

$$\binom{N}{n} = \frac{N!}{n!(N-n)!} = m \text{ (say)}$$

Calculate a statistic S (such as the mean, SD, median and mode) for each of these samples using the sample data $x_1, x_2, \ldots, x_n$ by $S = S(x_1, x_2, x_3, \ldots, x_n)$.

Sample number = 1, 2, 3, …, m
Statistic $S = S_1, S_2, S_3, \ldots, S_m$

Sampling distribution of the statistic is the set of values $\{S_1, S_2, \ldots, S_m\}$ of the statistic S obtained one for each sample. Thus, sampling distribution describes how a statistics will vary from one sample to the other of the same size. Though all the m samples are drawn from the given population, the members included in different samples are different. The variations in the values of the statistics attributed to chance are known as *sampling fluctuations*.

If the number of samples, each of the same size n, is infinitely large, then the probability distribution of the statistic is the sampling distribution of the statistic.

If the statistic S is the mean, variance or proportions then the corresponding distribution of the statistics is called the sampling distribution of mean, variance or proportions respectively.

Mean of the sampling distribution of S is

$$\overline{S} = \frac{1}{m}\sum_{i=1}^{m} S_i \tag{4.3}$$

Variance of the sampling distribution of S is

$$V(s) = \frac{1}{m}\sum_{i=1}^{m}(S_i - \overline{S})^2 \tag{4.4}$$

4.3.1 Standard Error (SE)

It is the SD of the sampling distribution of a statistic S. It gives an index for the precision of the estimate of the parameters. It decreases with the increase of the sample size n. It plays an important role in tests of hypothesis.

Sampling distribution of a statistic S enables us to find information about the corresponding population parameters.

4.3.2 Degrees of Freedom (υ)

The number of degrees of freedom (dof) of a statistic is a positive integer v which is equal to $(n - k)$, where n is the number of independent observations of the random sample and k is the number of population parameters which are calculated using the sample data. Thus, $v = n - k$, i.e., the difference between n the sample size and k the number of independent constraints imposed on the observations in the sample.

4.4 SAMPLING DISTRIBUTION OF MEANS (σ KNOWN)

The sampling distribution of means (SDM) $\overline{X}$ is the probability distribution of $\overline{X}$.

4.4.1 Finite Population of Size N

Consider a finite population of size N with mean μ and SD σ. Draw all possible samples of size n without replacement from this population. Let $\mu_{\bar{x}}$ and $\sigma_{\bar{x}}$ denote the mean and SD of the SDM. If $N > n$ then

$$\mu_{\bar{x}} = \mu \tag{4.5}$$

and

$$\sigma_{\bar{x}} = \frac{\sigma}{\sqrt{N}}\sqrt{\frac{N-n}{N-1}} \quad (4.6)$$

where $(N - n)/(N - 1)$ is known as the finite population correction factor.

4.4.2 Infinite Population

Suppose the samples are drawn from an infinite population or sampling is done with replacement, then

$$\mu_{\bar{x}} = \mu \quad (4.7)$$

and

$$\sigma_{\bar{x}} = \frac{\sigma}{\sqrt{n}} \quad (4.8)$$

The SE of mean $\sigma_{\bar{x}} = \frac{\sigma}{\sqrt{n}}$ measures the reliability of the mean as an estimate of the population mean μ.

Standardized sample mean $= Z = \frac{\bar{X} - \mu}{\sigma/\sqrt{n}}$ (4.9)

4.4.3 Non-normal Population (Large Sample)

Consider a population with unknown (non-normal) distribution with population mean μ and population variance σ, both finite. Let the population be finite or infinite. In case the population is finite, let its size be N which is at least twice the sample size n. Draw all possible samples of size n. Then the sampling distribution of $\bar{X}$ is approximately normally distributed with mean $\mu_{\bar{x}} = \mu$ and variance $\sigma^2_{\bar{x}} = \sigma^2/n$ provided the sample size $n \geq 30$.

4.4.4 Central Limit Theorem

If n is large, the sampling distribution of $\bar{X}$ is approximately normal with mean μ and variance σ^2/n regardless of the form of the population distribution.

The central limit theorem stated as follows, without proof, asserts this fact.

Theorem If $\bar{X}$ is the mean of a sample of size n drawn from a population with mean μ and finite variance σ^2, then the standardized sample mean is

$$Z = \frac{\bar{X} - \mu}{\sigma/\sqrt{n}}$$

which is a random variable whose distribution function approaches that of the standard normal distribution $N(z, 0, 1)$ as $n \to \infty$.

4.4.5 Normal Population (Small Sample)

Sampling distribution of $\bar{X}$ is normally distributed even for small samples of size $n < 30$ provided sampling is from normal population.

Example 4.1

A population consists of four numbers: 3, 4, 5 and 6. Consider all possible distinct samples of size 2 with replacement. Find

(a) Population mean

(b) Population SD

(c) SDM

(d) Mean of SDM

(e) SD of SDM

Verify (c) and (e) directly from (a) and (b) by using appropriate formula.

Solution

(a) Mean of population

$$\mu = \frac{\sum x_i}{n} = \frac{3+4+5+6}{4} = \frac{18}{4} = 4.5$$

(b) SD of Population

$$\sigma = \sqrt{\frac{\sum(x_i - \bar{x})}{n}} = \sqrt{\frac{\sum(X_i - \bar{X})^2}{n}}$$

$$= \sqrt{\frac{(3-4.5)^2 + (4-4.5)^2 (5-4.5)^2 + (6-4.5)^2}{4}}$$

$$= \sqrt{\frac{(-1.5)^2 + (-0.5)^2 + (0.5)^2 + (1.5)^2}{4}} = \sqrt{\frac{5}{4}}$$

$= 1.118033$

(c) Sampling with replacement (infinite population): Here N = population size = 4 and n = sample size = 2.

The total number of samples with replacement is $N^n = 4^2 = 16$. Let us list all possible samples of size 2 from population 3, 4, 5 and 6 with replacement.

We get

(3,3) (3,4) (3,5) (3,6)
(4,3) (4,4) (4,5) (4,6)
(5,3) (5,4) (5,5) (5,6)
(6,3) (6,4) (6,5) (6,6)

Now, we compute the statistic mean for each of these 16 samples.

The set of 16 means $\bar{X}$ of these 16 samples gives rise to the distribution of means of the samples known as the sampling distribution of means (SDM):

3	3.5	4	4.5
3.5	4	4.5	5
4	4.5	5	5.5
4.5	5	5.5	6

This SDM may also be arranged in the form of frequency distribution:

Sample mean	$\bar{X}_i$	3	3.5	4	4.5	5	5.5	6
Frequency	f_i	1	2	3	4	3	2	1

(d) The mean of these 16 means is known as the mean of SDM:

$$\mu_{\bar{x}} = \frac{1(3) + 2(3.5) + 3(4.0) + 4(4.5) + 3(5) + 2(5.5) + 1(6)}{16}$$

$$= \frac{72}{16} = 4.5$$

(e) Variance of the SDM

$$\sigma_{\bar{x}}^2 = \sqrt{\frac{\Sigma f_i(\bar{x}_i - \mu\bar{x})^2}{n}}$$

$$= \frac{1}{16}[1.(3-4.5)^2 + 2(3.5-4.5)^2$$

$$+3(4-4.5)^2 + 4(4.5-4.5)^2 + 3(5-4.5)^2$$

$$+2(5.5-4.5)^2 + 1(6-4.5)^2]$$

$$= \frac{10}{16} = 0.625$$

This gives SD of the SDM $\sigma_{\bar{x}} = 0.7905694$.

Verification

(i) $\mu = 4.5 = \mu_{\bar{x}}$

(ii) $\sigma_{\bar{x}} = 0.79056 \Rightarrow \frac{\sigma}{\sqrt{n}} = \frac{1.118033}{\sqrt{2}} = 0.79057$

Example 4.2

Solve Example 4.1 without replacement.

Solution

(a) $\mu = 4.5$

(b) $\sigma = 1.118083$

(c) Sampling without replacement implies finite population: Here N = population size = 4 and n = sample size = z.

The total number of samples without replacement is $\binom{N}{n} = \frac{4!}{(4-2)!2!} = 6.$

The six samples are (3,4), (3,5), (3,6), (4,5), (4,6) and (5,6). We compute the statistic mean for each of these samples as follows:

$\bar{X}_i$: 3.5 4 4.5 4.5 5 5.5

This SDM may also be arranged in the form of frequency distribution:

Sample mean	$\bar{X}_i$	3.5	4	4.5	5	5.5
Frequency	f_i	1	1	2	1	1

(d) Mean of the SDM

$$\mu_{\bar{x}} = \frac{3.5 + 4 + 2\,(4.5) + 5 + 5.5}{6} = \frac{27}{6} = 4.5$$

(e) Variance of the SDM

$$\sigma_{\bar{x}}^2 = \frac{1}{6}[(3.5-4.5)^2 + (4-4.5)^2$$

$$+ 2(4.5-4.5)^2 + (5-4.5)^2$$

$$+ (5.5\text{-}4.5)^2]$$

$$= \frac{2.5}{6} = 0.4166$$

This gives $\sigma_{\bar{x}} = \sqrt{0.4166} = 0.645497.$

Verification

(i) $\mu_{\bar{x}} = 4.5 = \mu$

(ii) $\sigma_{\bar{x}} = 0.4166 \Rightarrow \frac{\sigma}{\sqrt{N}}\sqrt{\frac{N-n}{N-1}}$

$$= \frac{1.11803}{\sqrt{4}} = \sqrt{\frac{4-2}{4-1}} = 0.4166$$

Example 4.3

Assume that the heights of 3000 male students at a college are normally distributed with mean 68 in. and SD 3 in. If 80 samples consisting of 25 students each are obtained, what would be the expected mean and SD of the resulting SDM if the sampling is done (a) with replacement and (b) without replacement?

Solution The numbers of samples of size $n = 25$ that could be obtained theoretically from a population of size $N = 3000$ are as follows:

(a) With replacement (infinite population): $N^n = 3000^{25}$

(b) Without replacement: $= \binom{N}{n} = \binom{3000}{25}$

These numbers are much larger than the number of samples which is 80.

We do not, therefore, get a true SDM, but only an experimental sampling distribution. Hence, the expected mean and SD would be close to those of theoretical distribution.

(a) $\mu_{\bar{x}} = \mu = 68$ in. and $\sigma_{\bar{x}} = \frac{\sigma}{\sqrt{n}} = \frac{3}{\sqrt{25}}$

$= \frac{3}{5} = 0.6$ in.

(b) $\mu_{\bar{x}} = 68$ in. and $\sigma_{\bar{x}} = \frac{\sigma}{\sqrt{n}} = \sqrt{\frac{N-n}{N-1}}$

$$= \frac{3}{\sqrt{25}}\sqrt{\frac{3000-25}{3000-1}} = 0.5975$$

$\sigma_{\bar{x}} \approx 0.6$

Example 4.4

In how many samples of Example 4.3 would you expect to find the mean (a) between 66.8 and 68.3 in. and (b) less than 66.4 in.?

Solution The mean $\bar{X}$ of a sample in standard units is given by $z = \frac{X - \mu_{\bar{x}}}{\sigma_{\bar{x}}} = \frac{\bar{X} - 68}{0.6}$.

(a) 66.8 in. standard units = (66.8 – 68.0)/0.6 = –2.0

68.3 in. standard units = (68.3 – 68.0)/0.6 = 0.5

Proportion of samples with means between 66.8 and 68.3 in.

= area under normal curve between $z = -2.0$ and $z = 0.5$

= area between $z = -2$ and $z = 0$ + area between $z = 0$ and $z = 0.5$

= 0.4772 + 0.1915 = 0.6687

∴ The expected number of samples = 80(0.6687) = 53 (Figure 4.1)

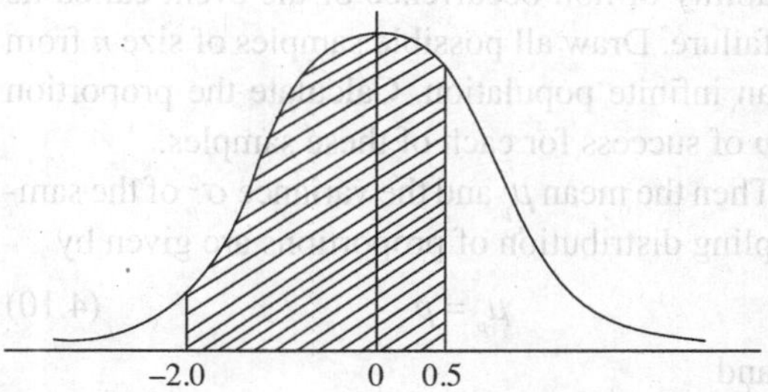

Figure 4.1 Example 4.4(a).

(b) 66.4 in. standard units = (66.4 – 68.0)/0.6 = –2.67

Proportion of samples with means less than 66.4 in.

= area under normal curve to the left of $z = -2.67$

= area to the left of $z = 0$ – area between $z = -2.67$ and $z = 0$

= 0.5 – 0.4962 = 0.0038

∴ The expected number of samples = 80 (0.0038) = 0.304 or zero (Figure 4.2).

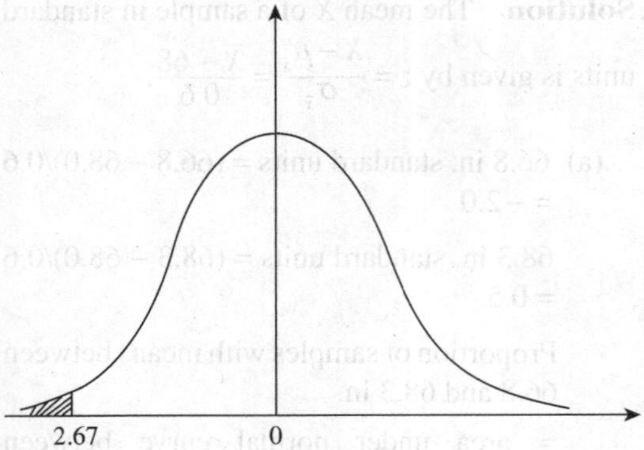

Figure 4.2 Example 4.4(b).

4·5 Sampling Distribution of Proportions

Let p be the probability of occurrence of an event called its success and $q = 1 - p$ be the probability of non-occurrence of the event called its failure. Draw all possible samples of size n from an infinite population. Calculate the proportion p of success for each of these samples.
Then the mean μ_p and the variance σ_p^2 of the sampling distribution of proportions are given by

$$\mu_p = p \qquad (4.10)$$

and

$$\sigma_p^2 = \frac{pq}{n} = \frac{p(1-p)}{n} \qquad (4.11)$$

where the population is binomially distributed and the sampling distribution of proportions is normally distributed whenever n is large.

For a finite population (with replacement) of size N,

$$\mu p = p \qquad (4.12)$$

and

$$\sigma_p^2 = \frac{pq}{n}\left(\frac{N-n}{N-1}\right) \qquad (4.13)$$

Example 4.5

Find the probability that in 120 tosses of a fair coin (a) between 40% and 60% will be heads.

Solution We consider the 120 tosses of the coin as a sample from the infinite population of all possible tosses of the coin. In this population, the probability of heads is $p = \frac{1}{2}$ and the probability of tails is $q = 1 - p = \frac{1}{2}$.

(a) We require the probability that the number of heads in 120 tosses will be between 40% and 60%:

$$P(40\%) = \frac{40}{100} \times 120 = 48$$

$$P(60\%) = \frac{60}{100} \times 120 = 72$$

We use normal approximation to the binomial distribution. Since the number of heads is a discrete variable, we ask for the probability that the number of heads lies between 47.5 and 72.5 (Figure 4.3).

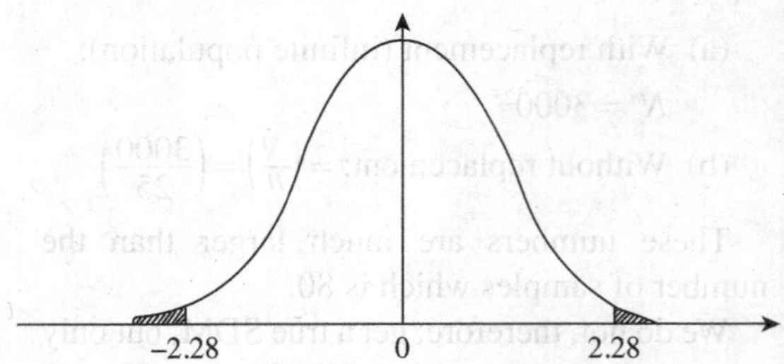

Figure 4.3 Example 4.5.

$$\mu = \text{expected number of heads} = np = 120\left(\frac{1}{2}\right)$$

$$= 60 \text{ and } \sigma = \sqrt{npq} = \sqrt{120 \times \frac{1}{2} \times \frac{1}{2}} = 5.48$$

$$47.5 \text{ in. standard units} = \frac{47.5 - 60}{5.48} = -2.28$$

$$72.5 \text{ in. standard units} = \frac{72.5 - 60}{5.48} = 2.28$$

The required probability is the area under normal curve between $z = -2.28$ and $z = 2.28$.

$$= 2 \text{ (area between } z = 0 \text{ and } z = 2.28)$$

$$= 2\,(0.4887) = 0.9774$$

4.6 Sampling Distribution of Differences and Sums

Suppose that we are given two populations. For each sample of size n_1 drawn from the first population, let us compute a statistic s_1. This gives a sampling distribution for s_1 whose mean and SD are denoted by μ_{s_1} and σ_{s_1} respectively. Similarly, for each sample of size n_2 drawn from the second population, let us compute a statistic s_2 whose mean and SD are μ_{s_2} and σ_{s_2} respectively.

Taking all possible combinations of these samples from the two populations, we can obtain a distribution of the two differences $s_1 - s_2$, which is called the *sampling distribution of differences* of the statistics. The mean and SD of this sampling distribution, denoted by $\mu_{s_1-s_2}$ and $\sigma_{s_1-s_2}$ respectively, are given by

$$\mu_{s_1-s_2} = \mu_{s_1} - \mu_{s_2} \tag{4.14}$$

and

$$\sigma_{s_1-s_2} = \sqrt{\sigma_{s_1}^2 + \sigma_{s_2}^2} \tag{4.15}$$

provided that the sample chosen do not in any way depend on each other, i.e., the samples are independent. In another words, the random variables s_1 and s_2 are independent.

If, for example, s_1 and s_2 are the sample means from two populations, denoted by $\bar{x}_1$ and $\bar{x}_2$ respectively, then the sampling distribution of the differences of means is given for infinitive populations with mean and SD, μ_1, σ_1 and μ_2, σ_2 respectively, by

$$\mu_{\bar{x}_1-\bar{x}_2} = \mu_{\bar{x}_1} - \mu_{\bar{x}_2} = \mu_1 - \mu_2 \tag{4.16}$$

and

$$\sigma_{\bar{x}_1-\bar{x}_2} = \sqrt{\sigma_{\bar{x}_1}^2 + \sigma_{\bar{x}_2}^2} = \sqrt{\frac{\sigma_1^2}{n_1} + \frac{\sigma_2^2}{n_2}} \tag{4.17}$$

This result holds for finite populations as well, if sampling is with replacement. The statement variable

$$z = \frac{(x_1 - x_2) - (\mu_1 - \mu_2)}{\sqrt{\frac{\sigma_1^2}{n_1} + \frac{\sigma_2^2}{n_2}}} \tag{4.18}$$

In that case, it is very nearly normally distributed if n_1 and n_2 are large ($n_1, n_2 \geq 30$).

Similar results can be obtained for sampling distributions of difference of proportions from two binomially distributed populations with parameters p_1, q_1 and p_2, q_2 respectively. We have

$$\mu_{p_1-p_2} = \mu_{p_1} - \mu_{p_2} = p_1 - p_2 \tag{4.19}$$

and

$$\sigma_{p_1-p_2} = \sqrt{\sigma_{p_1}^2 + \sigma_{p_2}^2} = \sqrt{\frac{p_1q_1}{n_1} + \frac{p_2q_2}{n_2}} \tag{4.20}$$

In sampling distribution of the sum of statistics s_1 and s_2, the mean and SD are given by

$$\mu_{s_1+s_2} = \mu_{s_1} + \mu_{s_2} \tag{4.21}$$

and

$$\sigma_{s_1} + {}_{s_2} = \sqrt{\sigma_{s_1}^2 + \sigma_{s_2}^2} \tag{4.22}$$

if the samples are independent.

Example 4.6

Let U_1 be a variable denoting any of the elements of the population 3, 7 and 8 and U_2 a variable denoting any of the elements of the population 2 and 4. Compute

(a) μ_{U_1} (b) μ_{U_2} (c) $\mu_{U_1-U_2}$

(d) σ_{U_1} (e) σ_{U_2} (f) $\sigma_{U_1-U_2}$

Solution

(a) μ_{U_1} = Mean of population of

$$U_1 = \frac{1}{3}(3+7+8) = 6$$

(b) μ_{U_2} = Mean of population of

$$U_2 = \frac{1}{2}(2+4) = 3$$

(c) The population consisting of the differences of any member of U_1 and any member of U_2 is

$3 - 2 = 1$ $\quad 7 - 2 = 5$ $\quad 8 - 2 = 6$

$3 - 4 = -1$ $\quad 7 - 4 = 3$ $\quad 8 - 4 = 4$

$\therefore \mu_{U_1-U_2}$ = mean of $(U_1 - U_2)$

$$= \frac{1+5+6+(-1)+3+4}{6}$$

$$= 3 = \mu_{U_1-U_2}$$

This proves the general result $\mu_{U_1-U_2} = \mu_{U_1} - \mu_{U_2}$.

(d) $\sigma^2_{U_1}$ = Variance of population of

$$U_1 = \frac{(3-6)^2 + (7-6)^2 + (8-6)^2}{3} = \frac{14}{3}$$

$$\Rightarrow \sigma_{U_1} = \text{SD of } U_1 = \sqrt{\frac{14}{3}} = 1.673320$$

(e) $\sigma^2_{U_2}$ = Variance of population of

$$U_2 = \frac{(2-3)^2 + (4-3)^2}{2} = 1$$

$$\Rightarrow \sigma_{U_2} = \text{SD of } U_2 = 1$$

(f) $\sigma^2_{U_1-U_2}$ = Variance of population of $(U_1 - U_2)$

$$= \frac{(1-3)^2+(5-3)^2(6-3)^2+(-1-3)^2}{6}$$

$$\frac{+(3-3)^2+(4-3)^2}{6}$$

$$= \frac{17}{3}$$

$$\Rightarrow \sigma_{U_1-U_2} = \text{SD of } (U_1 - U_2)$$

$$= \sqrt{\frac{17}{3}} = 2.380476$$

This proves the general result $\sigma_{U_1-U_2} = \sqrt{\sigma^2_{U_1} + \sigma^2_{U_2}}$.

Example 4.7

Let $\bar{X}_1$ and $\bar{X}_2$ be the average drying times of two types of oil paints with sample sizes $n_1 = n_2 = 18$. Find $P(\bar{X}_1 - \bar{X}_2 > 1)$ assuming that $\sigma_1 = \sigma_2 = 1$ and the mean drying times are equal for the two types of oil paints.

Solution We have

$$\sigma^2_{\bar{X}_1-\bar{X}_2} = \frac{\sigma_1^2}{n_1} + \frac{\sigma_2^2}{n_2} = \frac{1}{18} + \frac{1}{18} = \frac{1}{9}$$

$$P(\bar{X}_1 - \bar{X}_2 > 1) = P\left(z > \frac{1-(\mu_1-\mu_2)}{\bar{X}_1 - \bar{X}_2}\right)$$

$$= P\left(z > \frac{1}{\sqrt{1/9}}\right)$$

$$P(z > 3) = 1 - P(0 < z < 3) = 1 - 0.9987 = 0.0013$$

Example 4.8

Determine the expected number of random sample having their means

(a) Between 22.39 and 22.41

(b) Greater than 22.42

(c) Less than 22.37

(d) Less than 22.38 or greater than 22.41

Solution For the following data

N = size of population = 1500

n = size of samples = 36

Number of samples $N_s = 300$

Population mean $\mu = 22.4$

Population SD = 0.48

(a) $P(22.39 < \bar{X} < 22.41) = P(-1.26 < z < 1.26)$

Since $z = \frac{\bar{X}\ \mu}{\sigma/\sqrt{n}}$,

= 2(0.3962) = 0.7924, from the normal tables

Expected number of samples = total number of samples × probability

$$= N_s \times P(\bar{X}) = 300 \times 0.7924 = 238$$

(b) $P(\bar{X} > 22.42) = P(z > 2.53) = 0.00057$

Expected number of samples

$$N_s = 0.00057 \times 300 = 2$$

(c) $P(\bar{X} < 22.37) = P(z < -3.8) = 0.0001$

Expected number of samples

$$N_s = 300 \times 0.0001 = 0$$

(d) $P(\bar{X} < 22.38 \text{ and } \bar{X} > 22.41) = P(z < -2.53 \text{ and } z > 1.26) = 0.0057 + 0.1038 = 0.1095$

Expected number of samples
$N_s = 300 \times 0.1095 = 33$

Example 4.9

Find the probability that in 120 tosses of a fair coin (a) between 40% and 60% with be heads and (b) $\frac{5}{8}$ or more will be heads.

Solution

(a) $\mu_p = p = \frac{1}{2} = 0.50$ and $q = 1 - p = \frac{1}{2} = 0.5$

$$\sigma_p = \sqrt{\frac{pq}{N}} = \sqrt{\frac{\left(\frac{1}{2}\right)\left(\frac{1}{2}\right)}{120}} = 0.0456$$

$$40\% \text{ in standard units} = \frac{0.40 - 0.50}{0.0456} = -2.19$$

$$60\% \text{ in standard units} = \frac{0.60 - 0.50}{0.0456} = 2.19$$

The required probability is $P(-2.19 < z < 2.19)$

$$= 2F(2.19) - 1 = 0.9714$$

(b) $P\left(\geq \frac{5}{8}\right) = P\left(\geq \frac{0.6250 - 0.5000}{0.0456}\right)$

$$= P(\geq 2.74)$$

$$= 0.0040$$

Note $\frac{1}{2N} = \frac{1}{2 \times 120} = \frac{1}{240} = 0.00417$ is the correction factor since the proportion is actually a discrete variable.

Example 4.10

A normal curve has mean $\bar{x} = 20$ and the variance $\sigma^2 = 100$. Find the area (a) between $x = 26$ and $x = 38$ and (b) between $x = 15$ and $x = 40$.

Solution

Mean $\bar{x} = 20$
SD $\sigma = 10$
The normal variate $z = (x - \mu)/\sigma$

(a) For $x = 26$, we have $z = \frac{26 - 20}{10} = 0.6$

and for $x = 38$, we have $z = \frac{38 - 20}{10} = 1.8$

From the tables, we find

$$P(0.6) = P(0 \leq z \leq 0.6) = 0.2257$$

$$P(1.8) = P(0 \leq z \leq 1.8) = 0.4641$$

$\therefore$ Area under the curve from $x = 26$ to $x = 38$ is equal to the difference between these:

$$= P(1.8) - P(0.6) = 0.4641 - 0.2257$$

$$= 0.2384$$

(b) For $x = 15$, we have $z = \frac{15 - 20}{10} = -0.50$

and for $x = 40$, we have $z = \frac{40 - 20}{10} = 2$

Since both the z-values are on either side of the mean the area between them has to be taken as the sum of their tabulated values. From the tables showing the areas under normal curve from 0 to z, we find that the areas corresponding to $z = 0.50$ and $z = 2$ are 0.1916 and 0.4772 respectively.

$\therefore$ Required area $= 0.1916 + 0.4772 = 0.6688.$

Example 4.11

In a normal distribution, 7% of the items are under 35 and 89% are under 63. Determine the mean and the variance of the distribution.

Solution It is gives that 7% of items are under 35. This implies that 43% of items lie between the mean $\bar{x}$ and 35. From the tables of values of z, we find that the z-value that corresponds to the area 0.43 is $z = 1.48$. We have to take $z = -1.48$.

So, we have $z = \frac{35 - \bar{x}}{\sigma} = -1.48$

It is given that 89% of items are under 63. It implies that 39% of items lie between $x = \bar{x}$ and $x = 63$ for which $z = 1.23$.

$$\therefore z = \frac{63 - \bar{x}}{\sigma} = 1.23$$

The two equation for $\bar{x}$ and σ are $\bar{x} + 1.23\sigma = 63$ and $\bar{x} - 1.48\sigma = 35$.

Subtracting, we find $\sigma = \frac{28}{2.71} = 10.3321$ and substituting this value in $\bar{x} = 35 + 1.480$, we obtain $\bar{x} = 50.2315$.

Example 4.12

The income of a group of 10,000 persons was found to be normally distributed with mean Rs 750 per month and the SD is Rs 50. Show that about 95% of this group has income exceeding Rs 668 and only 5% has income exceeding Rs 832. Also, find the lowest income among the richest 100.

Solution Here $\bar{X} = 750$, $\sigma = 50$ and $x = 668$

$$\therefore z = \frac{668 - 750}{50} = -1.64$$

The area to the right of the ordinate at $z = -1.64$ is $0.4495 + 0.50 = 0.9495$.

$\therefore$ Expected number of persons getting above $668 = 95\%$ approximately

Also, the standard normal variance corresponding to 832 is $z = \frac{832 - 750}{50} = -1.64$.

The area to the right of the ordinate at $z = 1.64$ is $0.5000 - 0.4495 = 0.0505 = 5\%$ approximately.

Also, the number of persons getting Rs 832 and above is $10{,}000 \times 0.0505 = 505$.

4.7 Sampling Distribution of Means (σ Unknown): *t*-Distribution

In Sections 4.4 and 4.5, while dealing with problems of inference on a population mean and the difference between two population means, it was assumed that the population SD σ was known.

If σ is unknown, for large sample size $n \geq 30$, σ can be substituted by the sample SD s, calculated using the sample mean $\bar{x}$, by the following formula:

$$s^2 = \frac{1}{(n-1)} \sum_{i=1}^{n} (x_i - \bar{x})^2 \qquad (4.23)$$

In case the sample size n is small, $n < 30$, then the unknown σ can still be substituted by s, provided we assume that the sample is drawn from a normally distributed population.

Let $\bar{x}$ be the mean of a random sample of size n drawn from a normal population with mean μ and variance σ^2 then

$$t = \frac{\bar{x} - \mu}{s/\sqrt{n}} \qquad (4.24)$$

is a random variable having the t-distribution with $v = n - 1$ dof.

In 1908, W. S. Gosset[1] first published the probability distribution of t under the name 'Student'. So, the t-distribution is known as Student's t-distribution. In 1925, R. A. Fisher used t-distribution to test the regression coefficient.

Thus, for small samples, $n < 30$, and with σ not known, a natural statistic for inference on population mean M is

$$t = \frac{\bar{x} - \mu}{s/\sqrt{n}}$$

with the assumption that the sampling is from normal population. So, the above result is more general than the central limit theorem. Since, σ is not needed but less general than that since population is assumed to be normal.

The t-distribution curve is symmetric w.r.t. mean 0, unimodal, bell-shaped and is asymptotic on both sides of the t-axis (Figure 4.4). Thus, the t-distribution curve is similar to the normal curve. While the variance for normal distribution is unity, the variance for the t-distribution is

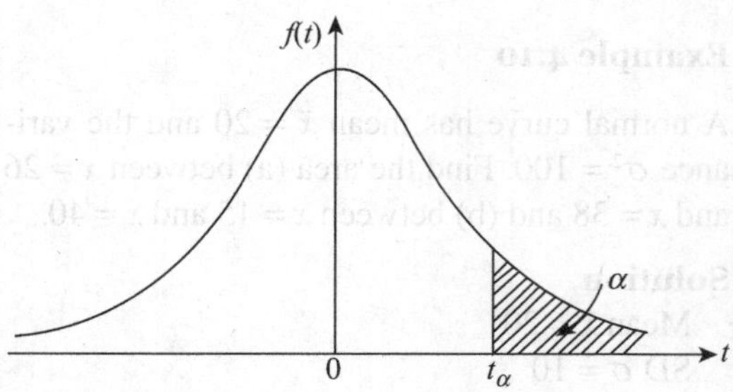

Figure 4.4 Critical value t_α for t-distribution.

[1]Wiiliam sealy Gosset, William sealy (1876-1937) is an English statistician.

greater than unity, since it depends on the is unity, the variance for the t-distribution is greater than unity, since it depends on the parameter v. So, the t-distribution is more variable. As $n \to \infty$, variance of the t-distribution approaches unity.

It follows therefore that as $v = (n - 1) \to \infty$, the t-distribution approaches the standard normal distribution. In fact, for $n \geq 30$, the standard normal distribution provides a good approximation to the t-distribution.

4.7.1 Critical Values of *t*-Distribution

Critical values of t-distribution are denoted by . It is a point such that the area under the curve to the right of t_2 is equal to α. Since the t-distribution is symmetric, it follows that

$$t_{1-\alpha} = -t_\alpha$$

It means that the t-value which has an area $1 - \alpha$ to its right and an area α to its left is equal to the negative t-value which has an area $- \alpha$ to its right and an area $1 - \alpha$ to its left (Figure 4.5).

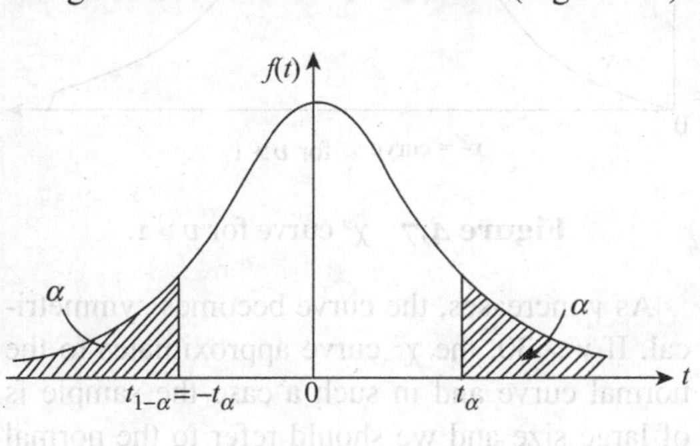

Figure 4.5 Symmetry property of t-distribution.

The critical values t_2 are tabulated for various values of the parameter v. In these tables, the LH column contains values of v while the column headings are areas α in the right-hand tail of the t-distribution, the entries are values of t_α.

Notes

1. In these tables, the areas are the column headings and the entries are the t-values. This is in contrast to what we find in normal tables, where the entries are areas and the column headings are the z-values.
2. Exactly 95% of the values of t-distribution with $v = n - 1$ dof lies between $t_{-0.02}$ and $t_{0.025}$.

The t-distribution is extensively used in tests of hypothesis about one mean or about equality of two means when σ is unknown.

Example 4.13

A chemical engineer claims that the population mean yield of a certain fetch process is 500 g/mL of raw material. To check this claim, he samples 25 batches each month. If the computer t-value falls between $- t_{0.05}$ and $t_{0.05}$, he is satisfied with his claim. What conclusion should be drawn from a sample that has a mean $\bar{x} = 518$ g/mL and a sample SD $s = 40$ g? Assume the distribution of yields to be approximately normal.

Solution From the tables, we find that $t_{0.05} =$ 1.711 for 24 dof.

$\therefore$ The manufacturer is satisfied with this claim if a sample of 25 batches yields a t-value between −1.711 and 1.711.

If $\mu = 500$ then

$$t = \frac{518 - 500}{40/\sqrt{25}} = 2.25$$

This value is well above 1.711. The probability of obtaining a t-value with $v = 24$, equal to or greater than 2.25 is approximately 0.02. If $\mu > 500$, the value of t computed from the sample would be more reasonable. Hence, the manufacturer is likely to conclude that the batch produces a better product than he thought.

Example 4.14

A company claims that the mean lifetime of tube lights is 500 h. Is the claim of the company acceptable if a random sample of 25 tube lights produced by the company has mean 518 h and SD 40 h?

Solution Here $\bar{x} = 518$ h, $n = 25$, $s = 40$ and $\mu = 500$.

$$\therefore t = \frac{\bar{x} - \mu}{s/\sqrt{n}} = \frac{518 - 500}{40/\sqrt{25}} = 2.25$$

Since $t = 2.25 < t_{0.01, v=24} = 2.492$, the claim of the company is acceptable.

Example 4.15

Find

(a) $t_{0.025}$ when $v = 14$
(b) $-t_{0.01}$ when $v = 10$
(c) $t_{0.995}$ when $v = 7$

Solution From the *t*-distribution tables, we have

(a) $t_{0.025} = 2.145$
(b) $-t_{0.01} = -2.764$
(c) $t_{0.995} = t_{1-0.005} = -t_{0.005} = -3.499$

$$\Rightarrow \therefore t_{1-\alpha} = t_{\alpha}$$

4.8 Chi-square (χ^2) Distribution

Let O_i and E_i ($I = 1, 2, 3, \ldots, n$) be the observed and expected frequencies of a class interval, then χ^2 is defined by

$$\chi^2 = \sum_{i-1}^{n} \frac{(O_i - E_i)^2}{E_i} \tag{4.25}$$

where $\sum O_i = \sum E_i = N \sum O_i$, the total frequency with dof is $n - 1$.

It describes the magnitude of deviation between the observed frequencies O_i and the expected frequencies E_i.

For large sample sizes, the sampling distribution of χ^2 can be closely approximated by a continuous curve called χ^2 distribution. It is defined by means of the following function:

$$y = ce^{-x^2/2}(\chi^2)^{(v-1)/2}$$

where v is the dof and c is a constant. In the case of binomial distribution, the dof is $n - 1$. For Poisson distribution, it is $n - 2$ and for normal distribution, the dof is $n - 3$. In fact, if we have $s \times t$ contingency table, then the dof is $(s - 1) \times (t + 1)$. χ^2 curve reduces to $y = ce^{-x^2/2}$ which is the right half of a normal curve shown in Figure 4.6.

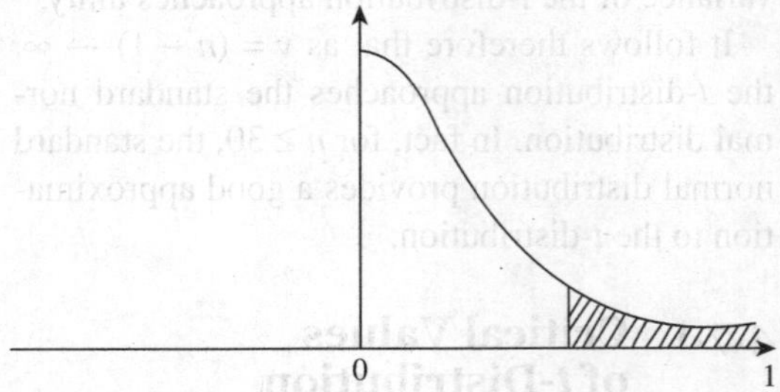

Figure 4.6 χ^2 curve for $\upsilon - 1$.

If $v > 1$, then χ^2 curve is tangential to the *x*-axis at the origin (Figure 4.7).

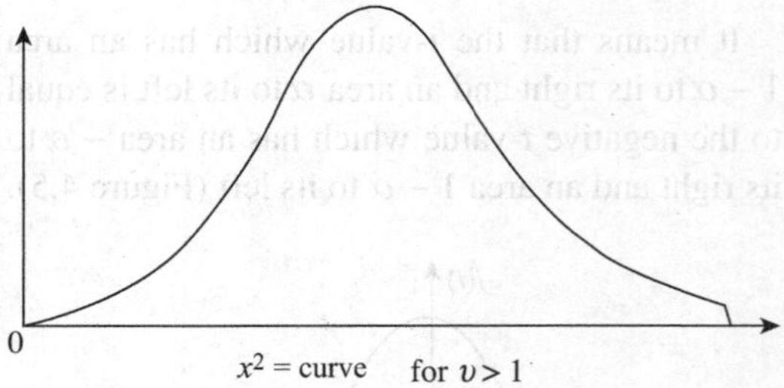

Figure 4.7 χ^2 curve for $\upsilon > 1$.

As v increases, the curve becomes symmetrical. If $v > 30$, the χ^2 curve approximates to the normal curve and in such a case the sample is of large size and we should refer to the normal distribution table.

The probability P that the value of χ^2 from a random sample will exceed χ_0^2 is given by

$$P = \int_{\chi_0^2}^{\infty} y\,dx$$

The value of χ^2 for dof from $v = 1$ to $v = 30$ have been tabulated for various convenient probability values. The table gives the values for the probability P that χ^2 exceeds a given value χ_0^2 (Figure 4.8).

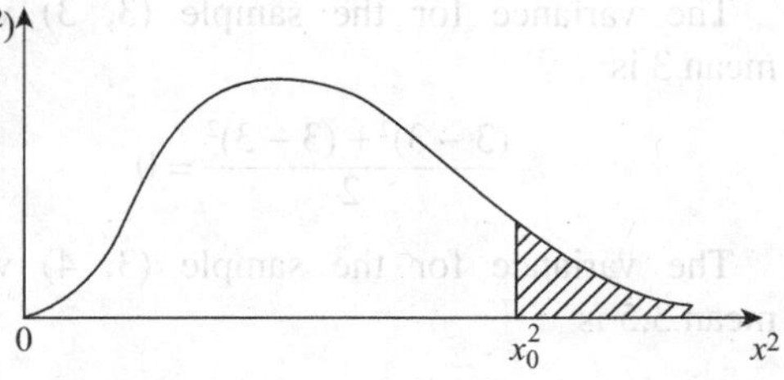

Figure 4.8 χ^2 distribution.

Hamlet discovered this χ^2 distribution in 1875. Karl Pearson rediscovered it independently in 1900 and applied it to test 'goodness-of-fit'.

The χ^2 test depends only on the set of observed and expected frequencies and on the dof. The χ^2 curve does not involve any parameter of the population. So, χ^2 distribution does not depend on the form of population. It is therefore called a non-parametric test or distribution-free test.

4.8.1 Properties of χ^2 Distribution

1. χ^2 distribution curve is not a normal curve. It lies completely in the first quadrant, since χ^2 varies from 0 to ∞. Thus, χ^2 distribution is not symmetrical.
2. It depends only on dof v.
3. It is a unimodal curve with its mode at $\chi^2 = k - 1$.
4. It is additive, i.e., if χ_1^2 and χ_2^2 are two independent distributions with dof v_1 and v_2 respectively, then $\chi_1^2 + \chi_1^2$ will be the χ^2 distribution of their sum with dof $(v_1 + v_2)$.

Here α denotes the area under the χ^2 distribution curve to the right of χ_α^2. Thus α denotes the probability that a random sample produces a χ^2 value $> \chi_\alpha^2$. So, χ_α^2 represents the χ^2 value such that the area under the χ^2 curve to its right is equal to α.

For various values of σ and v, the values of χ_α^2 are entered in tables.

In χ^2 table, the left-hand column contains values of v, the dof, the column headings are areas α; in the right-hand tail of χ^2 distribution curve, the table entries are values of χ^2.

4.8.2 χ^2 Test as a Test of Goodness-of-fit

The χ^2 test is used to test the deviations of the observed frequencies from the expected (theoretical) frequencies that they are significant or not. Thus, the test tells us how a set of observations fit a given distribution. Hence χ^2 test provides a test of goodness-of-fit for binomial, Poisson and normal distributions. If the calculated values of χ^2 are greater than tabular values, the fit is considered to be poor.

To apply χ^2 test, we first calculate χ^2. Then consulting the χ^2 table, we find the probability P corresponding to this calculated value of χ^2 for the given dof and apply the following hypotheses:

1. If $P < 0.005$, the observed value of χ^2 is significant at 5% level of significance.
2. If $P < 0.01$, the observed value of χ^2 is significant at 1% level of significance.
3. If $P > 0.05$, it is good fit and the value of χ^2 is not significant.

This implies that we accept the hypothesis if the calculated χ^2 is less than the tabulated value; otherwise, it is to be rejected.

4.8.3 Conditions for the Validity of χ^2 Test

We have pointed out that χ^2 test is used for large sample size. For the validity of χ^2 test as a test of goodness-of-fit with regard to significance of the deviation of the observed frequencies from the expected (theoretical) frequencies, the following conditions must be satisfied:

1. The sample observations should be independent.
2. The total frequency (the sum of the observed or expected frequencies) should be larger than 50.
3. No theoretical frequency should be less than 5 because χ^2 distribution cannot maintain continuity character if the frequency is less than 5.
4. Constraints on the frequencies, if any, should be linear.

4.9 Sampling Distribution of Variance σ^2

The theoretical sampling distribution of the sample variance for random samples from normal population is related to the uni-square distribution as follows:

Let s^2 be the variance of a random sample of size n, taken from a normal population having the variance σ^2. Then

$$\chi^2 = \frac{(x-1)s^2}{\sigma^2} = \sum_{i=1}^{n} \frac{(X_i - \overline{X})^2}{\sigma^2}$$

is a value of a random variable having the χ^2 distribution with $0 = n - 1$ dof.

Exactly 95% of χ^2 distribution lies between $\chi^2_{0.975}$ and $\chi^2_{0.025}$. When σ^2 is too small, χ^2 value falls to the right of $\chi^2_{0.025}$ and when σ^2 is too large, χ^2 falls to the left of $\chi^2_{0.975}$. Thus, when σ^2 is correct, χ^2 values fall to the left of $\chi^2_{0.975}$ or to the right of $\chi^2_{0.025}$.

Example 4.16

Find the mean and SD of sampling distribution of variances for the population 3, 4, 5 and 6 by drawing samples of size 2 (a) with replacement and (b) without replacement.

Solution

(a) With replacement:

The 16 samples with their corresponding means are

(3,3)	(3,4)	(3,5)	(3,6)
3	3.5	4	4.5
(4,3)	(4,4)	(4,5)	(4,6)
3.5	4	4.5	5
(5,3)	(5,4)	(5,5)	(5,6)
4	4.5	5	5.5
(6,3)	(6,4)	(6,5)	(6,6)
4.5	5	5.5	6

We compute the statistic variance for each of these 16 samples:

The variance for the sample (3, 3) with mean 3 is

$$\frac{(3-3)^2 + (3-3)^2}{2} = 0$$

The variance for the sample (3, 4) with mean 3.5 is

$$\frac{(3-3.5)^2 + (4-3.5)^2}{2} = 0.25$$

Similarly, computing the variance for each of the 16 samples we have

0	0.25	1	2.25
0.25	0	0.25	1
1	0.25	0	0.25
2.25	1	0.25	0

Thus, the SD of variances (with replacement) is

s^2	0	0.25	1	2.25
Frequency	4	6	4	2

Mean of the SD of variances

$$\mu_{s^2} = \frac{4(0) + 6(0.25) + 4(1) + 2(2.25)}{16}$$

$$= \frac{10}{16} = 0.625$$

Variance of the SD of variances

$$\sigma^2_{s^2} = \frac{1}{16}[4(0-0.625)^2 + 6(0.25-0.625)^2 + 2(2.25-0.625)^2]$$

$$= \frac{8.25}{16} = 0.515625$$

SD of SD of variances $\sigma^2_s = \sqrt{0.515625} = 0.718$

(b) Without replacement:

Samples	(3,4)	(3,5)	(3,6)	(4,5)	(4,6)	(5,6)
Means	3.5	4	4.5	4.5	5	5.5
Variances	0.25	1	2.25	0.25	1	0.25

Thus, the S.D. of variances (without replacement) is

σ_s^2	0.25	1	2.25
Frequency	3	2	1

Mean of the SD of variances μ_s^2

$$= \frac{3(0.25) + 2(1) + 1(2.25)}{6} = \frac{5}{6} = 0.8333$$

Variance of S.D. of variances

$$\sigma^2_{s^2} = \frac{2}{6}[3(0.25 - 0.8333)^2 + 2(1 - 0.8333)^2 + 1(2.25 - 0.8333)^2]$$

$$= \frac{3.08333}{6} = 0.51388$$

SD of SD of variances $\sigma_s^2 = \sqrt{0.51388}$

$$= 0.71686$$

4.10 Snedecor's F-Distribution

Let $x_1, x_2, \ldots, x_{n_1}$ and $y_1, y_2, \ldots, y_{n_2}$ be the values of two independent random samples drawn from two normal populations with equal variance σ^2. Let $\bar{x}$ and $\bar{y}$ be the sample means and suppose

$$s_1^2 = \frac{1}{n_1 - 1}\Sigma_{i=1}^{n_1}(x_1 - \bar{x})^2 \qquad (4.27)$$

$$s_2^2 = \frac{1}{n_2 - 1}\Sigma_{i=1}^{n_2}(y_i - \bar{y})^2 \qquad (4.28)$$

Then we define the statistic F by the relation

$$F = \frac{s_1^2}{s_2^2} \qquad (4.29)$$

The Snedecor's F-distribution is defined by the function

$$y = c\,F^{\left(\frac{\nu_1 - 2}{2}\right)}\left(1 + \frac{\nu_1}{\nu_2}F\right)^{-\left(\frac{\nu_1+\nu_2}{2}\right)} \qquad (4.30)$$

where the constant c depends on ν_1 and ν_2 is so chosen that the area under the curve is unity. The F-distribution is independent of the population variance σ^2 and depends only on ν_1 and ν_2, the numbers of dof of the samples. The F-curve is bell-shaped for $\nu_1 > 2$ as shown in Figure 4.9.

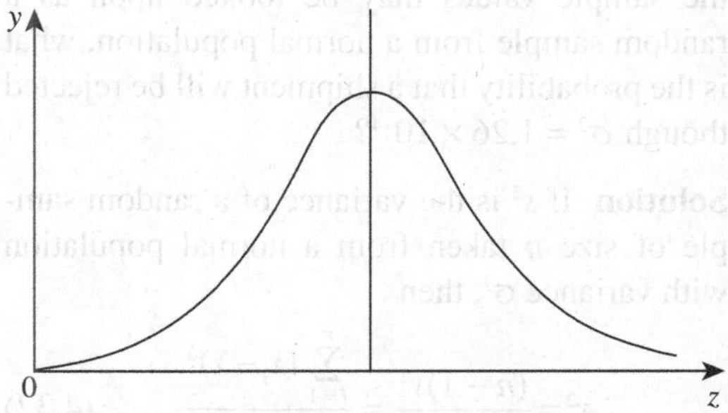

Figure 4.9 Snedecor's F-distribution curve.

Significant test is performed by means of Snedecors's F-table which provides 5% and 1% of points of significance for F. 5% of points of F means that the area under the F-curve, to the right of the ordinate at a value of F, is 0.05. Further, the F-table gives only single tail test. However, if we are testing the hypothesis that the population variances are same, the n we should use both tail areas under the F-curve and in that case F-table will provide 10% and 12% levels of significance.

4.11 Fisher's z-Distribution

Putting $F = e^{2z}$ in the F-distribution, we get

$$y = ce^{\nu_1 z}(\nu_1 e^{2z} + \nu_2) \qquad (4.31)$$

which is called the Fisher's z-distribution, where c is a constant depending upon ν_1 and ν_2 such that the area under the curve is unity. The curve for this distribution is more symmetrical than F-distribution.

Significance tests are performed from the z-table in a way similar to that of F-distribution.

Example 4.17

An optical firm purchases glass to be ground into lenses, and it is known from past experience that the variance of the refractive index of this kind of glass is 1.26×10^{-4}. As it is important that the various pieces of glass have nearly the same index of refraction, the firm rejects such a shipment if the sample variance of 20 pieces selected at random exceeds 2.00×10^{-4}. Assuming that the sample values may be looked upon as a random sample from a normal population, what is the probability that a shipment will be rejected though $\sigma^2 = 1.26 \times 10^{-4}$?

Solution If s^2 is the variance of a random sample of size n taken from a normal population with variance σ^2, then

$$x^2 = \frac{(n-1)s^2}{\sigma^2} = \frac{\sum_{i=1}^{n}(x_i - \bar{x})^2}{\sigma^2} \quad (4.32)$$

is a random variable having the χ^2 distribution with the parameter $v = n - 1$.

Here $n = 20$ so that $n - 1 = 19$

Sample variance of 20 pieces selected at random $= 2.00 \times 10^{-4}$

$$\sigma^2 = 1.26 \times 10^{-4}$$

Substituting these values into Eq. (4.32) for x^2 statistic, we get

$$x^2 = \frac{19(2.00 \times 10^{-4})}{1.26 \times 10^{-4}} = 30.2$$

From the relevant table for 19 dof, we find that

$$x^2_{0.05} = 30.1$$

Thus, the probability that a good shipment will be rejected erroneously is less than 0.05.

EXERCISES

1. If a sample of size 5 results in the sample values of 8, 5, 9, 6 and 2, find the sample mean.

   Ans: 6

2. In Question 1, find the sample variance s^2.

   Ans: 6

3. Assume that the heights of 3000 male students of a university are normally distributed with mean 68 in. and SD 3 in. If 80 samples consisting of 25 students each are obtained, what would be the mean and SD of the resulting sample of means if sampling is done (a) with replacement and (b) without replacement?

   Ans: (a) $\mu_{\bar{x}} = \mu = 68$ in. and $\sigma_{\bar{x}} = 0.6$in

   (b) $\mu_{\bar{x}} = \mu = 68$ in. and

   $\sigma^2 = \frac{3}{\sqrt{25}} \times \sqrt{\frac{3000 - 25}{3000 - 1}}$, which is slightly less than 0.6 in.

   (a) In Question 3, how many samples would you expect to find the mean (a) between 66.8 and 68.3 in. and (b) less than 66.4 in.?

   Ans: (a) 53 and (b) 0

4. The tools produced by a certain machine are found 2% defective. What is the probability that in a shipment of 400 such tools (a) 3% or more and (b) 2% or less will be found defective?

   Ans: (a) 0.1056 and (b) 0.5714

5. Consider ball bearings of a given brand weight 0.500z with a SD of 0.020z. What is the probability that 2 lots of 1000 ball bearing each will differ in weight by more than 20z?

   Ans: 0.0258

6. If 1 gallon can of paint covers on an average 513.3 sq. ft with a SD of 31.5 sq. ft, what is the probability that the sample mean area covered by a sample of 40 of these 1 gallon can will be anywhere from 510.0 to 520.0 sq. ft?

   Ans: $z_1 = -0.066$, $z_2 = 1.34$ and $P(-0.066 < z < 1.34) = 0.6533$

7. A fuse manufacturer claims that with a 30% overload, the fuses will blow in 13.30 min on the average. To test this claim, a sample of 25 of the fuses was subjected to a 30% overload, and the time it took them to blow had a mean of 11.70 min and a SD of

2.50 min. If it can be assumed that the data constitute a random sample from a normal population, do they tend to support or refute the claim of the manufacturer?

Ans: $t = -3.20 < -2.797$ and 0.005 is a very small probability, so we conclude that the data tend to refute the manufacturer's claim.

8. If two independent random samples of size $n_1 = 7$ and $n_2 = 13$ are taken from a normal population, what is the probability that the variance of the first sample will be at least thrice as large as that of the second sample?

Ans: From table $F_{0.05} = 3.00$ for $v_1 = 7.1 = 6$ and $v_2 = 13 - 1 = 12$; required probability is 0.05

9. A random sample of 100 is taken from an infinite population having the mean $\mu = 76$ and the variance $\sigma^2 = 256$. What is the probability that $\bar{x}$ will be between 75 and 78?

Ans: 0.6268

MULTIPLE CHOICE QUESTIONS

1. The size of the population is 2000 and the size of the sample is 200. Then the correction factor in the population is

(a) 0.5 (b) 0.6 (c) 0.8 (d) 0.9

Ans: (d)

$\left[\text{Hint: } N = 2000 \text{ and } n = 200 \Rightarrow \text{correction factor} = \frac{N-v}{\text{N}-1} = 0.9\right]$

2. The size of the sample is 25 and the population is 400 times the size of the sample. Then the correction factor of the population is

(a) 0.9999 (b) 0.9 (c) 0.999 (d) 0.99

Ans: (c)

3. A population consists of four numbers 1, 2, 3 and 4. Consider all possible distinct samples of size 2 which can be drawn without replacement from the population. Then the population mean is

(a) 3.5 (b) 2.5 (c) 3 (d) 2

Ans: (b)

4. In Question 3, SD of population is

(a) $\sqrt{\frac{5}{4}}\sqrt{(5/4)}$ (b) $\sqrt{\frac{5}{2}}\sqrt{(5/2)}$

(c) $\sqrt{5\frac{5}{4}}\sqrt{(5\,5/4)}$ (d) $\sqrt{\frac{15}{4}}\sqrt{(15/4)}$

Ans: (a)

5. In Question 3, the total number of samples without replacement is

(a) 4 (b) 3 (c) 6 (d) 10

Ans: (c)

6. In Question 3, the SE of means (without replacement) is (given that $\sigma^2 = 3.3$)

(a) 1.9 (b) 2 (c) 2.5 (d) 2.2

Ans: (b)

7. If a sample is taken from an infinite population and sample size is increased from 25 to 100. The effect of this on SE is

(a) divided by 4 (b) divided by 3
(c) divided by 2 (d) multiplied by 2

Ans: (c)

8. If $\mu = 30.5$, $n = 100$, $\bar{x} = 28.8$ and $\sigma = 8.35$, then $|z| =$

(a) 2.5 (b) 1.98 (c) 2.4 (d) 2.68

Ans: (d)

9. The sample of size 4 has values 25, 28, 26 and 25. Then variance of the sample is

(a) 3 (b) 1 (c) 2 (d) 4

Ans: (c)

10. If a sample of size 64 is taken from a population whose SD is 0.4, then the probable error is

(a) 0.118 (b) 0.337 (c) 0.216 (d) 0.5

Ans: (b)

FILL IN THE BLANKS

1. Population is also called as _____.

Ans: Universe

2. The number of possible samples of size n out N population units, without replacement is _____.

Ans: $\binom{N}{n}$

3. The number of possible samples of size n out of N population units, with replacement is _____.

Ans: N^n

4. The probability that any one sample of size n may be drawn out of N population units is _____.

Ans: $1/\binom{N}{n}$

5. Finite population correction factor is _____.

Ans: $[(N-n)/(N-1)]$

6. The name given to the parameter $(N-n)/(N-1)$, where N is the size of the population and n is the sample size, is _____.

Ans: Finite population correction factor

7. Formula for sample mean is _____.

Ans: $\bar{x} = \frac{1}{n}\sum_{i=1}^{n} x_i$

8. If $x_1, x_2, \ldots\ldots, x_n, x_1, x_2, \ldots, x_n$ are n observations, the formula for variance is _____.

Ans: $s^2 = \frac{1}{n-1}\sum_{i=1}^{n}(x_i - \bar{x})^2$

9. The probability that a specific item is included in a sample of size n drawn out of a population of size N is _____.

Ans: $\frac{1}{N}$

10. The discrepancy between sample estimate and population parameter is called a _____.

Ans: Sampling error

11. A population consisting of all real numbers is an example of _____.

Ans: An infinite population

12. The SD of all possible estimates from samples of fixed size is called a _____.

Ans: Standard error

5

Estimation Theory

5.1 INTRODUCTION

In Chapter 4, we have emphasized sampling properties of sample mean and sample variance and also displays of data in different forms. These presentations are meant to lay a foundation that permits statisticians to draw conclusions about the population parameters from experimental data. For example, the central limit theorem provides information about the distribution of the sample mean $\overline{X}$. The distribution involves the mean of population μ. Thus, any conclusions drawn concerning μ from an observed sample average must depend on knowledge of this sampling distribution. Similar remarks hold in respect of s^2 and σ^2. Now, we begin by outlining the purpose of statistical inference.

5.2 STATISTICAL INFERENCE

The theory of statistical inference consists of methods by which inferences or generalizations about a population are made. It can be divided into two major types:

1. Estimation of parameters
2. Testing of hypotheses

A study of either type of inferences about a population may lead to correct conjectures about the population.

The process of estimating a population parameter by using sample information is called *estimation* and the processes by which we can decide whether to accept or reject a set of hypotheses are called *basis of hypothesis*. There are two types of estimation procedures:

1. Point estimation
2. Interval estimation

5.3 POINT ESTIMATION

Based on sample data, a single number is calculated to estimate the population parameter. The rule or formula that describes this calculation is called the *point estimator* and the resulting number is called a *point estimate*.

Sampling distributions provide information on the basis of which we can select the best estimator. The sampling distribution of the point estimator should be quite close to the true value of the parameter to be estimated. That is, the estimator should not underestimate or overestimate the parameter of interest. Such an estimator is said to be unbiased.

An estimator of parameter is said to be unbiased if the mean of its distribution is equal to the true value of the parameter. Otherwise, the estimator is said to be biased.

Figure 5.1 shows sampling distribution for an unbiased estimator and a biased estimator.

Unbiased Estimator Definition A statistic $\hat{\theta}$ is called an unbiased estimator of the corresponding population parameter θ if

$$E(\hat{\theta}) = E(\text{Statistic}) = 0$$

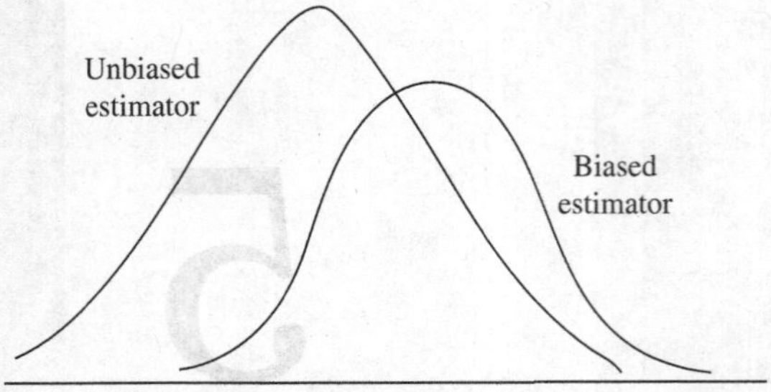

Figure 5.1 Sampling distributions for unbiased and biased estimators.

That is, the mean of the sampling distribution of estimator is equal to θ.

Maximum Error of Estimate E The sample mean estimate is seldom equal to the mean of population μ. Hence a point estimate is generally accompanied by a statement of error which gives the difference between the estimate and the quantity to be estimated, namely the estimator.

Thus, error $= \bar{x} - \mu$.

For large n, the random variable $\dfrac{\bar{x}-\mu}{\sigma/\sqrt{n}}$ is approximately equal to the normal variate. Now, the following inequality

$$-Z_{\alpha/2} \le \frac{\bar{x}-\mu}{\sigma/\sqrt{n}} \le Z_{\alpha/2} \text{ or } |\bar{x}-\mu| \le Z_{\alpha/2}\cdot\frac{\sigma}{\sqrt{n}}$$

is satisfied with probability $(1-\alpha)$.

Confidence interval for μ.

A $(1-\alpha)$ 100% confidence interval for is given by

$$\bar{x} - Z_{\alpha/2}\frac{\sigma}{\sqrt{n}} < \mu < \bar{x} + Z_{\alpha/2}\frac{\sigma}{\sqrt{n}}$$

So, the maximum error of estimate E with probability $(1-\alpha)$ is given by $E = Z_{\alpha/2}\dfrac{\sigma}{\sqrt{n}}$. Thus, in the point estimation of population mean μ with sample mean $\bar{x}$ for a large random sample ($n \ge 30$), we can assert with probability $(1-\alpha)$ that the error $|\bar{x}-\mu|$ will not exceed $Z_{\alpha/2}\dfrac{\sigma}{\sqrt{n}}$.

Sample Size When α, E and σ are known, the sample size n is given by $n = \left(\dfrac{Z_{\alpha/2}\sigma}{E}\right)^2$. When σ is unknown or sample size $n < 30$, the maximum error estimate E is $E = t_{\alpha/2}\dfrac{s}{\sqrt{n}}$ with $(1-\alpha)$ probability, where t distribution is with $(n-1)$ degrees of freedom (dof).

The second desirable characteristic of an estimator is that the spread (as measured by the variance) of the sampling distribution should be as small as possible.

Figure 5.2 shows the sampling distributions for two unbiased estimators, one with small variance and the other with larger variance.

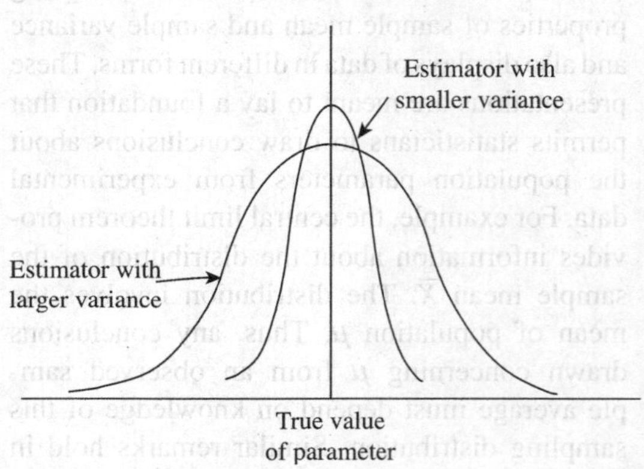

Figure 5.2 Sampling distributions for two unbiased estimators.

The distance between the estimate and the true value of the parameter is called the *error of estimation.*

We may assume that sample sizes are always large. Therefore, the unbiased estimators which we will be studying will have sampling distribution that can be approximated by a normal distribution because of central limit theorem. For unbiased estimators, the difference between the point estimator and the true value of the parameter will be less than 1.96 standard deviations or 1.96 standard errors. This quantity called the 95% margin of error provides a practical upper bound for the error of estimation (Figure 5.3).

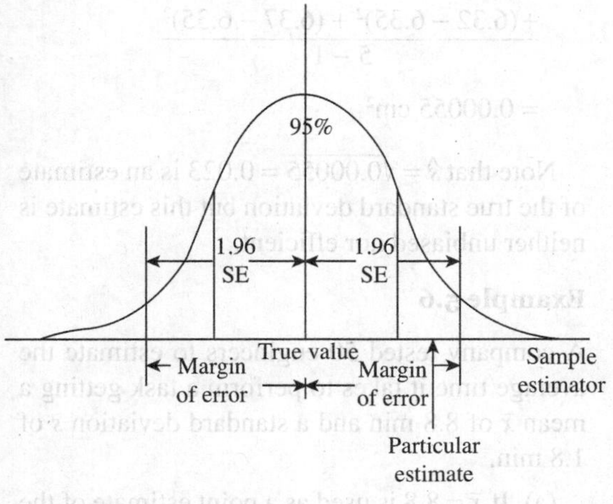

Figure 5.3 Sampling distribution of an unbiased estimators: SE, standard error.

Example 5.1

Show that S^2 is an unbiased estimator of the parameter σ^2.

Solution Consider

$$\sum_{i=1}^{n}(x_i-\bar{x})^2=\sum_{i=1}^{n}\left[(x_i-\mu)-(\bar{x}-\mu)\right]^2$$

$$=\sum_{i=1}^{n}(x_i-\mu)^2-2(\bar{x}-\mu)$$

$$\sum_{i=1}^{n}(x_i\ \mu)^2+n(\bar{x}-\mu)^2$$

$$=\sum_{i=1}^{n}(x_i-\mu)^2-n(\bar{x}-\mu)^2$$

Now,

$$E(S^2)=E\left[\frac{\sum_{i=1}^{n}(x_i-\bar{x})^2}{n-1}\right]$$

$$=\frac{1}{n-1}\left[\sum_{i=1}^{n}E(x_i-\mu)^2-nE(\bar{x}-\mu)^2\right]$$

$$=\frac{1}{n-1}\left(\sum\sigma_{xi}-n\sigma_{\bar{x}}^2\right)$$

But

$$\sigma_{xi}^2=\sigma^2 \text{ for } i=1,2,\ldots,n \text{ and } \sigma_{\bar{x}}^2=\frac{\sigma^2}{n}$$

$$\therefore E(S^2)=\frac{1}{n=1}\left(n\sigma^2-n\frac{\sigma^2}{n}\right)=\sigma^2$$

This proves that S^2 is an unbiased estimator.

Though S^2 is unbiased estimator of σ^2, S is a biased estimator of σ with the bias becoming insignificant for large samples.

This example shows why we divide by $n-1$ rather than n when the variance is estimated.

If we consider all possible unbiased estimators of some parameter θ, the one with the smallest variance is called the most efficient estimator of θ.

Example 5.2

If $\hat{\theta}_1$ and $\hat{\theta}_2$ are two unbiased estimators of θ, then $\hat{\theta}=\alpha\hat{\theta}_1+\beta\hat{\theta}_2$ is an unbiased estimator of θ, where α and β are constants such that $\alpha+\beta=1$.

Solution Suppose $\hat{\theta}_1$ and $\hat{\theta}_2$ are unbiased estimators of θ. This implies that

$$E(\hat{\theta}_1)=\theta \text{ and } E(\hat{\theta}_2)=\theta$$

$$\text{Now } E(\hat{\theta})=E(\alpha\hat{\theta}_1+\beta\hat{\theta}_2)$$

$$=\alpha E(\hat{\theta}_1)+\beta E(\hat{\theta}_2)$$

$$=\alpha\theta+\beta\theta=(\alpha+\beta)\theta=\theta$$

$$\Rightarrow E(\hat{\theta})=\theta \quad (\because \alpha+\beta=1)$$

$\therefore \hat{\theta}=\alpha\hat{\theta}_1+\beta\hat{\theta}_2$ is an unbiased estimator of θ.

Example 5.3

Let $x_1, x_2, \ldots, x_n$ be a random sample from a given population with mean μ and variance σ^2. Show that the sample mean $\bar{x}$ is an unbiased estimator of population mean μ, i.e. $E(\bar{x})=\mu$.

Solution We have

$$\bar{x}=\frac{1}{n}\sum_{i=1}^{n}x_i$$

$$\text{Now } E(\bar{x})=E\left(\frac{1}{n}\sum_{i=1}^{n}x_i\right)=\frac{1}{n}E\left(\sum_{i=1}^{n}x_i\right)$$

$$=\frac{1}{n}E(x_1+x_2+\cdots+x_n)$$

$$=\frac{1}{n}(E(x_1)+E(x_2)+\cdots+E(x_n))$$

$$=\frac{1}{n}(\mu+\mu+\cdots+\mu)\left[\begin{array}{l}\because E(x_i)=\mu\\ \text{for } i=1,2,\ldots,n\end{array}\right]$$

$$=\frac{1}{n}n\mu=\mu$$

$\therefore \bar{x}$ is an unbiased estimator of μ.

Example 5.4

Give examples of estimators (or estimates) which are (a) unbiased and efficient, (b) unbiased and inefficient and (c) biased and inefficient.

Solution Assume that the population is normally distributed. Then

(a) The sample mean $\hat{X}$ and the modified sample variance $\hat{S}^2 = \frac{n}{n-1}S^2$ are two examples of unbiased and efficient estimates.

(b) The sample median is an unbiased but inefficient estimate of the population mean since mean of its sampling is inefficient when compared with $\hat{X}$.

(c) The sample standard deviation S, the modified standard deviation $\hat{S}$ and the mean deviation are examples of biased and inefficient estimates for evaluating the population standard deviation σ.

Example 5.5

Samples of 5 measurements of the diameter of a sphere are recorded as 6.33, 6.37, 6.36, 6.32 and 6.37 cm. Under the assumption that the measured diameter is normally distributed, find unbiased and efficient estimates of

(a) True mean
(b) True variance

Solution

(a) An unbiased and efficient estimate of the true mean (population mean) is

$$\bar{x} = \frac{\sum x}{n} = \frac{6.33 + 6.37 + 6.36 + 6.32 + 6.37}{5}$$

$$= 6.35 \text{ cm}$$

(b) An unbiased and efficient estimate of the true variance (population variance)

$$\hat{S}^2 = \frac{n}{n-1}S^2 = \frac{\sum(x-\bar{x})^2}{n-1}$$

$$= \frac{(6.33-6.35)^2 + (6.37-6.35)^2 + (6.36-6.35)^2}{5-1}$$

$$\frac{+(6.32-6.35)^2 + (6.37-6.35)^2}{5-1}$$

$$= 0.00055 \text{ cm}^2$$

Note that $\hat{s} = \sqrt{0.00055} = 0.023$ is an estimate of the true standard deviation but this estimate is neither unbiased nor efficient.

Example 5.6

A company tested 50 engineers to estimate the average time it takes to perform a task getting a mean $\bar{x}$ of 8.8 min and a standard deviation s of 1.8 min.

(a) If $\bar{x} = 8.8$ is used as a point estimate of the actual average time required to perform the task, find the maximum error with 95% confidence.

(b) Construct 99% confidence intervals for the true average time it takes to do the task.

(c) With what confidence can we assert that the sample mean does not differ from the true mean by more than 15 s?

Solution Here $\bar{x} = 8.8$ and $s = 1.8$. For 95%, $Z_{\alpha/2} = 1.96$, where $n = 50$.

(a) Maximum error of estimate $E = Z_{\alpha/2} \times \frac{\sigma}{\sqrt{n}}$

$$= 1.96 \times \frac{1.8}{\sqrt{50}} = 0.4989, \text{ assuming } \sigma = s = 1.8$$

(b) For 99% confidence, $Z_{\alpha/2} = 2.575$

$$\therefore E = Z_{\alpha/2} \times \frac{\sigma}{\sqrt{n}} = 2.575 \times \frac{1.8}{\sqrt{50}} = 0.6555$$

99% confidence interval limits are

$$x \pm Z_{\alpha/2}\frac{\sigma}{\sqrt{n}} = \bar{x} \pm E = 8.8 \pm 0.6555$$

$\therefore$ Confidence interval is (8.1445, 9.4555).

(c) We have to find $Z_{\alpha/2}$ such that $E = Z_{\alpha/2} \times \frac{\sigma}{\sqrt{n}}$

$$15 \text{ s} = \frac{1}{4} \text{ min} = Z_{\alpha/2} \times \frac{1.8}{\sqrt{50}}$$

$$\Rightarrow Z_{\alpha/2}\frac{1}{4} \times \frac{\sqrt{50}}{1.8} = 0.9821$$

From normal tables, the area corresponding to $Z_{\alpha/2} = 0.9821$ is 0.3365. Then the area is between $Z_{-\alpha/2}$ and $Z_{\alpha/2}$. Thus we can assert with 67.3% confidence.

Example 5.7

The mean and standard deviation of a population are 225 and 278 respectively. What can we assert with 95% confidence about the maximum error if $\bar{x} = 225$ and $n = 100$?

Solution Mean of the population $\mu = 225$

Standard deviation of the population $\sigma = 278$

Sample size $n = 100$

Formula for maximum error $E = Z_{\alpha/2} \times \frac{\sigma}{\sqrt{n}}$

The value of $Z_{\alpha/2}$ for 95% confidence = 1.96

Maximum error $E = Z_{\alpha/2} \times \frac{\sigma}{\sqrt{n}}$

$$= 1.96 \times \frac{278}{\sqrt{100}} = 54.488$$

Example 5.8

Construct 95% confidence interval for the true mean for Example 5.7.

Solution 95% confidence interval

$$= \left(\bar{x} - Z_{\alpha/2} \frac{\sigma}{\sqrt{n}}, \bar{x} + Z_{\alpha/2} \frac{\sigma}{\sqrt{n}}\right)$$

We have computed that $Z_{\alpha/2} \frac{\sigma}{\sqrt{n}} = 54.488$

(in Example 5.7.)

It is given that $\bar{x} = 225$.

∴ The required confidence interval is

$$(225 - 54.488, 225 + 54.488)$$
$$= (170.512, 279.488)$$

Example 5.9

An industrial engineer intends to use the mean of a random sample of size $n = 144$ to estimate the average mechanical aptitude of assembly line workers in a large industry. If, on the basis of experience, the engineer can assume that $\sigma = 5.8$ for such data, what can he assert with probability 0.99 about the maximum size of his error?

Solution Maximum error E is given by

$$E = Z_{\alpha/2} \frac{\sigma}{\sqrt{n}}$$

Here we have $n = 144$, $\sigma = 5.8$ and

$$Z_{0.005} = 2.575$$

∴ We have

$$E = 2.575 \times \frac{5.8}{\sqrt{144}} = 1.24$$

Thus, the engineer can assert with probability 0.99 that his error will be at most 1.24.

Example 5.10

A research worker wants to determine the average time it takes a mechanic to rotate the tyres of a car and he wants to be able to assert with 95% confidence that the mean of the sample is off by at most 0.25 min. If he can presume from his past experience that $\sigma = 0.8$ min. what is the sample size to be taken?

Solution Formula to determine sample size is

$$n = \left(\frac{Z_{\alpha/2}\sigma}{E}\right)^2$$

Here $E = 0.25$, $\sigma = 0.8$ and $Z_{0.025} = 1.96$

$$\therefore n = \left(\frac{1.96 \times 0.8}{0.25}\right)^2 = 39.3 \cong 40$$

since n is to be an integer.

Hence the research worker will have to time 40 mechanics performing the task of rotating tyres.

5.4 Interval Estimation

An interval estimator is a rule for calculating two numbers a and b, say, to create an interval that we are fairly certain contains the parameter of interest. By 'fairly certain', we mean 'with high probability'. We measure this probability using the confidence coefficient $1 - \alpha$.

The probability that a confidence interval will contain the estimated parameter is called the *confidence coefficient* denoted by $1 - \alpha$.

For example, those who conduct experiments often construct 95% confidence intervals. This means that the confidence coefficient or the probability that the interval will contain the estimated parameter is 0.95. We can increase or decrease the amount certainly by changing the confidence coefficient. Some values used in this respect are 0.90, 0.95, 0.98 and 0.99.

Constructing a Confidence Interval When the sampling distribution of a point estimator is approximately normal, an interval estimator or confidence interval can be constructed using the following procedure. Let us assume that the confidence coefficient is 0.95, for simplicity (Figure 5.4).

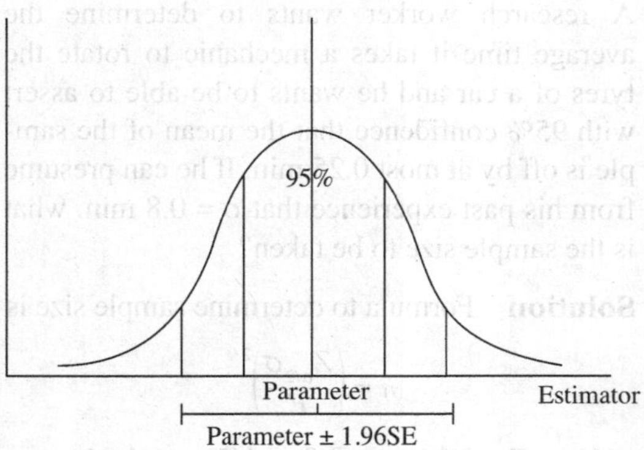

Figure 5.4 Construction of a confidence interval: SE, standard error.

1. Of all the possible values of the estimator that we might select, 95% of them will be in the interval: parameter ±1.96 standard error.
2. Since the value of the parameter is unknown, consider constructing the interval estimator ±1.96 standard error which has the same width as the above interval but has a variable centre.
3. How often this interval will work properly and enclose the parameter of interest can be seen from Figure 5.5.

The first two intervals work properly. The parameter, indicated by thick line, is contained

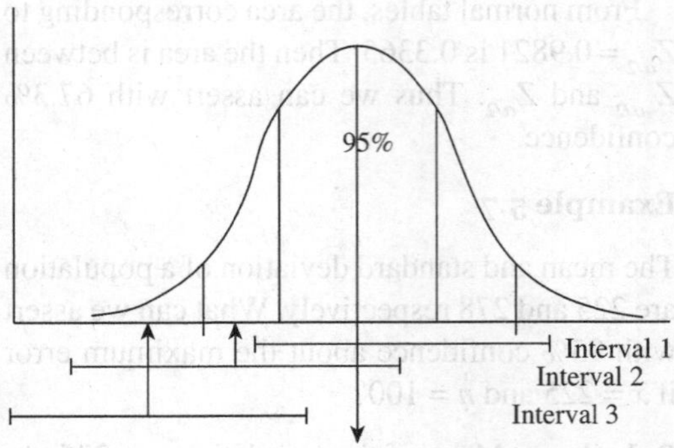

Figure 5.5 Some 95% confidence intervals.

within both intervals. The third interval does not work, since it does not enclose the parameter.

A (1 − α) 100% large sample confidence interval for point estimator is given by $\pm Z_{\alpha/2} \times$ (standard error of the estimator).

where $Z_{\alpha/2}$ is the z-value with an area $\alpha/2$ in the right tail of a standard normal distribution. Table 5.1 gives values of z commonly used for confidence levels.

Table 5.1 Values of z commonly used for confidence intervals.

Confidence Coefficient $(1-\alpha)$	α	$\alpha/2$	$Z_{\alpha/2}$
0.90	0.10	0.050	1.645
0.95	0.05	0.025	1.960
0.98	0.02	0.10	2.230
0.99	0.01	0.005	2.580

Example 5.11

A random sample of size 100 is taken from a population with $\sigma = 5-1$. Given that the sample mean is $\bar{x} = 21.6$, construct a 95% confidence interval for the population mean μ.

Solution Formula for confidence interval is

$$= \left(\bar{x} - Z_{\alpha/2}\frac{\sigma}{\sqrt{n}}, \bar{x} + Z_{\alpha/2}\frac{\sigma}{\sqrt{n}}\right)$$

Here sample mean $\bar{x} = 21.6$, sample size $= n = 100$ and $\sigma = 5.1$ (standard deviation).

The value of $Z_{\alpha/2}$ corresponding to 95% confidence is 1.96.

$$Z_{\alpha/2} = 1.96$$

$$\therefore \bar{x} - Z_{\alpha/2}\frac{\sigma}{\sqrt{n}} = 21.6 - \frac{1.96 \times 5.1}{\sqrt{100}}$$

$$= 21.6 - 0.9996 = 20.6004$$

$$\bar{x} + Z_{\alpha/2}\frac{\sigma}{\sqrt{n}} = 21.6 + \frac{1.96 \times 5.1}{\sqrt{100}}$$

$$= 21.6 - 0.9996 = 22.5996$$

Hence, 95% confidence interval for the population mean μ is (20.6004, 22.5996).

Example 5.12

A sample of size 10 and standard deviation 0.03 is taken from a population. Find the maximum error with 99% confidence.

Solution Here sample size = n = 10 and standard deviation $s = 0.03$.

The value of $t_{\alpha/2}$ corresponding to 99% confidence with dof $v = 9$ is $t_{\alpha/2} = 3.25$

$$\text{Maximum error } E = t_{\alpha/2}\frac{s}{\sqrt{n}} = 3.25 \times \frac{0.03}{\sqrt{10}}$$

$$= 0.0308$$

Example 5.13

What is the maximum error one can expect to make with probability 0.9, when using the mean of a random sample of size $n = 64$ to estimate the mean of a population with $\sigma^2 = 2.56$?

Solution The formula for maximum error is

$$E = Z_{\alpha/2}\frac{\sigma}{\sqrt{n}}$$

The value of $Z_{\alpha/2}$ corresponding to 90% confidence is 1.645.

$$Z_{\alpha/2} = 1.645$$

Here standard deviation $\sigma = \sqrt{2.56} = 1.6$ and sample size $n = 64$.

$$\therefore E = 1.645 \times \frac{1.6}{\sqrt{64}} = 1.645 \times 0.2 = 0.329$$

Hence maximum error with 90% confidence is 0.329.

Example 5.14

The dean of a college wants to use the mean of a random sample to estimate the average amount of time students take to get from one class to the next and he wants to be able to assert with 99% confidence that the error is at most 0.25 min. If it can be presumed from experience that $\sigma = 1.40$ min, how large a sample will have to be taken?

Solution Formula for finding the sample size is

$$n = \left(\frac{Z_{\alpha/2}\,\sigma}{E}\right)^2$$

Here maximum error E = 0.25 min and standard deviation $\sigma = 1.40$ min.

The value of $Z_{\alpha/2}$ corresponding to 99% confidence is $Z_{\alpha/2} = 2.575$.

Substituting these values in the formula

$$n = \left(\frac{Z_{\alpha/2}\,\sigma}{E}\right)^2$$

$$= \left(2.575 \times \frac{1.40}{0.25}\right)^2 = 207.94 \cong 208$$

since n is to be an integer.

Example 5.15

A random sample of 10 ball bearings produced by a company have a mean diameter of 0.496 cm with a standard deviation of 0.002 cm. Find the maximum error estimate E and 95% confidence interval for the actual mean diameter of all ball bearings produced by the company assuming that the sampling is from a normal population.

Solution Since the sample size n = 10 is less than 30, we have to use t-distribution which relates to small sampling.

At 95% confidence, $t_{\alpha/2} = 2.262$

Also, $n = 10$ and $\sigma = 0.002$

$\therefore$ Maximum error estimate at 95% confidence is

$$E = t_{\alpha/2}\frac{\sigma}{\sqrt{n}} = 2.262 \times \frac{0.002}{\sqrt{10}} = 1.4306 \times 10^{-3}$$

$$= 0.00143$$

Since $t_{0.025}$ with $n - 1 = 10 - 1 = 9$ dof $= 2.262$

95% confidence interval limits are $\bar{x} \pm t_{\alpha/2}\frac{\sigma}{\sqrt{n}}$

$= 0.496 \pm 2.262 \times \frac{0.002}{\sqrt{10}} = (0.49457, 0.49743)$

5.5 Bayesian Estimation

5.5.1 Introduction

Personal or subjective probability is a new concept introduced in the probability theory through Bayesian methods.

In general, the parameters that are to be estimated are unknown constants. But in Bayesian methods, they are considered as random variables.

To estimate the mean of a population, μ is treated as a random variable whose distribution indicates the strong feelings or assumption of a person about the possible value of μ.

5.5.2 Bayesian Estimation

Let μ_0 and σ_0 be the mean and standard deviation, respectively, of such a subjective prior distribution.

Combining the prior feelings about the possible values of μ with direct sample evidence, the posterior distribution of μ in Bayesian estimation is approximated by normal distribution with

$$\mu_1 = \frac{n\bar{x}\sigma_0^2 + \mu_0\sigma^2}{n\sigma_0^2 + \sigma^2}$$

and

$$\sigma_1 = \sqrt{\frac{\sigma^2\sigma_0^2}{n\sigma_0^2 + \sigma^2}}$$

Here μ_1 is known as the mean and σ_1 the standard deviation of the posterior distribution.

In computing μ_1 and σ_1, we assume that σ^2 is known.

When σ^2 is unknown, σ^2 is replaced by sample variance s^2 provided the sample is large for which $n \geq 30$.

5.5.3 Bayesian Interval for μ

A $(1 - \alpha)$ Bayesian interval for μ is given by

$$\mu_1 - Z_{\alpha/2}\sigma_1 < \mu < \mu_1 + Z_{\alpha/2}\,\sigma_1$$

Example 5.16

Let a sample of size $n = 20$ from a normal distribution with unknown mean μ and variance $\sigma^2 = 9$. Assume that the prior distribution for μ is normal with mean $\mu_0 = 2$ and variance $\sigma_0^2 = 4$. If the sample mean $\bar{x} = 1.5$, then find the mean and standard deviation of posterior distribution approximated by Bayesian formulas.

Solution Bayesian formula for

$$\mu = \frac{n\bar{x}\,\sigma_0^2 + \mu_0\sigma^2}{n\sigma_0^2 + \sigma^2}$$

Here $n = 20, \bar{x} = 1.5, \sigma_0^2 = 4, \mu_0 = 2$ and $\sigma^2 = 9$

$$\therefore \mu_1 = \frac{20 \times 1.5 \times 4 + 2 \times 9}{20 \times 4 + 9} = \frac{138}{89} = 1.55$$

Bayesian formula for $\sigma_1 = \sqrt{\frac{\sigma_0^2\,\sigma^2}{n\sigma_0^2 + \sigma^2}}$

$$= \sqrt{\frac{4 \times 9}{20 \times 4 + 9}} = \sqrt{\frac{36}{89}} = 0.6360$$

EXERCISES

1. A computer company tested 40 engineers to estimate the average time it takes to assemble a certain computer component. The mean is 12.73 min and the standard deviation is 2.06 min.

(a) If $\bar{x} = 12.73$ is used as a point estimate of the actual average time required to perform the task, determine the maximum error with 99% confidence.

(b) Construct 98% confidence intervals for the true average time it takes to do the work.

Ans: (a) $E = 0.8387$ and (b) 12.73 ± 0.7589

$\Big[$Hint: (a) $E = Z_{\alpha/2} \times \frac{\sigma}{\sqrt{n}} = 2.575 \times \frac{2.06}{\sqrt{40}}$

$= 0.8387$

(b) $\bar{x} = 12.73$, $\sigma = 2.06$, $Z\frac{\sigma}{z} = 2.575$ and $n = 40$

For 98% confidence interval,

$E = 2.33 \times \frac{2.06}{\sqrt{40}} = 0.7589\Big]$

2. In Question 1, with what confidence can we assert that the sample mean does not differ from the true mean by more than 30 s?

Ans: We can assert with 87.4% of confidence.

[Hint: From the normal tables, the area corresponding to $Z_{\alpha/2}$ = 1.535 is 0.437. Area between $Z_{-\alpha/2}$ and $Z_{\alpha/2}$ is 2(0.437) = 0.874.]

3. Given the population standard deviation as 0.3, find (a) 95% and (b) 99% confidence intervals with the mean lead concentration recovered from a sample of lead measurements in 36 different locations is 2.6 g/mL.

Ans: (a) $2.5 < \mu < 2.7$ and (b) $2.47 < \mu < 2.73$

$\Big[$Hint: $\bar{x} \pm Z_{\alpha/2} \times \frac{\sigma}{\sqrt{n}} = 2.6 \pm (1.96 \text{ or } 2.575)$

$\frac{0.3}{6}\Big]$

4. In Question 3, for 99% confidence, if our estimate of μ is off by less than 1%, how large a sample should be chosen?

Ans: $n = \left(2.575 \times \frac{2.3}{0.01}\right)^2 = 5967.5625 = 5968$

5. The mean weight loss of $n = 16$ grinding balls after a certain length of time in mill slurry is 3.42 g with a standard deviation of 0.68 g. Construct a 99% confidence interval for the true mean weight loss of such grinding balls.

Ans: (2.92, 3.92)

$\bar{x} = \left(\frac{0.68}{\sqrt{16}}\right)$

[Hint: $n=16$, $\bar{x}=3.42$, $s=0.68$ and $t_{0.005} = 2.947$; for $n-1=15$ dof, $3.42 \pm 2.947 = (2.92, 3.92)$].

6. Sample size is 10 for a normal distribution with unknown mean μ and variance $\sigma^2 = 4$. Assume that the prior distribution for μ is normal with mean $\mu_0 = 0$ and variance $\sigma_0^2 = 1$. If the sample mean is 0.75, then find the mean and standard deviation of posterior distribution estimated by Bayesian formulas.

Ans: 0.5357 and 0.5345

FILL IN THE BLANKS

1. If $\hat{\theta}$ is an unbiased estimator of θ, then $\hat{\theta}^2$ is a/an _____ estimator of θ.

Ans: Biased

2. If $\hat{\theta}_1$ and $\hat{\theta}_2$are two unbiased estimators of θ then $\hat{\theta} =$ _____ is a/an _____ estimator of θ.

Ans: $\alpha\hat{\theta}_1 + \beta\hat{\theta}_2$ $(\alpha + \beta)$ and unbiased

3. In sampling from an $N(\mu, \sigma^2)$ population, the sample mean is a _____ estimator of μ.

Ans: Consistent

4. If x_i $(i = 1, 2, \ldots, n)$ is a random sample from a normal population $N(\mu, 1)$, then $t = \frac{1}{n}\sum_{i=1}^{n} x_i^2$ is a/an _____ estimator of _____.

Ans: Unbiased and $\mu^2 + 1$

5. If the maximum error with the probability 0.95 is 1.2 and the standard deviation of the population is 10, then the sample size is _____.

Ans: 267

$\Big[$Hint: $n = \left(Z_{\alpha/2} \times \frac{\sigma}{E}\right)^2 = \left(1.96 \times \frac{10}{1.2}\right)^2$

$= 266.77 \cong 267\Big]$

6. If $n = 16$, $s = 2$ and $\bar{x} = 18$, then 99% confidence interval for mean is _____.

Ans: (16.5265, 19.4735)

$\left[\text{Hint: } n = 16 < 30\text{; therefore, small sample to use } (1 - a)\ 100\% \text{ confidence interval for } \mu \text{ is } \left(\bar{x} - t_{\alpha/2} \times \frac{s}{\sqrt{n}}, \bar{x} + t_{\alpha/2} \times \frac{s}{\sqrt{n}}\right) = (16.5265, 19.4735) \quad \therefore t_{\alpha/2} = t_{0.005} = 2.947 \text{ for } n - 1 = 15 \text{ dof.}\right]$

7. A sample of size 100 is taken whose standard deviation is 5. The maximum error with probability 0.95 is _____.

Ans: 0.98

$\left[\text{Hint: } E = Z_{\alpha/2} \times \frac{\sigma}{\sqrt{n}} = 1.96 \times \frac{5}{\sqrt{100}} = 0.98\right]$

8. A sample size is 25. The maximum error with 0.95 probability is 0.1. Then the standard deviation is _____.

Ans: $\sigma = 0.2551$

$\left[\text{Hint: } E = Z_{\alpha/2} \times \frac{\sigma}{\sqrt{n}} = 1.96 \times \frac{\sigma}{\sqrt{25}} = 0.1 \Rightarrow \sigma = \frac{0.1 \times 5}{1.96} = 0.2551\right]$

9. If $n = 16$, $s = 2$ and $\bar{x} = 18$, then 95% confidence interval for mean is _____.

Ans: (16.9345, 19.0655)

$\left[\text{Hint: } t_{\alpha/2} = t_{0.025} = 2.131 \text{ for } n - 1 = 15 \text{ dof} \Rightarrow 18 \pm 2.131 \times \frac{2}{\sqrt{16}} = (16.9345, 19.0655)\right]$

10. If there are 5 defective items among 4000, one-sided 99% confidence interval of proportion is _____.

Ans: 0.0033

6

Inferences Concerning Means and Proportions

6.1 Introduction

In practical situations, statistical inference can involve either estimating a population parameter or making decisions about the value of the parameter. In the Chapter 5, we have considered estimation. Now we take up a study of tests of hypotheses. In tests of hypotheses, a postulate or a statement about a parameter of the population is tested for its validity.

Statistical decisions are conclusions arrived at about the population parameters on the basis of a random sample from the population.

6.2 Statistical Hypotheses

Statistical hypothesis is a guess made about the parameter or parameters of population distributions.

Suppose, e.g., we want to decide that a coin is not fair. We formulate the hypothesis that it is fair, i.e. $P(\text{H}) = 0.5$—the probability of heads occurring is 0.5. Similarly, if we want to decide that one procedure is better than another, we make a hypothesis that there is no difference between the two procedures.

6.2.1 Null Hypothesis

A null hypothesis (NH) is a statistical hypothesis formulated for reaching the decision to accept a stated hypothesis or to reject it or sometimes not to make a decision yet but to take another sample. An NH is denoted by H_0.

6.2.2 Alternative Hypothesis

Any hypothesis that differs from a given NH is called an *alternative hypothesis* (AH), e.g. the NH is $P(\text{H}) = 0.5$ then AH is $P(\text{H}) = 0.7$, i.e. $P(\text{H}) \neq 0.5$ or $P(\text{H}) > 0.5$. An AH is denoted by H_1.

6.3 Tests of Hypotheses and Significance

A test of hypothesis is a process to determine whether to accept or reject the NH H_0. This test determines whether the observed samples differ significantly from the expected results. Acceptance of hypothesis only indicates that the data would not give sufficient evidence to refute the hypothesis, whereas rejection is a firm conclusion where the sample evidence refutes it.

A test of hypothesis is also called a test of significance or decision rule. When the NH is accepted, it means that the result is significant.

6.4 Type I and Type II Errors

If we reject a hypothesis when it happens to be true, we say that a *Type I* error has been made. If, on the other hand, we accept a hypothesis when it should be rejected, then we say that a *Type II* error has been made. In either case, a wrong decision or error in judgement has occurred.

In order to have good tests of hypotheses, they must be designed to have minimum errors of decision. The only way to reduce both types of errors is to increase the sample size which may or may not be possible. Table 6.1 illustrates types of errors in test of hypothesis.

Table 6.1 Types of errors in test of hypothesis.

Decision	Unknown Truth	
Accept H_0	H_0 is true True decision $p = 1 - \alpha$	H_1 is true. Type II error $p = \beta$
Accept H_1	Type I error $p = \alpha$	True decision $p = 1 - \beta$

6.5 Levels of Significance

In testing a given hypothesis, the maximum probability with which we would be willing to risk a Type I error is called the *level of significance* (LOS) of the test and is denoted by α. This probability is often specified before any sample is drawn so that results obtained will not influence our decision.

In practice, an LOS of $\alpha = 0.05$ or 0.01 is used. If, e.g., a 0.05% or 5% LOS is chosen in designing a test of hypothesis, then there are about 5 chances in 100 that we would reject the hypothesis when it should be accepted. That is, whenever the NH is true, we are about 95% confident that we would make the right decision. In such cases, we say that the hypothesis has been rejected at 0.05 LOS, which means that we could be wrong with a probability 0.05.

An LOS is also known as the size of the test.

Thus, α = probability of committing Type I error

$$\Rightarrow P(\text{reject } H_0/H_0) = \alpha$$

and

$$\beta = \text{probability of committing Type II error}$$

$$\Rightarrow P(\text{accept } H_0/H_1) = \beta$$

The power of the test is computed as $1 - \beta$.

Notes

1. When the size of the sample is increased, the probabilities of committing both Type I and Type II errors, i.e. α and β, are decreased.
2. α and β are known as producer's risk and consumer's risk respectively.
3. When both α and β are small, then the test procedure is good and hence the chance of making right decision is increased.

Simple Hypothesis It is a statistical hypothesis that completely specifies an exact parameter. An NH H_0 is always a simple hypothesis stated as an equality specifying an exact value of the parameter which includes any value not stated by an AH H_1.

Examples

1. NH H_0: $\mu = \mu_0$ (population mean is equal to a specified constant μ_0)
2. NH H_0: $\mu_1 - \mu_2 = \delta$ (difference between the sample means is equal to a constant δ)

Composite Hypothesis It is a hypothesis stated in terms of several possible values in the form of inequalities.

An AH is a composite hypothesis involving statements in the form of inequalities <, > or ≠.

Examples

1. AH H_1: $\mu > \mu_0$
2. AH H_1: $\mu < \mu_0$
3. AH H_1: $\mu \neq \mu_0$

6.6 Statistical Test of Hypothesis Procedure

The hypothesis test procedure consists of the following steps:

1. Decide upon an NH H_0 and an AH H_1 (which is usually the hypothesis we want to test).
2. Select a test statistic (common test statistics are the sample mean $\bar{X}$ and the sample variance S^2).
3. Choose α, the LOS of the test (α is usually chosen as 0.05 or 0.01 but any value between 0 and 1 can be selected).
4. Choose a rejection region, called the critical region (CR) for the test, i.e. choose a set of

possible test statistic values such that if H_0 is true then the probability that the value of the test statistic will fall in the rejection region is α.

(a) For one-tailed test (OTT), a CR is often chosen to be all the numbers greater or smaller than a critical value.

(b) For a two-tailed test (TTT), another popular choice is the set of all numbers either less than a left critical value or greater than a right critical value.

The complement of the rejection region is called the *acceptance region*.

5. Calculate the test statistic of a random sample from the population. If this value falls in the CR then reject H_0 and accept H_1, otherwise accept H_0.
6. Calculate the *p*-value of the test. It is the probability test. If H_0 is true, a value at least as extreme as the observed test statistic will be observed. Then the decision rule for accepting or rejecting is "reject H_0 if the *p*-value is less than α". The *p*-value provides useful information as to whether H_0 is to be accepted or rejected.

Example 6.1

A software development company has verified that the number of lines of code per programmer per week X has a normal distribution with mean 280 and standard deviation 25. The company decides to use a statistical test to determine whether the new program they have used has led to better productivity. A random sample obtained from 100 programmers yields $\bar{X} = 290$ lines per week. Determine if at 5% LOS whether the mean of X has increased.

Solution We use the NH that there has been no change.

$$H_0: \mu = \mu_0 = 280 \text{ lines/week}$$

Hoping that μ has increased, let

$$H_1: \mu > 280 \text{ lines/week}$$

$\bar{X}$ is normally distributed with mean $\mu = 280$ and standard deviation $\sigma = 25$.

Also, sample size $n = 100$

Standard normal random variable is given by

$$Z = \frac{\bar{X} - \mu}{\frac{\sigma}{\sqrt{n}}} = \frac{\bar{X} - 280}{\frac{25}{\sqrt{100}}} = \frac{\bar{X} - 280}{2.5}$$

Since $Z_{0.05} = 1.645$, the rejection region is all $\bar{X}$ such that

$$\frac{\bar{X} - 280}{2.5} > 1.645 \text{ or}$$

$$\bar{X} > 280 + 2.5 \times 1.645$$

$$= 284.1125$$

Since $\bar{X} = 290$, we reject H_0 and conclude that $\mu > 280$.

Suppose X, as it has a normal distribution with mean 285 and standard deviation 25 (Figure 6.1).

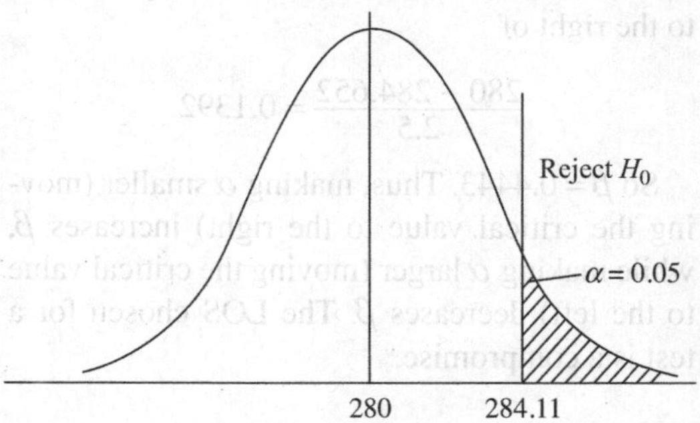

Figure 6.1 $\bar{X}$ has normal distribution with mean $\mu = 280$ and standard deviation $\sigma = 25$.

Then the H_1 is simple and the test can be as shown in Figure 6.2. Here β, the probability of accepting H_0 when H_1 is true, is the area under the density function for $\bar{X}$ (when H_1 is true) to the left of 284.11. Since for this calculation we assume that H_1 is true, β is also the area of the tail of a standard normal density to the right of the value.

$$Z = \frac{285 - 284.11}{2.5} = 0.356$$

Consulting the normal table, we find $\beta = 0.3404$. Hence the power of the test is 0.6596.

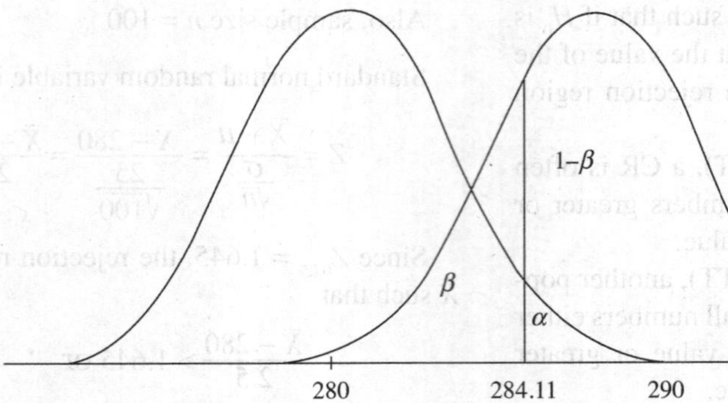

Figure 6.2 (Example 6.1).

Suppose we reduce α to 0.01 (one per cent), the CR becomes $\overline{X} > 280 + 2 \times (2.326) = 284.652$.

Since $Z_{0.01} = 2.326$, this means that β is the area under the standard normal density function to the right of

$$\frac{280 - 284.652}{2.5} = 0.1392$$

So $\beta = 0.4443$. Thus, making α smaller (moving the critical value to the right) increases β, while making α larger (moving the critical value to the left) decreases β. The LOS chosen for a test is a compromise.

6.7 Reasoning of Statistical Test of Hypothesis

The reasoning used in statistical test of hypothesis is similar to the process in a court trial. In trying a person for theft, the court must decide between innocence and guilt. The accused is assumed to be innocent. The prosecution collects and presents all available evidence to contradict the innocent hypothesis and obtains a conviction. If there is sufficient evidence against innocence, the court rejects the innocence hypothesis and declares the accused guilty. The court will find him not guilty if the prosecution does not present sufficient evidence to prove the guilt of the accused. Note that this does not prove that the accused is in fact innocent but only shows that there has not been enough evidence to conclude that the accused has been guilty.

6.7.1 Critical Region

From the sample data, a test statistic S has been calculated. In any test of hypothesis, it is used to accept or reject the NH of the test. Consider the area under the probability curve of the sampling distribution of the test statistic, which follows some given distribution. This area under the probability curve is divided into two regions: region of rejection and region of acceptance. In the region of rejection, which is also called the region of significance or CR, the NH is rejected. In the region of acceptance, which is also called the region of non-significance or non-CR, the NH is accepted.

The area of the CR is equal to the LOS α. It is to be noted that the CR always lies on the tail regions of distributions. Whether the CR lies on the left side tail or the right side tail or both the tails depends upon the nature of the AH.

6.7.2 Critical Region or Significant Value

For a given LOS α, the value of the test statistic S_α that divides the area under the probability curve into CR and non-CR is called the CR or significant value.

One-tailed Test

Right One-tailed Test Suppose that the AH is $H_1: \mu > \mu_0$ or $H_1: \sigma_1^2 > \sigma_2^2$ type. Then the entire CR lies to the right side of the right-side tail of the probability density curve as shown in Figure 6.3.

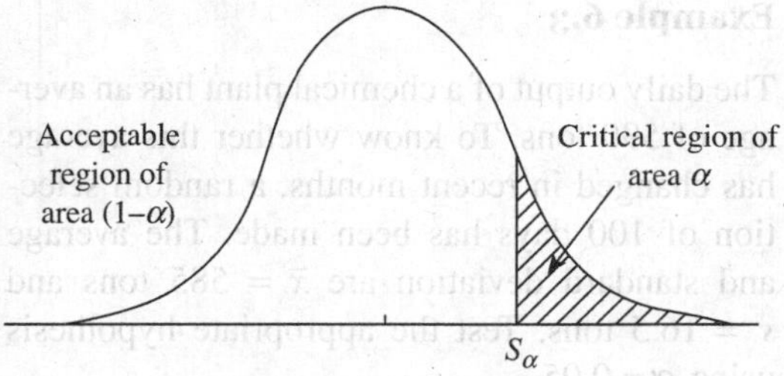

Figure 6.3 Right one-tailed test.

In this case the test of hypothesis is known as the right one-tailed test (ROTT).

Left One-tailed Test Suppose the AH is H_1: $\mu < \mu_0$ or H_1: $\sigma_1^2 < \sigma_2^2$ type. Then the entire CR lies to the left side of the left-side tail of the probability density curve as shown in the Figure 6.4. In this case the test of hypothesis is known as the left one-tailed test (LOTT).

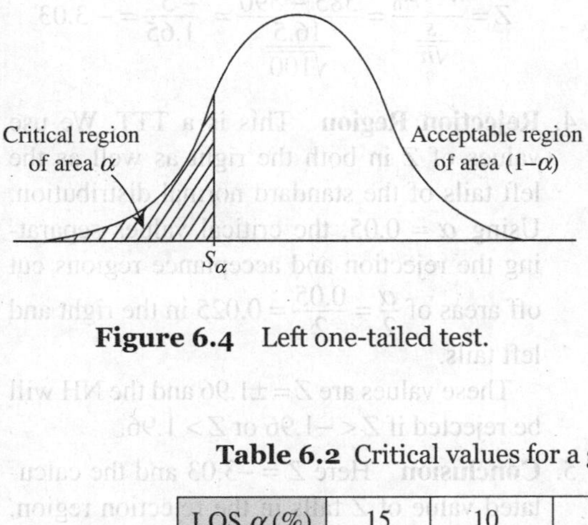

Figure 6.4 Left one-tailed test.

Two-tailed Test Suppose that the AH is H_1: $\mu \neq \mu_0$ or H_1: $\sigma_1^2 \neq \sigma_2^2$ type. Then the CR lies on the right of the right tail and the left side of the left-side tail of the probability density curve such that CR of area $\frac{1}{2}\alpha$ lies to the right side of the right-side tail and the CR of area $\frac{1}{2}\alpha$ lies on the left side of the left-side tail as shown in Figure 6.5. In this case the test of hypothesis is known as the two-tailed test (TTT).

Refer to Table 6.2 for critical values for a given LOS α for ROTT, LOTT and TTT.

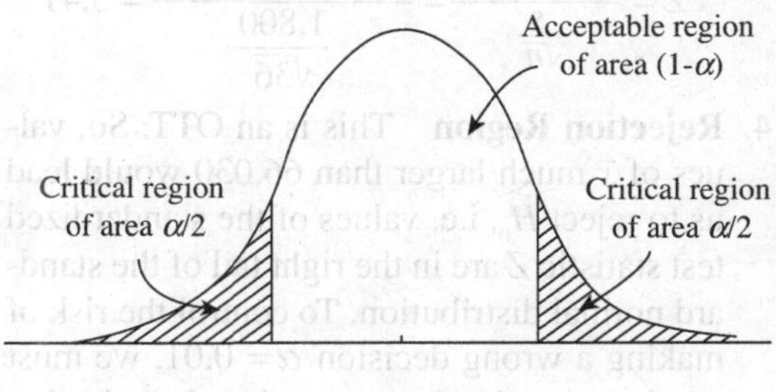

Figure 6.5 Two-tailed test.

Example 6.2

The monthly salaries of women professionals average to Rs 66,030. Do men in similar positions have average monthly salaries that are higher than those for women? A random sample of n = 36 men in professional positions shows $\bar{x}$ = Rs 67,053 and s = Rs 1,800. Test the appropriate hypothesis using α = 0.01.

Table 6.2 Critical values for a given LOS α for ROTT, LOTT and TTT.

LOS α (%)	15	10	5	4	1	0.5	0.2
LOS α	0.15	0.1	0.05	0.04	0.01	0.005	0.002
$(-Z_{\alpha/2}, Z_{\alpha/2})$ for TTT	(–1.44, 1.44)	(–1.645, 1.645)	(–1.96, 1.96)	(–2.06, 2.06)	(–2.58, 2.58)	(–2.81, 2.81)	(–3.08, 3.08)
$-Z_\alpha$ for LOTT	–1.04	–1.28	–1.645	–2.6	–2.33	–2.58	–2.88
Z_α for ROTT	1.04	1.28	1.645	2.6	2.33	2.58	2.88

Notes: TTT, two-tailed test; ROTT, right one-tailed test and LOTT, left one-tailed test.

Solution We would like to show that the average monthly earnings of men are higher than Rs 66,030, the women's average. If μ is the average monthly earnings of men, we can set out the formal test of hypothesis in following steps:

1. NH H_0: $\mu = 66{,}030$
2. AH H_1: $\mu > 66{,}030$
3. **Test Statistic** Using the sample information with σ as an estimate of the population standard deviation, we calculate

$$Z = \frac{\bar{x} - 66{,}030}{\frac{s}{\sqrt{n}}} = \frac{67{,}053 - 66{,}030}{\frac{1{,}800}{\sqrt{36}}} = 3.41$$

4. **Rejection Region** This is an OTT. So, values of $\bar{x}$ much larger than 66,030 would lead us to reject H_0, i.e. values of the standardized test statistic Z are in the right tail of the standard normal distribution. To control the risk of making a wrong decision $\alpha = 0.01$, we must set the critical value separating the rejection region from the acceptance region in such a way that the right tail is exactly $\alpha = 0.01$. This value is found from table to be 2.33 as shown in Figure 6.6. The NH H_0 will be rejected if the observed value of the test statistic Z is greater than 2.33.
5. **Conclusion** Compare the observed value of the test statistic $Z = 3.41$ with the value necessary for rejection of $Z = 2.33$. Since the observed value of the test statistic falls in the rejection region, we can reject H_0 and conclude that the average monthly earnings of men are higher than the average of women's earnings. The probability that we have made an incorrect decision is $\alpha = 0.01$.

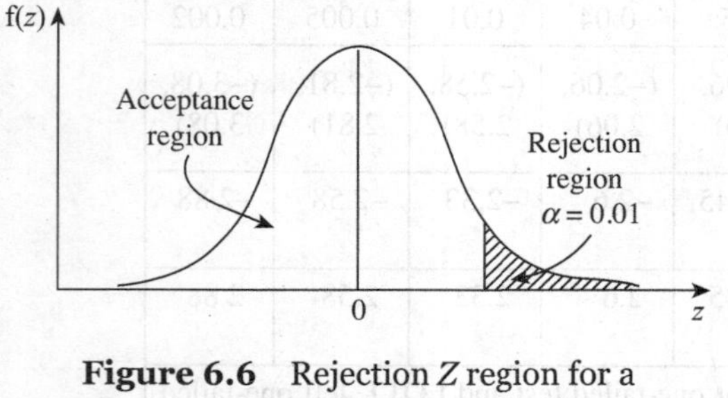

Figure 6.6 Rejection Z region for a right-tailed test with $\alpha = 0.01$.

Example 6.3

The daily output of a chemical plant has an average of 590 tons. To know whether this average has changed in recent months, a random selection of 100 days has been made. The average and standard deviation are $\bar{x}$ = 585 tons and s = 16.5 tons. Test the appropriate hypothesis using $\alpha = 0.05$.

Solution We want to show that the average output in recent months has improved and is higher than 590 tons. If μ is the average output, we may set out the formal test of hypothesis as follows:

1. NH H_0: $\mu = 590$
2. AH H_1: $\mu \neq 590$
3. **Test Statistic** So, the point estimate for μ is $\bar{x}$. So the test statistic is

$$Z = \frac{\bar{x} - \mu_0}{\frac{s}{\sqrt{n}}} = \frac{585 - 590}{\frac{16.5}{\sqrt{100}}} = \frac{-5}{1.65} = -3.03$$

4. **Rejection Region** This is a TTT. We use values of Z in both the right as well as the left tails of the standard normal distribution. Using α = 0.05, the critical values separating the rejection and acceptance regions cut off areas of $\frac{\alpha}{2} = \frac{0.05}{2} = 0.025$ in the right and left tails.

   These values are $Z = \pm 1.96$ and the NH will be rejected if $Z < -1.96$ or $Z > 1.96$.
5. **Conclusion** Here $Z = -3.03$ and the calculated value of Z falls in the rejection region. We can reject the NH H_0 that μ = 590 and conclude that it has changed. The probability of rejecting H_0 when H_0 is true and $\alpha = 0.05$, a fairly small probability. Hence, we can be confident that the decision is correct.

6.7.3 Calculating the *p*-Value

In Examples 6.1–6.3, the decision to reject or accept H_0 was made by comparing the computed value of the test statistic with a critical value of Z based on the LOS α of the test. However, different LOS may lead to different conclusions. For example, if in a right-tailed test the test statistic is $Z = 2.03$, we can reject H_0 at the 5% (or 0.05) LOS because the test statistic exceeds $Z = 1.645$. On the other hand, we cannot reject H_0 at the 1% (0.01) LOS because the test statistic is less than $Z = 2.33$ (Figure 6.7). To avoid ambiguity in drawing conclusions, it is preferred to use a variable LOS called the *p*-value for the test.

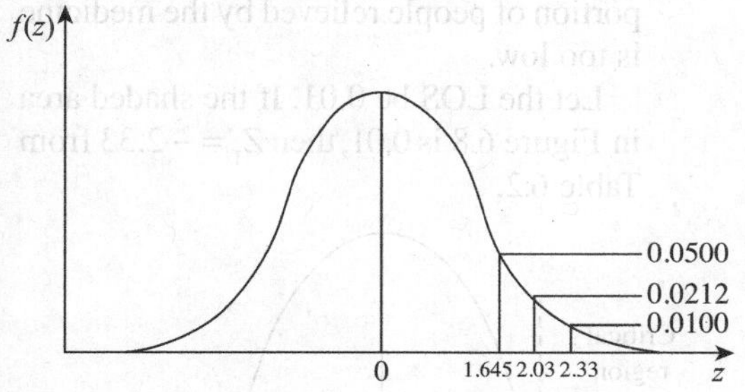

Figure 6.7 Variable rejection regions.

***p*-Value Definition** The *p*-value or observed LOS of a statistical test is the smallest value of α for which H_0 can be rejected. It is the actual risk of committing a Type I error, if H_0 is rejected based on the observed value of the test statistic. The *p*-value measures the strength of H_0.

In the right-tailed test with observed test statistic $Z = 2.03$, the smallest critical value we can use and still reject H_0 is $Z = 2.03$. For this critical value, the risk of incorrect decisions is

$$P(Z \geq 2.03) = 1 - 0.9788 = 0.0212$$

This probability is the *p*-value for the test. Note that it is actually the area to the right of the calculated value of the test statistic.

A small *p*-value indicates that the observed value of the statistic lies far away from the hypothetical value of μ. This gives strong evidence that H_0 is false and should be rejected. Large *p*-values indicate that the observed test statistic is not far from the hypothetical mean and does not support rejection of H_0.

If the *p*-value is less than or equal to a pre-assigned LOS α, then NH H_0 can be rejected and we can report that the results are statistically significant at level α.

In the previous case, by choosing $\alpha = 0.05$ as LOS, H_0 can be rejected because the *p*-value is less than 0.05. On the other hand, if we choose $\alpha = 0.01$ as LOS, the *p*-value 0.0212 is not small enough to permit rejection of H_0. The results are significant at the 5% level but not at 1% level. This is reported in professional journals as significant ($p < 0.05$).

Example 6.4

In Example 6.3, we want to know whether the daily output which averaged 590 tons has changed in recent months. With the data calculated, compute the *p*-value for this TTT of hypothesis. Use the *p*-value to draw conclusions regarding the statistical test.

Solution The rejection region for this TTT of hypothesis is found in both tails of the normal probability distribution. Since the observed value of the test statistic is $Z = -3.03$, the smallest rejection region that we can use and still reject H_0 is $|Z| > 3.03$. For this rejection region, the value of α is the *p*-value:

$$p\text{-Value} = P(Z > 0.03) + P(Z < -0.03)$$
$$= (1 - 0.9988) + 0.0012 = 0.0024$$

Note that the two-tailed *p*-value is actually twice the tail area corresponding to the calculated value of the test statistic. If this *p*-value = 0.0024 is less than or equal to the pre-assigned LOS α, H_0 can be rejected. For this test, we can reject H_0 at either the 1% or 5% LOS.

6.7.4 Hypothesis Concerning One Mean (σ Known)

Example 6.5

It is claimed that the thermal conductivity of a certain kind of cement brick is 0.340. To test this, a test sample of size $n = 35$ has been taken with 0.05 LOS. From the information gathered from similar studies, the standard deviation is 0.010.

Solution We have the following:

1. NH H_0: $\mu = 0.340$
2. AH H_1: $\mu \neq 0.340$
3. LOS $\alpha = 0.05$

The AH H_1 is two-sided here since we want to reject the NH H_0 if the mean of the determinations is significantly less than or greater than 0.340.

Now the standardized statistic for test concerning mean σ known is

$$Z = \frac{\bar{x} - \mu_0}{\sigma/\sqrt{n}}$$

CRs for testing $\mu = \mu_0$ and normal population and σ known are given in Table 6.3.

Table 6.3 Critical values.

AH H_1	Reject NH H_0
$\mu < \mu_0$	If $Z < -Z_\alpha$
$\mu > \mu_0$	If $Z > Z_\alpha$
$\mu \neq \mu_0$	If $Z < -Z_{\alpha/2}$ or $Z > Z_{\alpha/2}$

If $\alpha = 0.05$, the critical values of the criteria are −1.645 and 1.645 for one-sided alternatives and −1.96 and 1.96 for the two-sided alternatives.

If $\alpha = 0.01$, the critical values are −2.33 and 2.33,

for one-sided alternatives, and the critical values are − 2.575 and 2.575,

for two-sided alternatives, and they are as shown in Table 6.3.

Example 6.6

A manufacturer of a patent medicine claimed that it was 90% effective in relieving an allergy for a period of 8 h. In a sample of 200 people who had the allergy, the medicine provided relief for 160 people.

(a) Determine whether the manufacturer's claim is legitimate by using 0.01 as the LOS.

(b) Find the p-value of the test.

Solution

(a) Let p denote the probability of getting relief from the allergy by using the medicine. In the test of hypothesis, the various steps are as follows:

1. NH $H_0 : p = 0.9$ and the claim is correct.
2. AH $H_1 : p < 0.9$ and the claim is incorrect.

We choose an OTT because we are interested in determining whether the proportion of people relieved by the medicine is too low.

Let the LOS be 0.01. If the shaded area in Figure 6.8 is 0.01, then $Z_1 = -2.33$ from Table 6.2.

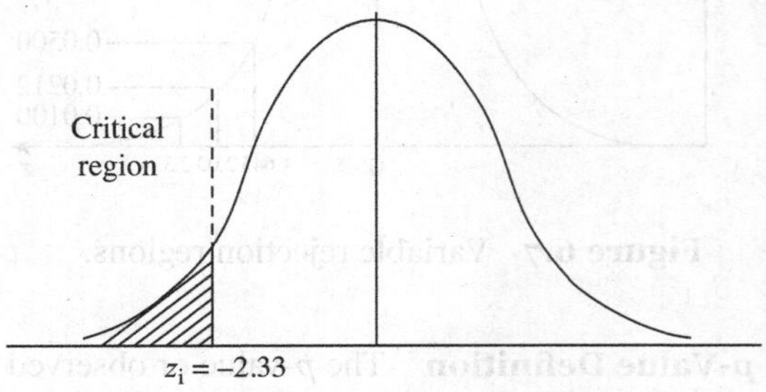

Figure 6.8 Critcal region of example 6.6.

Decision Rule

1. The claim is not legitimate if $Z < -2.33$ (in this case, we reject H_0).
2. Otherwise, the claim is legitimate, and the observed results are due to chance (in this case, we accept H_0).

Let H_0 be true. Then

$$\mu = np = 200 \times 0.9 = 180$$

$$\sigma = \sqrt{npq} = \sqrt{200 \times 0.9 \times 0.1} = 4.23$$

We write 160 in standard units.

$$\frac{x - \mu}{\sigma} = \frac{160 - 180}{4.23} = -4.73$$

This is much less than − 2.33.

By our decision rule, we conclude that the claim is not legitimate and the sample results are highly significant.

(b) The p-value of the test is $P(Z \leq -4.73) = 0$. This shows that the claim is almost certainly false.

That is, if H_0 is true, it is almost certain that a random sample of 200 allergic suffers who used the medicine would include more than 160 people who found relief.

Example 6.7

The mean lifetime of a sample of 100 bulbs produced by a company is computed as 1570 h with a standard deviation of 120 h. If μ is the mean lifetime of all the bulbs produced by the company, test the hypothesis $\mu = 1600$ h against the AH $\mu \neq 1600$ h using a LOS of as (a) 0.05 and (b) 0.01. Also (c) find *p*-value of the test.

Solution We have to decide between the two hypotheses

1. NH $H_0 : \mu = 1600$
2. AH $H_1 : \mu \neq 1600$

Here we have to use a TTT because the inequality $\mu \neq 1600$ includes values both greater than and smaller than 1600.

(a) For a TTT at a LOS of 0.05, we have the following decision rule:

1. Reject H_0 if the Z score of the sample mean is outside the range −1.96 to 1.96.
2. Accept H_0 (or withhold any decision) otherwise.

The statistic under consideration is the sample mean $\bar{X}$. The sampling distribution of X has a mean $\mu_{\bar{x}} = \mu$ and standard deviation $\sigma_{\bar{x}} = \sigma/\sqrt{n}$, where μ and σ are the mean and standard deviation of the population of all bulbs produced respectively.

Under the hypothesis H_0, we have $\mu = 1600$. Also $\sigma_{\bar{x}} = \sigma/\sqrt{n} = \dfrac{120}{\sqrt{100}} = 12$, using the sample standard deviation as an estimate of σ.

$$\text{Now } Z = \frac{\bar{x} - \mu}{\sigma/\sqrt{n}} = \frac{1570 - 1600}{12} = -2.50$$

It lies outside the range −1.96 to 1.96.

$\therefore$ We reject H_0 at a 0.05 LOS.

(b) Suppose the LOS is 0.01, the range −1.96 to 1.96 in the decision rule of part (a) is replaced by − 2.58 to 2.58. Since the Z score of − 2.50 lies inside this range, we accept H_0 (or withhold any decision) at a 0.01 LOS.

(c) The *p*-value of the TTT is $P(Z \leq -2.50) + P(Z \geq 2.50) = 0.0124$.

This is the probability that a mean lifetime of less than 1570 h or more than 1630 h would occur by chance if H_0 were true.

Example 6.8

In Example 6.7, test the hypothesis $\mu = 1600$ h, using a LOS of (a) 0.05 and (b) 0.01. Also (c) find the *p*-value of the test.

Solution We have to decide between the two hypotheses:

1. NH $H_0 : \mu = 1600$
2. AH $H_1 : \mu < 1600$

An OTT has to be used here.

(a) The LOS is 0.05 and is shown in Figure 6.9 as the shaded region.

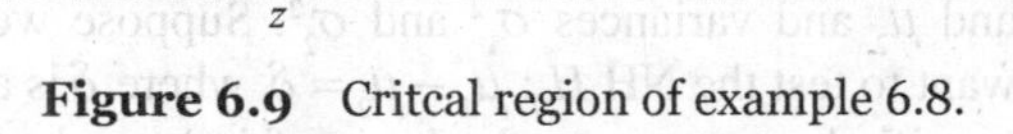

Figure 6.9 Critcal region of example 6.8.

It has an area of 0.05. The corresponding value of Z is − 1.645. We adopt the decision rule:

1. Reject H_0 if $Z < -1.645$
2. Accept H_0 (or withhold any decision) otherwise

(b) Let the LOS be 0.01.Then Z-value is −2.33. So, we adopt the decision rule:

1. Reject H_0 if $Z < -2.33$
2. Accept H_0 (or withhold any decision) otherwise

As in (a), the Z score is -2.50. This is less than -2.33. So, we reject H_0 at a 0.01 LOS. Note that this decision is not the same as that reached in Example 6.8 (b) using a TTT.

It follows that decision relating to a given hypothesis H_0 based on OTT and TTT are not always in agreement. This is, of course, to be expected because we are testing H_0 against a different alternative in each case.

(c) The p-value of the test is $P(Z < 1570) = 0.0062$.

This is the probability that a mean lifetime of less than 1570 h would occur by chance in case H_0 is true.

6.8 Inference Concerning Two Means

6.8.1 Introduction

There arise many statistical problems in which we have to take decisions about the relative size of the means of two or more populations. We shall therefore consider below tests concerning the difference between two means.

6.8.2 Inference Concerning Two Means Procedure

Consider two populations having the means μ_1 and μ_2 and variances σ_1^2 and σ_2^2. Suppose we want to test the NH H_0: $\mu_1 - \mu_2 = \delta$, where δ is a specified constant, on the basis of independent random samples of size n_1 and n_2 respectively.

Similar to the tests concerning one mean, we consider tests of this NH against each of the AH as follows:

1. H_1:$\mu_1 - \mu_2 < \delta$
2. H_1:$\mu_1 - \mu_2 > \delta$
3. H_1:$\mu_1 - \mu_2 \neq \delta$

The test itself will depend on the difference between the sample means: $\overline{X}_1 - \overline{X}_2$.

If both samples are from normal populations with known variances, it can be based on the statistic

$$Z = \frac{(\overline{X}_1 - \overline{X}_2) - \delta}{\sigma_{(\overline{X}_1 - \overline{X}_2)}}$$

which is a random variable having the standard normal distribution.

Here $\sigma_{(\overline{X}_1 - \overline{X}_2)}$ is the standard deviation of the sampling distribution of the difference for random samples from infinite populations which may be obtained using the following theorem:

Theorem If the distribution of two independent random variables have the means μ_1 and μ_2 and the variances σ_1^2 and σ_2^2, then the distribution of their sum (or difference) has the mean $\mu_1 + \mu_2$ (or $\mu_1 - \mu_2$) and the variance $\sigma_1^2 + \sigma_2^2 \left(\sigma_1^2 - \sigma_2^2\right)$

To find the variance of the difference between the means of two independent random samples of size n_1 and n_2 from infinite population, note that the variances of the two means are

$$\sigma_{\overline{X}_1}^2 = \frac{\sigma_1^2}{n_1} \text{ and } \sigma_{\overline{X}_2}^2 = \frac{\sigma_2^2}{n_2}$$

where σ_1^2 and σ_2^2 are the variances of the respective populations. By the above theorem, we have

$$\sigma_{(\overline{X}_1 - \overline{X}_2)}^2 = \frac{\sigma_1^2}{n_1} + \frac{\sigma_2^2}{n_2}$$

The test statistic can be written as

$$Z = \frac{(\overline{X}_1 - \overline{X}_2) - \delta}{\sqrt{\dfrac{\sigma_1^2}{n_1} + \dfrac{\sigma_2^2}{n_2}}}$$

which is a random variable having the standard normal distribution. In deriving this result, we have assumed that the sampling is from normal populations. However, the above statistic can be used when the samples are large enough so that we can apply the central limit theorem and approximate σ_1 and σ_2 with S_1 and S_2, i.e. when n_1 and $n_2 \geq 30$.

$$\therefore Z = \frac{(\overline{X}_1 - \overline{X}_2) - \delta}{\sqrt{\frac{S_1^2}{n_1} + \frac{S_2^2}{n_2}}}$$

The CRs for testing the NH $\mu_1 - \mu_2 = \delta$ are given in Table 6.4. Table 6.4 gives the CRs for testing $\mu_1 - \mu_2 = \delta$. Normal populations and σ_1 and σ_2 are known or large samples with size n_1 and $n_2 \geq 30$.

Table 6.4 Decision rule for alternative hypotheses.

AH H_1	Reject NH H_0
$\mu_1 - \mu_2 < \delta$	If $Z < -Z_\alpha$
$\mu_1 - \mu_2 > \delta$	If $Z > Z_\alpha$
$\mu_1 - \mu_2 \neq \delta$	If $Z < -Z_{\alpha/2}$ or $Z > Z_{\alpha/2}$

Although δ can be any constant, it is to be noted that in majority of the problems if value is zero, we test the NH of no difference, i.e. H_0: $\mu_1 = \mu_2$.

Example 6.9

It has been claimed that the resistance of electric wire can be reduced by more than 0.050 ohm by alloying. To test this claim, 32 values obtained for standard wire yielded $\overline{X}_2 = 0.136$ ohm and $s_1 =$ 0.004 ohm and 32 values obtained for alloyed wire yielded $x_2 = 0.083$ ohm and $s_2 = 0.005$ ohm. At the 0.05 ohm LOS, does this support the claim?

Solution We follow the following steps:

1. NH H_0: $\mu_1 - \mu_2 = 0.050$
2. AH H_1: $\mu_1 - \mu_2 > 0.050$
3. LOS $\alpha = 0.050$
4. **Criterion** Reject the NH H_0
5. **Calculation** If $Z > 1.645$, then

$$\therefore Z = \frac{(\overline{X}_1 - \overline{X}_2) - \delta}{\sqrt{\frac{S_1^2}{n_1} + \frac{S_2^2}{n_2}}}$$

$$= \frac{(0.136 - 0.083) - 0.050}{\sqrt{\frac{(0.004)^2}{32} + \frac{(0.005)^2}{32}}}$$

$$= 2.65$$

6. **Conclusion** Since $Z = 2.65$ is greater than 1.645, the NH must be rejected.

The claim is supported by the data. Also, the p-value is 0.004. So, the evidence for alloying is very strong. Only 4 in 1000 times would Z be at least 2.65 if the mean difference is 0.05.

Example 6.10

A company claims that the light bulbs are superior to those of its main competitor. If a study showed that a sample of $n_1 = 40$ of its bulbs has a mean lifetime of 647 h of continuous use with a standard deviation of 27 h, while a sample of $n_2 = 40$ bulbs made by the competitor had a mean lifetime of 638 h of continuous use with a standard deviation of 31 h. Does this support the claim at 0.05 LOS?

Solution We follow the following steps:

1. NH H_0: $\mu_1 - \mu_2 = 0$
2. AH H_1: $\mu_1 - \mu_2 > 0$
3. LOS $\alpha = 0.05$
4. **Criterion** Reject the NH H_0
5. **Calculation** If $Z > 1.645$, then

$$\therefore Z = \frac{(\overline{X}_1 - \overline{X}_2)\ \delta}{\sqrt{\frac{S_1^2}{n_1} + \frac{S_2^2}{n_2}}} = \frac{647 - 638}{\sqrt{\frac{27^2}{40} + \frac{31^2}{40}}} = 1.38$$

6. **Conclusion** Since $Z = 1.38$ does not exceed 1.645, the NH cannot be rejected. That is, the observed difference between the two sample means is not significant.

Also, the p-value $= 0.0838$ (Figure 6.10) so the evidence against equal means is not very strong.

$$\therefore p\text{-Value} = 0.0838 \text{ for } Z = 1.38$$

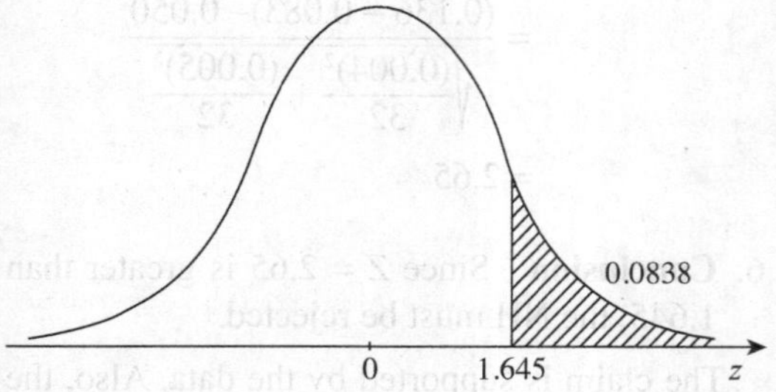

Figure 6.10 Large samples test for example 6.10.

EXERCISES

1. A machine runs on an average 125 h/year. A random sample of 49 machines has an annual average use of 126.9 h with standard deviation of 8.4 h. Does this support the claim that machines are used on an average more than 125 h/year at 0.05 LOS?

   Ans: H_0: $\mu = 125$, H_1: $\mu > 125$ and LOS $\alpha = 0.05$; CR $Z > Z_\alpha = Z_{0.05} \Rightarrow Z = \dfrac{\overline{X} - \mu}{\sigma/\sqrt{n}} = \dfrac{126.9 - 125}{8.4/\sqrt{49}} = 1.58$. Accept H_0 if $Z = 1.58 < 1.64 = Z_{0.05}$. Cannot believe that machine works more than 125h/year.

2. The breaking strength of cables produced by a manufacturer has mean 1800 lb and standard deviation 100 lb by a new technique. It is claimed that the breaking strength can be increased. To test this claim, a sample of 50 cables is tested and it is found that the mean breaking strength is 1850 lb.

   (a) Can we support the claim at a 0.01 LOS?
   (b) What is the p-value of the test?

   Ans: (a) H_0: $\mu = 1800$, H_1: $\mu > 1800$ and OTT at LOS $\alpha = 0.01$; $s = 3.55 > 2.33 = Z_{0.01}$. Results are highly significant, so claim should be supported.

   (b) $P(Z \geq 3.55) = 0.0002$

3. A tracking firm suspects the claim that the average lifetime of certain tyres is at least 28,000 miles. To check the claim the firm puts 40 of these tyres on its trucks and gets a mean lifetime of 27,463 miles with a standard deviation of 1,348 miles. What can we conclude if the probability of Type I error is to be at most 0.01?

   Ans: H_0: $\mu \geq 28{,}000$, H_1: $\mu < 28{,}000$ and at LOS $\alpha \leq 0.01$. H_0 must be rejected if $Z < -2.33$ where $\dfrac{\overline{X} - \mu}{\sigma/\sqrt{n}} = \dfrac{27{,}463 - 28{,}000}{1{,}348/\sqrt{40}} = -2.52 < -2.33 = Z_{0.01}$. Firm's suspicion is confirmed.

4. A random sample of 6 steel beams has a mean compressive strength 58,392 psi with standard deviation 648 psi and LOS $\alpha = 0.05$. Test whether the true average strength of the steel from which the sample is taken is 58,000 psi. Assume normality.

   Ans: Accept NH H_0

   [Hint: NH H_0: $\mu = 58{,}000$; AH H_1: $\mu \neq 58{,}000$ and LOS $\alpha = 0.05$. Test statistic $t = \dfrac{\overline{X} - \mu}{\sigma/\sqrt{n}}$, where $\overline{X} = 58{,}392$, $\sigma = 648$ and $n = 6$. Therefore, computed value of $t = \dfrac{58{,}392 - 58{,}000}{648/\sqrt{6}} = 1.49$. So, $t_{\alpha/2} = t_{0.025} = 2.57$. Since $t < t_{0.025}$, the decision is to accept NH H_0.]

FILL IN THE BLANKS

1. If $p = 0.9$, $n = 64$, $\sigma^2 = 2.56$ and $Z_{\alpha/2} = 1.645$, then maximum error $E =$ _____.

   Ans: 13,290

   $\left[\text{Hint: } E = Z_{\alpha/2} \times \dfrac{\sigma}{\sqrt{n}} = 1.645 \times \dfrac{\sqrt{2.56}}{\sqrt{64}} = 13290\right]$

2. If maximum error $E = 0.49$, $\sigma = 2$ and $Z_{\alpha/2} = 1.96$, then sample size $n =$ _____.

Ans: 64

$$\left[\text{Hint: } n = \left(Z_{\alpha/2} \times \frac{\sigma}{E}\right)^2 = \left(1.96 \times \frac{2}{0.49}\right)^2 = 64\right]$$

3. If NH H_0: $\mu = 800$ and AH H_0: $\mu \neq 800$, then the CR is _____ if LOS $\alpha = 0.5\%$.

Ans: $-2.81 < Z < 2.81$

4. If $\overline{X} = 788$, $n = 30$, $\mu = 800$ and $\sigma = 40$, then test statistic $Z =$ _____.

Ans: -1.643

$$\left[\text{Hint: } Z = \frac{\overline{X} - \mu}{\sigma/\sqrt{n}} = \frac{788 - 800}{40/\sqrt{30}} = -1.643\right]$$

5. If $Z = -1.643 > -2.81 = Z_{\alpha/2} = Z_{0.0025,}$ then decision is _____.

Ans: Accept NH H_0

6. The CR for LOS $\alpha = 1\%$ is _____.

Ans: $-2.58 < Z < 2.58$

7. The decision in Question 6 if $Z = -1.643 > -2.58 = Z_{\alpha/2} = Z_{0.0025}$ is _____.

Ans: Accept NH H_0

8. If $\overline{X} = 31.45$, $\mu = 32.3$, $\sigma = 1.6$ and $n = 40$, then $Z = \dfrac{\overline{X} - \mu}{\sigma/\sqrt{n}} =$ _____.

Ans: -3.33

9. If $\overline{X}_1 = 292.5$, $\overline{X}_2 = 266.10$, $\sigma_1 = 15.6$, $\sigma_2 = 18.2$, $\delta = 70$, $n_1 = 60$ and $n_2 = 60$, then the test statistic Z for two means is _____.

Ans: -14.09

$$\left[\text{Hint: } Z = \frac{\overline{X}_1 - \overline{X}_2 - \delta}{\sqrt{\frac{\sigma_1^2}{n_1} + \frac{\sigma_2^2}{n_2}}} = \frac{292.5 - 266.1 - 70}{\sqrt{\frac{15.6^2}{60} + \frac{18.2^2}{60}}}\right.$$

$$\left. = \frac{-43.6}{\sqrt{4.056 + 5.521}} = \frac{-43.6}{3.095} = -14.09\right]$$

10. If $n_1 = 10, n_2 = 10, \overline{X}_1 = 121, \overline{X}_2 = 112, \sigma_1 = 8,$ $\sigma_2 = 8$ and $\partial = 0$, then $Z = \dfrac{\overline{X}_1 - \overline{X}_2 - \delta}{\sqrt{\frac{\sigma_1^2}{n_1} + \frac{\sigma_2^2}{n_2}}}$ $=$ _____.

Ans: 2.52

$$\left[\text{Hint: } Z = \frac{\overline{X}_1 - \overline{X}_2 - \delta}{\sqrt{\frac{\sigma_1^2}{V_1} + \frac{\sigma_2^2}{n_2}}} = \frac{121 - 112 - 0}{\sqrt{\frac{8^2}{10} + \frac{8^2}{10}}}\right.$$

$$\left. = \frac{9}{\sqrt{6.4+6.4}} = \frac{9}{3.577} = 2.52\right]$$

7

Tests of Significance

7.1 Introduction

In Chapter 6, we have considered large sample statistical estimation and hypotheses testing techniques which depend on the central limit theorem to justify normality of the estimators and the test statistics. They apply when the samples are large ($n \geq 30$).

Situations arise where large sampling is not possible. Consider, e.g., populations such as synthetic diamonds, satellites, aeroplanes, super computers and nuclear reactors which involve heavy expenditure. In such cases, the size of the sample is small ($n < 30$) and we have to consider statistical techniques that are suitable for them.

7.2 Test for One Mean (Small Sample)

7.2.1 Student's *t*-Distribution

While discussing sampling distribution, we noted the following points:

1. When the original sampled population is normal, $\overline{X}$ and $Z = \dfrac{\overline{X} - \mu}{\sigma/\sqrt{n}}$ both have normal distributions for any sample size n.
2. When the original sampled population is not normal, $\overline{X}$ and $Z = \dfrac{\overline{X} - \mu}{\sigma/\sqrt{n}}$ both have approximately normal distributions, if the sample size n is large.

But when the sample size n is small, the statistic $Z = \dfrac{\overline{X} - \mu}{s/\sqrt{n}}$ does not have normal distributions. In such a case, there are two ways to proceed:

1. We use an empirical approach. We draw repeated samples and compute $Z = \dfrac{\overline{X} - \mu}{s/\sqrt{n}}$ for each sample. The relative frequency distribution that we construct using these values will approximate the shape and locations of the sampling distribution.
2. We use a mathematical approach to derive the actual density function or curve that describes the sampling distribution.

The second approach was used by an English statistician W. S. Gosset in 1908. He derived a complicated formula for the density function of $t = \dfrac{\overline{X} - \mu}{s/\sqrt{n}}$ for random samples of size n from a normal population. He published his results under the pen name 'Student' and hence the statistic has been known as Student's *t*-distribution. It has the following characteristics:

1. It is mound shaped and symmetric about $t = 0$ just like Z.
2. It is more variable than z. It has heavier tails. That is the t-curve approaches the horizontal axis more slowly when compared with the normal curve. This is due to the fact that t-statistic involves two random quantities $\overline{X}$, and s whereas the Z-statistic involves only the sample mean $\overline{X}$ (Figure 7.1).

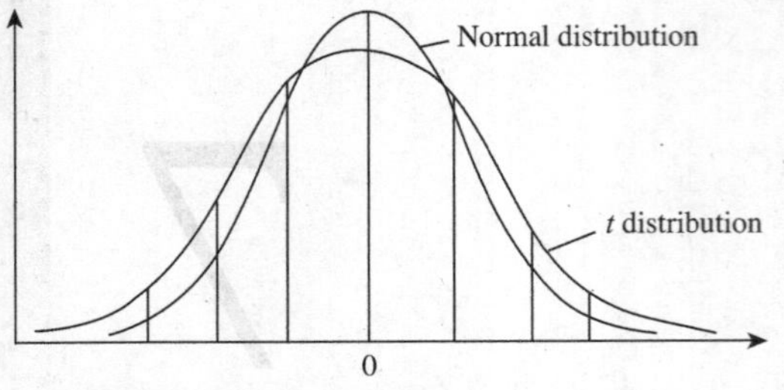

Figure 7.1 Standard normal Z and t-distributions.

3. The shape of the t-distribution depends on the sample size n. As n becomes larger, the variability of t decreases because the estimate of s of σ is based on more and more information. Consequently, when n is infinitely large, the t and Z distributions are identical.

The sample variance s^2 is given by the formula

$$s^2 = \frac{\sum (x_i - \bar{X})^2}{n-1}$$

or by

$$s^2 = \frac{\sum x_i^2 - \frac{(\sum x_i)^2}{n}}{n-1}$$

where $\sum x_i^2$ is the sum of the squares of the individual sum of the squares of the individual measurements and $(\sum x_i)^2$ is the square of the sum of the individual measurements.

The divisor $(n-1)$ in the formula is called the number of degrees of freedom (dof) associated with s^2. It determines the shape of the t-distribution. The term 'degrees of freedom' refers to the number of independent deviations in s^2 that are available for estimating σ^2. The dof may be different for different applications and may specify the correct t-distribution to be used.

For calculating critical values or p-values for the t-statistic, the table of probabilities for the standard normal Z-distribution is not useful. Instead, we have to use the t-distribution table.

For a t-distribution with 5 dof, the value of t that has area 0.05 to its right is found in row 5 in the column $t_{0.05}$. For this particular t-distribution, the area to the right of $t = 2.015$ is 0.05. Only 5% of all values of the t-statistic will exceed this value.

Example 7.1

Suppose that we have a sample of size $n = 10$ from a normal distribution. Find the value of t such that only 1% of all values of t will be smaller than this.

Solution The dof that give us the correct t-distribution are $n - 1 = 10 - 1 = 9$. The required t-value must be in the lower portion of the t-distribution with area = 0.01, i.e. 1%, to its left (Figure 7.2). Since the t-distribution is symmetric w.r.t. 0, this value is the negative of the value on the right-hand side with area 0.01 to its right, i.e. $-t_{0.1} = -2.821$.

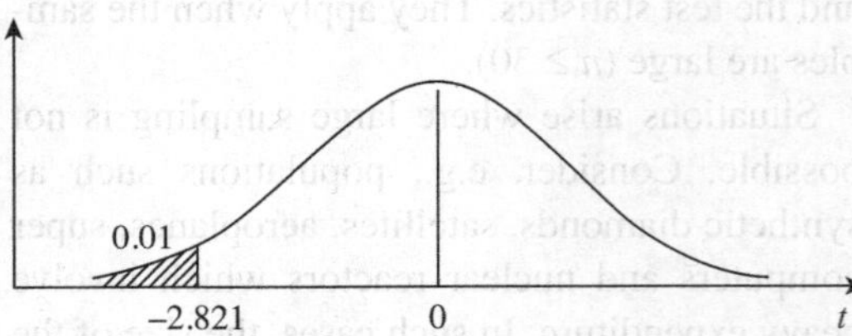

Figure 7.2 t-distribution for example 7.1.

Assumptions Relating to Student's *t*-Distribution The critical values of t permit us to make reliable inferences only if we follow all the rules. This implies that the sample must meet the following requirements specified by the t-distribution:

1. The sample must be randomly selected.
2. The population from which sampling is done must be normally distributed.

7.2.2 Small Sample Inferences Concerning a Population Mean

We have considered large sample inferences in Section 7.2.1. As with large sample inference, small sample inference may involve either estimation or hypothesis testing. Instead of Z-statistic and normal distribution, we will use a different sample statistic, namely, $t = \dfrac{\overline{X} - \mu}{s/\sqrt{n}}$ and a different sampling distribution, namely, the Student's t-distribution with $(n-1)$ dof.

Small Sample Hypothesis Test for μ

1. Null hypothesis (NH) H_0: $\mu = \mu_0$
2. Alternative hypothesis (AH) H_1: $\mu < \mu_0$
   (a) *One-tailed test*: H_1: $\mu > \mu_0$ or H_1: $\mu < \mu_0$
   (b) *Two-tailed test*: H_1: $\mu \neq \mu_0$
3. **Test Statistic:** $t = \dfrac{\overline{X} - \mu_0}{s/\sqrt{n}}$
4. **Rejection Region:** Reject H_0 when
   (a) *One-tailed test:* $t > t_\alpha$ or $t < -t_\alpha$ when the AH is H_1: $\mu < \mu_0$ or p value $< \alpha$
   (b) *Two-tailed test:* $t > t_{\alpha/2}$ or $t < -t_{\alpha/2}$

The critical values of t, t_α and $t_{\alpha/2}$ based on $(n-1)$ dof are shown in Figure 7.3. These values are found in the t-distribution table.

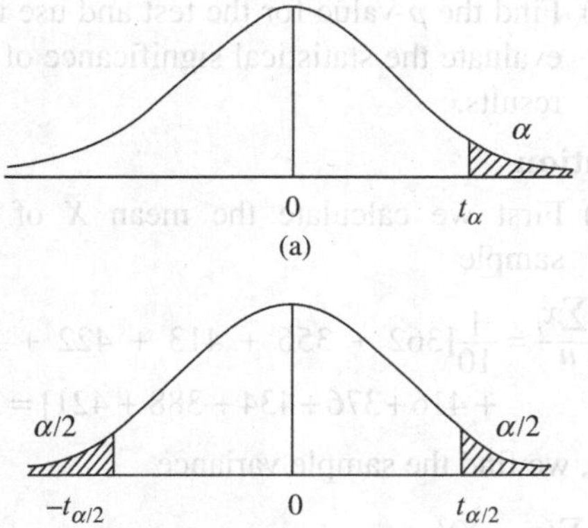

Figure 7.3 Critical values of t, t_α and $t_{\alpha/2}$ based on $(n-1)$ dof.

Small Sample $(1 - \alpha)$ 100% Confidence Interval for μ

It is

$$\left(\overline{X} - t_{\alpha/2}\frac{s}{\sqrt{n}}, \overline{X} + t_{\alpha/2}\frac{s}{\sqrt{n}}\right)$$

where $\dfrac{s}{\sqrt{n}}$ is the estimated standard error of $\overline{X}$. It is called the standard error of the mean.

Two Ways to Conduct a Test of Hypothesis They are as follows:

1. **Critical Value Approach** Based on the critical value of the sampling distribution of the test statistic, we set up rejection region. If the test statistic falls within the rejection region we can reject H_0.
2. ***p*-Value Approach** Based on the observed value of the test statistic, calculate the p-value. If it is smaller than the LOS α, we can reject H_0.

Note Example 7.2 uses the first approach and Example 7.3 uses the second approach.

Example 7.2

A new process of producing synthetic diamonds can be operated at a profitable level only if the average weight of the diamonds is greater than 0.5 carat. To test the profitability of the process, 6 diamonds are produced with weights 0.45, 0.60, 0.52, 0.49, 0.58 and 0.54 carat respectively. Do the 6 measurements present sufficient evidence to indicate that the average weight of the diamonds produced by the process is in excess of 0.5 carat?

Solution The mean of the population of the diamonds produced by the new process is $\mu = 0.5$.

Mean of the weights of the six diamonds is

$$\overline{X} = \frac{1}{n}\sum x_i = \frac{1}{6}[0.45 + 0.60 + 0.52 + 0.49 + 0.58 + 0.54]$$

$$= \frac{3.18}{6} = 0.53$$

$$\sum(x_i - \overline{X})^2 = (0.45 - 0.53)^2 + (0.60 - 0.53)^2 + (0.52 - 0.53)^2 + (0.49 - 0.53)^2 + (0.58 - 0.53)^2 + (0.54 - 0.53)^2$$

$$= (-0.08)^2 + (0.07)^2 + (-0.01)^2 + (-0.04)^2 + (0.05)^2 + (0.01)^2$$

$$= 0.0064 + 0.0049 + 0.0001 + 0.0016 + 0.0025 + 0.0001$$

$$= 0.0156$$

Sample size $n = 6$

Sample variance

$$s^2 = \frac{\Sigma(x_i - \overline{X})^2}{n-1} = \frac{0.0156}{6-1} = 0.00312$$

Sample standard deviation

$$s = \sqrt{0.00312} = 0.05586$$

We now carry out the formal test of hypothesis in steps as follows:

1. NH H_0: $\mu = 0.5$
2. AH H_1: $\mu > 0.5$
3. **Test Statistic:**

$$t = \frac{\overline{X} - \mu_0}{s/\sqrt{n}} = \frac{0.53 - 0.50}{0.05586/\sqrt{6}} = 1.3155$$

The test statistic provides evidence for either rejecting or accepting H_0 depending on how far from the centre of the t-distribution does the test statistic lie.

4. **Rejection Region:** Let us choose a 5% LOS, i.e. $\alpha = 0.05$. The right-side tail rejection region is found using the critical values of t from the t-distribution table.

Number of dof $= n - 1 = 6 - 1 = 5$

So, we can reject H_0 if $t > t_{0.05} = 2.015$, as shown in Figure 7.4.

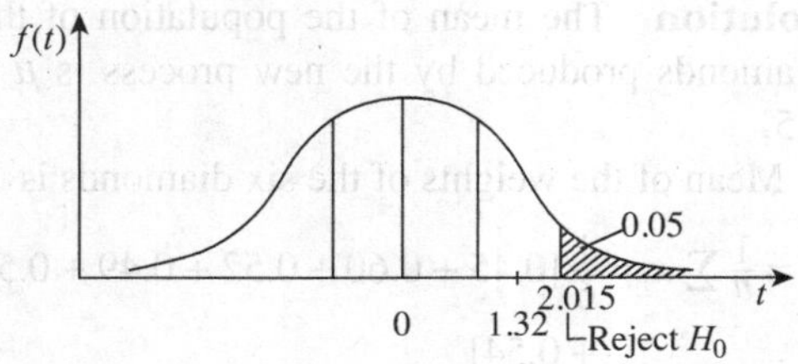

Figure 7.4 Rejection region for example 7.2.

5. **Conclusion:** Here the calculated value of test statistic is $t = 1.3155$. It does not fall within the rejection region. Therefore, we cannot reject H_0. Hence the data do not give sufficient evidence to indicate that the mean diamond weight exceeds 0.5 carat.

6. **Remarks:** The conclusion to accept H_0 requires the difficult calculation of β. This is the case of Type II error. To avoid this, we choose not to reject H_0. Towards this end, we consider a 95% lower one-sided confidence bound for μ as follows:

$$\bar{x} - t_\alpha\left(\frac{s}{\sqrt{n}}\right) = 0.53 - 2.015 \times \frac{0.05586}{\sqrt{6}}$$

$$= 0.53 - 0.04595 = 0.484$$

Thus, a 95% lower bound for μ is $\mu > 0.484$. The range of possible values includes mean diamond weights both smaller and greater than 0.5. This confirms the failure of our test to show that μ exceeds 0.5.

Example 7.3

Labels on one-gallon cans of paint indicate the drying time and the area that can be covered in one coat. A manufacture of paints claims that the brand of paint they manufacture will cover 420 sq. ft per 1 gallon. To test this, a random sample of 10 one-gallon cans of paints was tested. The actual areas painted in sq. ft are 362, 356, 413, 422, 372, 416, 376, 434, 388 and 421.

(a) Do the data present sufficient evidence to indicate that the average differs from 425 sq. ft?

(b) Find the p-value for the test and use it to evaluate the statistical significance of the results.

Solution

(a) First we calculate the mean $\overline{X}$ of the sample

$$\overline{X} = \frac{\Sigma x_i}{n} = \frac{1}{10}[362 + 356 + 413 + 422 + 372 + 416 + 376 + 434 + 388 + 421] = 396$$

Next, we find the sample variance

$$s^2 = \frac{\Sigma(x_i - \overline{X})^2}{n-1}$$

$$= \frac{1}{9}[(362-396)^2 + (356-396)^2 + (413-396)^2 + (422-396)^2 + (372-396)^2 + (416-396)^2 + (376-396)^2 + (434-396)^2 + (388-396)^2 + (421-396)^2]$$

$$= \frac{1}{9}[(-34)^2 + (-40)^2 + (17)^2 + (26)^2 + (-24)^2 + (20)^2 + (-20)^2 + (38)^2 + (-8)^2 + (25)^2]$$

$$= \frac{7230}{9} = 803.3333$$

Now, the standard deviation is

$$s = \sqrt{s^2} = 28.343$$

We calculate the test statistic

$$t = \frac{\overline{X} - \mu_0}{s/\sqrt{n}} = \frac{396 - 420}{28.343/\sqrt{10}} = 2.678$$

The p-value for the test is the probability of observing a value of the t-statistic as contradictory to NH. For the above set of data, we have obtained $t = -2.678$. This is a two-tailed test. The p-value is the probability that either $t \le -2.678$ or $t \ge 2.678$.

From the t-distribution tables, we observe that it gives values of t corresponding to right-side tail areas. Since in the present case dof $= n - 1 = 9$, we read the areas 'α' in row 9 corresponding to the values of t:

t_α	0.100	0.050	0.025	0.010	0.005
α	1.383	1.833	2.262	2.821	3.250

The five critical values for various tail areas are also shown in Figure 7.5.

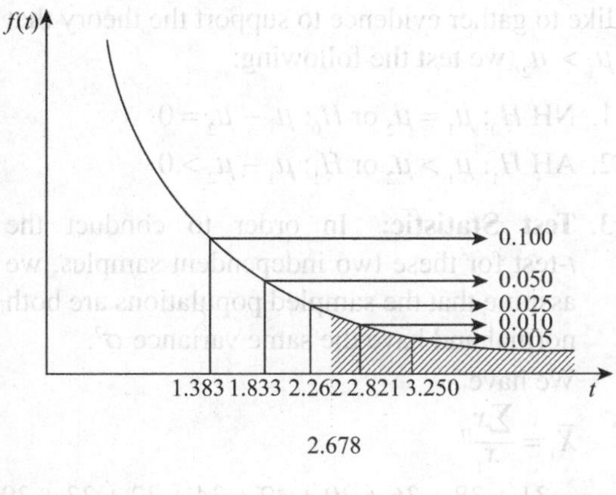

Figure 7.5 p-value for example 7.3.

The value $t = 2.678$ falls between 2.262 and 2.821 which correspond to $t_{0.025}$ and $t_{0.010}$ respectively.

This area represents only half of the p-value. We have $0.01 < \frac{1}{2}$ (p-value) $< 0.025 \Leftrightarrow 0.02 < p\text{-value} < 0.05$, so reject H_0 at the 5% level but not at the 2% or 1% level. Therefore, the p-value is less than 0.1.

For this test of hypothesis, H_0 is rejected at the 5% LOS. There is sufficient evidence to indicate that the average coverage differs from 420 sq. ft.

(b) **Limits for the actual average coverage** A 95% confidence interval gives the limits for μ as

$$\left(\overline{X} - t_{\alpha/2}\left(\frac{s}{\sqrt{n}}\right), \overline{X} + t_{\alpha/2}\left(\frac{s}{\sqrt{n}}\right)\right)$$

$$= \left(396 - 2.262\left(\frac{20.343}{\sqrt{10}}\right), 396 + 2.262\left(\frac{20.343}{\sqrt{10}}\right)\right)$$

$$= (375.73, 416.27)$$

We can estimate that the average coverage of area by one gallon of paint lies in the interval (375.73, 416.27). If we increase the size of the sample, a shorter interval can be obtained.

Observe that the upper limit of the interval 416.27 is close to 420 sq. ft, the coverage claimed on the label. The observed value of t is -2.678 which is slightly greater than the left critical value of $t_{0.05} = -2.821$, making the p-value slightly more than 0.05.

7.3 Test for Two Means

7.3.1 Small Sample Test Concerning Difference Between Two Means

Let n_1 and n_2 be the two sides of two samples, which are small (i.e. n_1 and $n_2 < 30$). Suppose that these samples are drawn from two normal populations with population variances σ_1^2 and σ_2^2 unknown but equal (i.e. $\sigma_1 = \sigma_2 = \sigma$). Then the pooling variance σ^2 is given by

$$\sigma^2 = \frac{\sum(x_{1i} - \overline{X}_1)^2 + \sum(x_{2i} - \overline{X}_2)^2}{x_1 + x_2^2}$$

$$= \frac{(n_1 - 1)s_1^2 + (n_2 - 1)s_2^2}{n_1 + n_2 - 2}$$

where $\overline{X}$, s_1^2 and $\overline{X}$, s_2^2 are the mean and variance of two samples of sizes n_1 and n_2 respectively. In a test concerning the difference between the means for small samples, the t-test statistic is given by

$$t = \frac{(\bar{X}_1 - \bar{X}_2) - \delta}{\sigma_{\bar{X}_1 - \bar{X}_2}}$$

with $(n_1 + n_2 - 2)$ dof. This test is also known as two-sample pooled t-test. The above formula can be rewritten as

$$t = \frac{(\bar{X}_1 - \bar{X}_2) - \delta}{\sqrt{(n_1 - 1)s_1^2 + (n_2 - 1)s_2^2}} \sqrt{\frac{n_1 n_2 (n_1 + n_2 - 2)}{n_1 + n_2}}$$

with $(n_1 + n_2 - 2)$ dof.

Test of Hypothesis Concerning the Difference between Two Means Based on Independent Random Samples

1. NH H_0: $\mu_1 - \mu_2 = \delta$ H_0:$\mu_1 - \mu_2 = \delta$ where δ is some specified difference that you wish to test. In many tests it may be assumed that there is no difference between μ_1 and μ_2 so that $\delta = 0$.
2. AH
   (a) *One-tailed test*: H_1: $\mu_1 - \mu_2 > \delta$ or H_1: $\mu_1 - \mu_2 < \delta$
   (b) *Two-tailed test*: H_1: $\mu_1 - \mu_2 \neq \delta$
3. **Test Statistic:**

$$t = \frac{\bar{X}_1 - \bar{X}_2 - \delta}{\sqrt{s^2\left(\frac{1}{n_1} + \frac{1}{n_2}\right)}}$$

   where

$$s^2 = \frac{(n_1 - 1)s_1^2 + (n_2 - 1)s_2^2}{n_1 + n_2 - 2}$$

4. **Rejection Region:** Reject H_0 when
   (a) *One-tailed test*: $t > t_\alpha$ or $t < -t_\alpha$ when the AH is H_1: $\mu_1 - \mu_2 < \delta$ or when p value $< \alpha$
   (b) *Two-tailed test*: $t > t_{\alpha/2}$ or $t < -t_{\alpha/2}$

   The critical values of t, t_α and $t_{\alpha/2}$ are based on $(n_1 + n_2 - 2)$ dof.

Small Sample $(1 - \alpha)$ 100% Confidence Interval for $(\mu_1 - \mu_2)$ Based on Independent Random Samples

$$(\bar{X}_1 - \bar{X}_2) \pm \sqrt{s^2\left(\frac{1}{n_1} + \frac{1}{n_2}\right)}$$

where s^2 is the pooled estimate of σ^2.

Example 7.4

In a manufacturing company, a one-month training programme is conducted to each employee. A new training programme has been developed. To test the new programme, 2 groups of 9 employees each were assigned a job, the first group trained under the old system and the second trained under the new programme. Their performance timings in minutes are recorded as follows:

Old programme	31	38	36	30	42	34	32	33	39
New programme	34	33	26	36	31	28	27	29	35

Do the job execution times present sufficient evidence to indicate that the mean time is less for the new training programme?

Solution Let μ_1 and μ_2 be the mean time to execute the job after the old and new training programmers respectively. Then, as we would like to gather evidence to support the theory that $\mu_1 > \mu_2$, we test the following:

1. NH H_0: $\mu_1 = \mu_2$ or H_0: $\mu_1 - \mu_2 = 0$
2. AH H_1: $\mu_1 > \mu_2$ or H_1: $\mu_1 - \mu_2 > 0$
3. **Test Statistic:** In order to conduct the t-test for these two independent samples, we assume that the sampled populations are both normal and have the same variance σ^2.

   We have

$$\bar{X}_1 = \frac{\sum x_{1i}}{x_1}$$

$$= \frac{31 + 38 + 36 + 30 + 42 + 34 + 32 + 33 + 39}{9}$$

$$= \frac{315}{9} = 35$$

$$\bar{X}_2 = \frac{\sum x_{2i}}{x_2}$$

$$= \frac{34 + 33 + 26 + 36 + 31 + 28 + 29 + 35}{9}$$

$$= \frac{279}{9} = 31$$

$$s_1^2 = \frac{\sum(x_{1i} - \overline{X}_1)^2}{n_1 - 1}$$

$$= \frac{1}{8}[(-4)^2 + 3^2 + 1^2 + (-5)^2 + 7^2 + (-1)^2 + (-3)^2 + (-2)^2 + 4^2]$$

$$= \frac{1}{8}(16 + 9 + 1 + 25 + 49 + 1 + 9 + 4 + 16)$$

$$= \frac{130}{8} = 16.25$$

$$s_2^2 = \frac{\sum(x_{2i} - \overline{X}_2)^2}{n_2 - 1}$$

$$= \frac{1}{8}[(+3)^2 + 2^2 + (-5)^2 + 5^2 + 0 + (-3)^2 + (-4)^2 + (-2)^2 + 4^2]$$

$$= \frac{1}{8}(9 + 4 + 25 + 25 + 0 + 9 + 16 + 4 + 16)$$

$$= \frac{108}{8} = 13.50$$

Next, we have the standard deviations of the two samples as

$$s_1 = \sqrt{s_1^2} = \sqrt{16.25} = 4.0311$$

$$s_2 = \sqrt{s_2^2} = \sqrt{13.50} = 3.6742$$

There is no significant variation between these figures and we may assume that the two distributions are of the same shape.

We now calculate the pooled estimate of the common variance as

$$s^2 = \frac{(n_1 - 1)s_1^2 + (n_2 - 1)s_2^2}{n_1 + n_2 - 2}$$

$$= \frac{8(4.0311)^2 + 8(3.6742)^2}{9 + 9 - 2} = 14.8748$$

We then calculate the test statistic t given by

$$t = \frac{\overline{X}_1 - \overline{X}_2}{\sqrt{s^2\left(\frac{1}{n_1} + \frac{1}{n_2}\right)}} = \frac{35 - 31}{\sqrt{14.8748\left(\frac{1}{9} + \frac{1}{9}\right)}}$$

$$= \frac{4 \times 3}{\sqrt{14.8748 \times 2}} = 2.2$$

Here the number of dof is $n_1 + n_2 - 2 = 9 + 9 - 2 = 16$.

4. **Rejection region:** The alternative hypothesis $H_1: \mu_1 > \mu_2$ or $H_1: \mu_1 - \mu_2 > 0$ implies that we have to use a one-tailed test on the right-side tail of the t-distribution with 16 dof. The appropriate critical value for a rejection region with $\alpha = 0.05$ from the Student's t-distribution table can be found and H_0 will be rejected if $t > 1.746$. The observed value of test statistic t is 2.2. Comparing this with the critical value $t_{0.05} = 1746$ we can reject H_0. There is sufficient evidence to indicate that the new training programme is superior at the 5% LOS (Figure 7.6).
5. **Conclusion:** Since the observed test statistic value $t = 2.2$ is greater than the critical value $t_{0.025} = 2.120$, we can conclude that the new training programme excels at the 2.5% LOS.

Example 7.5

Find the p-value that would be reported for the statistical test of Example 7.4.

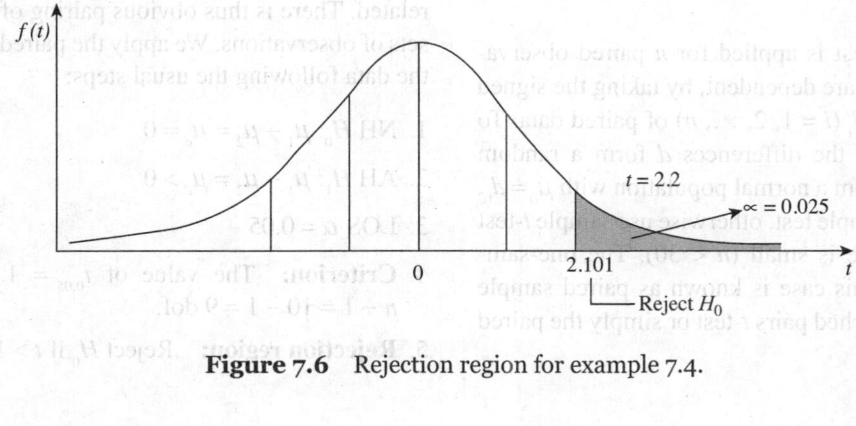

Figure 7.6 Rejection region for example 7.4.

Solution The observed value of t for this single-tailed test is $t = 2.2$ for a t-statistic with 16 dof. The observed value of $t = 2.2$ lies between $t_{0.025} = 2.120$ and $t_{0.01} = 2.583$ and the tail area to the right of $t = 2.2$ is between 0.01 and 0.025. The p-value for this test would be reported as $0.01 < p\text{-value} < 0.025$.

Since the p-value is less than 0.025, most researches would report the results as significant.

7.3.2 Paired Sample *t*-Test

Let μ_1 and μ_2 be the unknown means of two samples of the same size n drawn from two normal distributions. We would like to test NH that the means are equal $\mu_1 = \mu_2$ against the AH $\mu_1 = \mu_2$ (say). The variances need not be known but are assumed to be equal.

Suppose each value of the first sample corresponds precisely to one value of the other because corresponding values result from the same person, animal or thing (paired comparison). For example, two measurements of the same thing by two different methods or two measurements from the two eyes of the same person or animal. More generally, they may result from pairs of similar individuals or things, e.g. identical twins and pairs of used front tyres from the same car.

In such cases, we have to form the differences of corresponding values and test the hypothesis that the population corresponding to the differences has mean zero. If the two samples are not independent, we cannot use the previous method.

Paired t-test is applied for n paired observations, which are dependent, by taking the signed differences d_i $(i = 1, 2, \ldots, n)$ of paired data. To test whether the differences d form a random sample is from a normal population with $\mu_0 = d_0$, use large sample test, otherwise use sample t-test if the sample is small ($n < 30$). The one-sample test in this case is known as paired sample t-test or matched pairs t-test or simply the paired t-test.

The test statistic for the paired t-test is

$$t = \frac{\bar{d} - \mu_d}{s_d / \sqrt{n}} \quad t = \frac{\bar{d} - \mu_d}{sd / \sqrt{n}}$$

with $v = n - 1$ dof

where

$$\bar{d} = \frac{\Sigma d_i}{n} \text{ is the mean}$$

$$s_d^2 = \frac{\Sigma\left(d_i - \bar{d}\right)^2}{n-1}$$ is the variance and the positive square root $|s_d|$ is the standard deviation of the differences d_i $(i = 1, 2, \ldots, n)$

Example 7.6

The following are the average weekly losses of worker-hours due to accidents in 10 industrial units before and after introduction of certain safety measures

Before	43	66	109	35	56
After	36	50	96	37	53

Before	85	36	38	49	42
After	78	28	48	41	42

Use 0.05 LOS to test whether the safety measures are effective.

Solution In this problem, the independent sample test is not applicable because the data in the first row and that in the second row are correlated. There is thus obvious pairing of the two sets of observations. We apply the paired t-test to the data following the usual steps:

1. NH H_0: $\mu_1 - \mu_2 = \mu_d = 0$
2. AH H_1: $\mu_1 - \mu_2 = \mu_d > 0$
3. LOS $\alpha = 0.05$
4. **Criterion:** The value of $t_{0.05} = 1.833$ for $n - 1 = 10 - 1 = 9$ dof.
5. **Rejection region:** Reject H_0 if $t > 1.833$

6. **Calculation:** $d_i = (7, 16, 13, -2, 3, 7, 8, -10, 8, 0)$

Mean of differences

$$\bar{d} = \frac{\Sigma d_i}{n} = \frac{1}{10}(7+16+13-2+3+7+8-10+8+0)$$

$$= \frac{50}{10} = 5$$

$$\text{Variance } s_d^2 = \frac{\Sigma(d_i - \bar{d})^2}{n-1} = \frac{1}{9}\,[2^2 + 11^2 + 8^2 + (-7)^2 + (-2)^2 + 2^2 + 3^2 + (-15)^2 + 3^2 + (-5)^2]$$

$$= \frac{1}{9}(4 + 121 + 64 + 49 + 4 + 4 + 9 + 225)$$

$$= \frac{514}{9} = 57.1111$$

Standard deviation $s_d = \sqrt{s_d^2} = \sqrt{57.1111} = 7.5571$

7. **Test Statistic:**

$$t = \frac{\bar{d} - \mu_d}{s_d/\sqrt{n}} = \frac{5-0}{7.5571/\sqrt{10}} = 2.0922$$

8. **Conclusion:** The observed t-statistic value is $t = 2.0922$. This figure exceeds the critical value of t, viz. $t_{0.05} = 1.833$ at $v = n - 1 = 10 - 1 = 9$ dof. Hence, we conclude that the industrial safety measures adopted are effective and the evidence is strong to this effect.

9. ***p*-Value:** The observed t-statistic value viz. $t = 2.09$ lies between $t_{0.05} = 1.833$ and $t_{0.025} = 2.262$ at 9 dof and the tail area to the right of $t = 2.09$ is between 0.025 and 0.05. The p-value for this test would be reported as $0.025 < p\text{-value} < 0.05$. Since the p-value is less than 0.05, most researches would report the results as significant.

Example 7.7

In Example 7.6, find a 90% confidence interval for the mean improvement in lost worker-hours.

Solution Here $n = 10$, mean of the differences $\bar{d} = 5.0$. Also, the standard deviation $s_d = 7.56$. Since $t_{0.05} = 1.833$ at 9 dof, the 90% confidence interval for μ_d, the mean improvement, is

$$\bar{d} - t_\alpha\left(\frac{s_d}{\sqrt{n}}\right) < \mu_d < \bar{d} + t_\alpha\left(\frac{s_d}{\sqrt{n}}\right)$$

$$5.0 - 1.833\left(\frac{7.56}{\sqrt{10}}\right) < \mu_d < 5.0 + 1.833\left(\frac{7.56}{\sqrt{10}}\right)$$

$$0.62 < \mu_d < 9.38$$

We conclude that 90% confidence interval for the mean improvement in lost worker hours per week is lost, on an average, after the implementation of the safety measures in the industrial unit.

7.4 Test of Hypothesis

7.4.1 Test of Hypothesis: One Proportion (Small Sample)

Tests of hypothesis concerning proportions are required in many areas.

Examples

1. A politician is interested in knowing what fraction of the voters will favour him in the next election.
2. All manufacturing companies are concerned about the proportion of defective items while marketing.
3. A gambler wishes to know what proportion of outcomes would be favourable to him. Consider the problem of testing the hypothesis that the proportion of success in a binomial experiment is equal to a given value. The steps for testing an NH about a proportion against various alternatives using the binomial probabilities are the following:

1. NH $H_0: p = p_0$
2. AH $H_1: p < p_0$
   (a) *One-tailed test:* $H_1: p > p_0$ or $H_1: p < p_0$
   (b) *Two-tailed test:* $H_1: p \neq p_0$
3. Choose an LOS
4. **Test Statistic:** Binomial variable x with $p = p_0$
5. **Computations:** Find x, the number of success and compute the appropriate p-value

6. **Conclusion:** Draw appropriate conclusions based on the *p*-value

Example 7.8

A builder claims that heat pumps are installed in 70% of all homes being constructed today. Would you agree with this claim if a random survey of new homes shows that 8 out of 15 had heat pumps installed? Use a 0.10 LOS.

Solution

1. NH $H_0: p = 0.7$
2. AH $H_1: p \neq 0.7$
3. LOS $\alpha = 0.10$
4. **Test statistic:** Binomial variable x with $p = 0.7$ and $n = 15$
5. **Computations:** $x = 8$ and $np_0 = 15(0.7) = 10.5$. From the table of binomial probability sums, we find that the computed value of p is $p = 2P(x \le 8 \text{ when } p = 0.7) = 2\sum_{x=0}^{8} b(x:15, 0.7) = 0.2622 > 0.10$
6. **Conclusion:** Do not reject H_0. We conclude that there is insufficient reason to doubt the builder's claim.

7.4.2 Test of Hypothesis: One Proportion (Large Sample)

For large samples, normal curve approximation with parameters $\mu = np_0$ and $\sigma^2 = np_0q_0$ is usually preferred for large n. This gives accurate results as long as p_0 is not extremely close to 0 or 1. For using normal approximation, the Z-value for testing $p = p_0$ is given by

$$Z = \frac{x - np_0}{\sqrt{np_0q_0}}$$

which is a value of the standard normal variable Z. Hence, for a two-tailed test at the LOS α, the critical region is $Z < -Z_{\alpha/2}$ and $Z > Z_{\alpha/2}$. For one-sided alternative $p < p_0$, the critical region is $Z < -Z_\alpha$ and for the alternative $p > p_0$, the critical region is $Z > Z_\alpha$.

Example 7.9

A commonly prescribed drug for relieving nervous tension is believed to be only 60% effective. Experimental results with a new drug administered to a random sample of 100 adults who had been suffering from nervous tension show that 70 received relief. Is this sufficient evidence to conclude that the new drug is better? Use a 0.05 LOS.

Solution

1. NH $H_0: p = 0.6$
2. AH $H_0: p > 0.6$
3. LOS $\alpha = 0.05$
4. Critical region $Z > 1.645$
5. **Computations:** Here $x = 70$ and $n = 100$

$$np_0 = 100 \times 0.6 = 60$$

$$Z = \frac{x - np_0}{\sqrt{np_0q_0}} = \frac{70 - 60}{\sqrt{100 \times 0.6 \times 0.4}} = 2.04$$

$$p = P(Z > 2.04) < 0.025$$

6. **Conclusion:** Reject H_0. We conclude that the new drug is superior.

7.4.3 Test of Hypothesis: Two Proportions

Let A and B be two distinct populations and suppose each member of these populations belongs to two mutually exclusive classes depending on whether it possesses an attribute C (success) or not (failure).

Table 7.1 Success and failure in random samples of sizes n_1 and n_2 from two populations A and B.

	Class with Attribute C (Success)	Class Without C (Failure)	Total
Sample from A	x_1	$n_1 - x_1$	n_1
Sample from B	x_2	$n_2 - x_2$	n_2
Total	$x_1 + x_2$	$n_1 + n_2 - x_1 - x_2$	0

Let x_1 and x_2 be the number of items possessing attribute C in random samples of sizes x_1 and x_2 drawn from the two population A and B respectively (Table 7.1).

Then $p_1 = \frac{x_1}{n_1}$ and $p_2 = \frac{x_2}{n_2}$ are the sample proportions.

Let p_1 and p_2 be the population proportions of A and B respectively. To determine whether the proportion of items having attribute C (success) is same in both the population tests, the steps are as follows:

1. NH $H_0: p_1 = p_2$ or $p_1 - p_2 = 0$
2. Against AH $H_1: p_1 < p_2$
   (a) *One-tailed test*: $H_1: p_1 > p_2$ or $H_1: p_1 < p_2$
   (b) *Two-tailed test*: $H_1: p_1 \neq p_2$

For large samples when n_1 and $n_2 \geq 30$, p_1 and p_2 are asymptotically normally distributed and hence the sampling distribution of differences in proportions $(p_1 - p_2)$ will be approximately normally distributed with mean $\mu_{p_1-p_2} = 0$ and standard deviation is given by

$$\sigma_{p_1-p_2} = \sqrt{\hat{p}(1-\hat{p})\left(\frac{1}{n_1}+\frac{1}{n_2}\right)}$$

Here, an unbiased pooled estimate of the population proportion $\hat{p}$ is

$$\hat{p} = \frac{n_1p_1 + n_2p_2}{n_1 + n_2} = \frac{x_1 + x_2}{n_1 + n_2}$$

obtained by pooling together the data from both the samples.

$\therefore$ Z-statistic for testing $p_1 = p_2$ is given by

$$Z = \frac{p_1 - p_2}{\sigma_{p_1-p_2}} = \frac{\frac{x_1}{n_1} - \frac{x_2}{n_2}}{\sqrt{\hat{p}(1-\hat{p})\left(\frac{1}{n_1}+\frac{1}{n_2}\right)}}$$

Using the critical points of the standard normal curve, the critical regions are determined depending on the appropriate AH.

Example 7.10

If 120 out of 200 patients suffering from a certain disease are cured by allopathy and 240 out of 500 patients are cured by homeopathy, is there reason enough to believe that allopathy is better than homeopathy in curing the disease? Use $\alpha = 0.05$ LOS.

Solution Let p_1 and p_2 be the proportions of patients cured by allopathy and homeopathy respectively.

1. NH $H_0: p_1 = p_2$
   There is no difference between allopathy and homeopathy.
2. AH $H_1: p_1 > p_2$
   Allopathy is better than homeopathy.
3. LOS $\alpha = 0.05$
   Reject H_0 if critical region $Z > 1.645$
4. **Computations:** By a right-side one-tailed test,

$$p_1 = \frac{x_1}{n_1} = \frac{120}{200} = 0.60$$

$$p_2 = \frac{x_2}{n_2} = \frac{240}{500} = 0.48$$

$$\text{Pooled proportion } \hat{p} = \frac{x_1 + x_2}{n_1 + n_2} = \frac{120 + 240}{200 + 500} = 0.51$$

$$\therefore Z = \frac{\frac{x_1}{n_1} - \frac{x_2}{n_2}}{\sqrt{\hat{p}(1-\hat{p})\left(\frac{1}{n_1}+\frac{1}{n_2}\right)}} = \frac{0.60 - 0.48}{\sqrt{(0.51)(0.49)\left(\frac{1}{200}+\frac{1}{500}\right)}} = 2.9$$

$$p = p\,(Z > 2.9) = 0.0019$$

5. **Conclusion:** Reject H_0 and agree that the proportion of patients following allopathy is higher than the proportion following homeopathy, i.e. allopathy is better than homeopathy in curing particular disease.

7.4.4 Test of Hypothesis for Several Proportions

Consider K as the binomial populations with parameters $p_1, p_2, \ldots, p_k$ to test whether the population proportions of these K populations are all equal. Consider the following:

Table 7.2 Success and failures of all samples.

	Sample 1	Sample 2	...	Sample K	Total
Success	x_1	x_2	...	x_k	x
Failure	$n_1 - x_1$	$n_2 - x_2$	...	$n_k - x_k$	$n - x$
Total	n_1	n_2	...	n_k	n

1. NH H_0: $p_1 = p_2 = \cdots = p_k$
2. Against AH H_1: $p_i \neq p_j$ for some i, j
3. **Computations:** Let us draw K independent random samples of sizes $n_1, n_2, \ldots, n_k$, one from each of the K populations. Let $n_1, n_2, \ldots, n_k$ denote the number of items possessing the attribute (success) Table 7.2.

   Here n denotes the total number of trails and x denotes the total number of successes for all the samples put together. The expected cell frequencies e_{ij} are calculated by

$$e_{1j} = \frac{n_j x}{n}$$

$$e_{2j} = \frac{n_j(n-x)}{n}$$

$$\vdots$$

4. **Test Statistic:** Difference among the proportions is given by

$$\chi^2 = \sum_{i=1}^{2} \sum_{j=1}^{k} \frac{(o_{ij} - e_{ij})^2}{e_{ij}}$$

   where o_{ij} are the observed cell frequencies and e_{ij} are the expected cell frequencies.
5. **Conclusion:** Reject H_0 if $\chi^2 > \chi^2_\alpha$ with $(K-1)$ dof.

Example 7.11

In a shop study, a set of data was collected to determine whether or not the proportion of defectives produced by workers was the same for the day, evening or night shift worked (Table 7.3).

Table 7.3 Number of products produced in a day for all three shifts.

Shift	Day	Evening	Night
Defectives	45	55	70
Non-defectives	905	890	870

Use a 0.025 LOS to determine if the proportion of defectives is the same for all three shifts.

Solution Let p_1, p_2 and p_3 represent the true proportions of defectives for the day, evening and night shifts respectively.

1. NH H_0: $p_1 = p_2 = p_3$
2. AH H_1: p_1, p_2 and p_3 are not all equal.
3. LOS $\alpha = 0.025$
4. **Critical Region:** $\chi^2 > 7.378$ for $v = 2$ dof
5. **Computations:** Corresponding to the observed frequencies $o_1 = 45$ and $o_2 = 55$, we find the following expected frequencies:

$$e_1 = \frac{950 \times 750}{2835} = 57.0 \text{ and}$$

$$e_2 = \frac{945 \times 170}{2835} = 56.7$$

   All other expected frequencies are found by subtraction and are displayed in Table 7.4.

$$\chi^2 = \frac{(45-57.0)^2}{57.0} + \frac{(55-56.7)^2}{56.7} + \frac{(70-56.3)^2}{56.3} + \frac{(905-893.0)^2}{893.0} + \frac{(890-888.3)^2}{888.3}$$

Table 7.4 Observed and expected frequencies.

Shift	Day	Evening	Night	Total
Defectives	45 (57.0)	55 (56.7)	70 (56.3)	170
Non defectives	905 (893.0)	890 (888.3)	870 (883.7)	2665
Total	950	945	940	2835

$$+\frac{(870-883.7)^2}{883.7}=6.29$$

and

$$p = 0.04$$

6. **Conclusion:** We do not reject H_0 at $\alpha = 0.025$. With the above *p*-value computed, it would certainly be improper to conclude that the proportion of defectives produced is the same for all shifts.

7.5 Analysis of R × C Tables (Contingency Tables)

A classification in which attributes are divided into more than two classes is called a manifold classification. Let an attribute *A* be divided into *r* classes $A_1, A_2, \ldots, A_r$ and another attribute *B* be divided into *c* classes $B_1, B_2, \ldots, B_c$. Then the various cell frequencies can be expressed in the form of a table called an $r \times c$ (*r* by *c*) manifold contingency table. In this table, A_i is the number of items possessing the attribute A_i with $i = 1, 2, \ldots, r$ and B_j is the number of items having attribute B_j with $j = 1, 2, \ldots, c$. Also, o_{ij}, called the observed frequencies, denote the number of items possessing both the attributes A_i and B_j (Table 7.5). Here the total frequency is

$$N=\sum_{i=1}^{r}A_i=\sum_{j=1}^{c}B_j$$

Notes RT, row totals and CT column totals. They are also known as marginal frequencies.

Thus $r \times c$ table is expressed in a matrix form with *r* as rows and *c* as columns containing $m \times n$ $(r \times c)$ cells with cell frequencies o_{ij}.

These tables are essentially in the following two kinds of problems:

1. Test for independence
2. Test for homogeneity

7.5.1 Test for Independence

In this kind of problems, *c* samples from one population with each item are classified w.r.t. two

Table 7.5 $r \times c$ table.

A/B	B_1	B_2	.....	B_j	.....	B_c	Row Total
A_1	o_{11}	o_{12}	....	o_{1j}	...	o_{1c}	RT_1
A_2	o_{21}	o_{22}	....	o_{2j}	...	o_{2c}	RT_2
⋮	⋮	⋮	⋮	⋮	⋮	⋮	⋮
A_i	o_{i1}	o_{i2}	....	o_{ij}	...	o_{ic}	RT_i
⋮	⋮	⋮	⋮	⋮	⋮	⋮	⋮
A_r	o_{r1}	o_{r2}	....	o_{rj}	...	o_{rc}	RT_r
Column Total	CT_1	CT_2	....	CT_j	....	CT_c	$N=\sum_{i=1}^{r}A_i=\sum_{j=1}^{c}B_j$

qualitative attributes. The row totals and column totals are not fixed, but random. Only the grand total N is fixed. The NH consists of testing whether the two attributes are independent. Then

p_{ij} = (probability of getting a value belonging to ith row) × (probability of getting a value belonging to jth column)

The AH is that the two attributes are not independent, i.e. the attributes are dependent.

7.5.2 Test for Homogeneity

In this kind of problem, samples from several c populations are considered with each trial permitting more than two possible outcomes. Here both the marginal frequencies, i.e. row and column totals, are fixed beforehand. To test whether an attribute is common to all the populations, i.e. to determine whether the c populations are homogeneous w.r.t. an attribute, consider the following:

1. NH H_0: $p_{i1} = p_{i2} = \ldots = p_{ic}$, for $i = 1, 2, \ldots, r$, i.e. probability of obtaining an observation in the rth row is the same for each column, $\sum_{i=1}^{r} p_{ij} = 1$ for each column.
2. The AH of p's are not all equal for at least one row (i.e., non homogeneous). In either of the problems the expected cell frequencies denoted by e_{ij} are calculated by

   e_{ij} = (total observed frequencies in the jth column) × (total observed frequencies in the rth row) ÷ (total of all cell frequencies)

   = (column total × row total)/grand total
3. **Test Statistic:** Statistic for analysis of $r \times c$ tables is

   $$\chi^2 = \sum_{i=1}^{r} \sum_{j=1}^{c} \frac{(o_{ij} - e_{ij})^2}{e_{ij}} \text{ with } (r-1)\times(c-1) \text{ dof}$$
4. **Conclusion:** Reject H_0 if $\chi^2 > \chi^2_\alpha$ with $(r-1)(c-1)$ dof

7.6 Goodness-of-Fit Test: χ^2 Distribution

To determine if a population follows a specified theoretical distribution such as normal, binomial or Poisson distribution, the χ^2 test is used to ascertain how closely the actual distribution approximates the assumed theoretical distribution. This test which is carried out to see how good a fit is between the observed frequencies o_i from the sample and the expected frequencies e_i from the theoretical distribution is known as goodness-of-fit test. This test judges whether the sample is drawn from a certain hypothetical distribution, i.e. whether the observed frequencies follow a postulated distribution or not.

The test statistic χ^2 is a measure of the discrepancy between the observed and expected frequencies

Statistic for test of goodness-of-fit is

$$\chi^2 = \sum_{i=1}^{k} \frac{(o_i - e_i)^2}{e_i} \tag{7.1}$$

Here o_i and e_i are the observed and expected frequencies of the ith cell or class interval such that $\sum o_i = \sum e_i = N$ = total frequency.

K is the number of cells or class intervals, in the given frequency distribution.

χ^2 is a random variable which is closely approximated with v dof.

Degrees of Freedom for χ^2 Distribution

Let K be the number of terms in the Eq. (7.1) for χ^2. Then the dof for χ^2 are as follows:

1. $v = K - 1$ if e_i can be calculated without having estimate population parameters from sample statistics.
2. $v = K - 1 - m$ if e_i can be calculated only by estimating m number of population parameters from sample statistics.

Binomial Distribution p is the parameter and $m = 1$, then

$$v = K - 1 - m = K - 1 - 1 = K - 2$$

Poisson Distribution λ is the parameter and $m = 2$, then

$$v = K - 2$$

Normal Distribution μ and σ are two parameters and $m = 2$, then

$$v = K - 1 - m = K - 1 - 2 = K - 3$$

TEST FOR GOODNESS-OF-FIT

1. NH H_0: Good fit exists between the theoretical distribution and given data (observed frequencies).
2. AH H_1: No good fit.
3. LOS α (given).
4. **Critical region:** Reject H_0 if $\chi^2 > \chi^2_\alpha$ with v dof, i.e. the theoretical distribution is a poor fit.
5. **Computation:** Calculate χ^2 using formula.
6. **Conclusion:** Accept H_0 if $\chi^2 > \chi^2_\alpha$, i.e. the theoretical distribution is a good fit to the data.

Conditions for Validity of χ^2 Test Sample size n should be large $n \geq 50$. If individual frequencies o_i (or e_i) are small, i.e. $o_i < 10$, then combine neighbouring frequencies so that combined frequency $o_i \geq 10$. The number of classes K should be in several, within the range $4 \leq K \leq 16$.

Example 7.12

Fit a Poisson distribution to the following data and test the goodness-of-fit at a 0.5 LOS:

x	0	1	2	3	4	5	6	7
$f(x)$	305	366	210	80	28	9	2	1

Solution

1. NH H_0: Poisson distribution is a good fit.
2. AH H_1: Poisson distribution is not a good fit.
3. LOS $\alpha = 0.05$.
4. **Critical Region:** $\chi^2 < \chi^2_{0.5}$ with 4 dof.
5. **Test Statistic:** $\chi^2 = \sum_{i=1}^{n} \frac{(o_i - e_i)^2}{e_i}$
6. **Computations:** To find the mean of the distribution μ

$$\mu = \frac{\sum x f(x)}{\sum f(x)}$$

$$= \frac{0 \times 305 + 1 \times 366 + 2 \times 210 + 3 \times 80}{305 + 366 + 210 + 80 + 28 + 9 + 2 + 1}$$

$$\frac{+ 4 \times 28 + 5 \times 9 + 6 \times 2 + 7 \times 1}{305 + 366 + 210 + 80 + 28 + 9 + 2 + 1}$$

$$= \frac{1202}{1001} = 1.2$$

Recurrence formula for the Poisson distribution is

$$P(x+1) = \frac{\mu}{x+1} P(x)$$

Expected frequency $f(x) = NP(x)$ where $N = 1001$

$$f(x+1) = \frac{\mu}{x+1} f(x)$$

$$f(0) = 1001 \times e^{-1.2} = 1001 \times 0.3 = 300.3$$

$$f(1) = \frac{1.2}{1} f(0) = 1.2 \times 300.3 = 360.36$$

$$f(2) = \frac{1.2}{2} f(1) = 0.6 \times 360.36 = 216.216$$

$$f(3) = \frac{1.2}{3} f(2) = 0.4 \times 216.216 = 86.486$$

$$f(4) = \frac{1.2}{4} f(3) = 0.3 \times 86.486 = 25.946$$

$$f(5) = \frac{1.2}{5} f(4) = 0.24 \times 25.946 = 6.227$$

$$f(6) = \frac{1.2}{6} f(5) = 0.20 \times 6.227 = 1.245$$

$$f(7) = \frac{1.2}{7} f(6) = \frac{1.2}{7} \times 1.245 = 0.213$$

Taking $o_5 + o_6 + o_7 = 12$ (sum of the last three) so that $N = 1001$. We construct Table 7.6.

$n = 6$

$$v = n - 2 = 6 - 2 = 4 \text{ dof}$$

$$\chi^2_{0.5}(4) = 9.488$$

$$\chi^2 = \frac{16}{301} + \frac{25}{361} + \frac{36}{216} + \frac{49}{87} + \frac{4}{26} + \frac{4}{10} = 1.41$$

$$\chi^2 (= 1.41) < \chi^2_{0.5}(4)(= 9.488)$$

Table 7.6 Table of observed and expected frequencies.

x	Observed Frequency	Expected Fequency	$o_i - e_i$	$(o_i - e_i)^2$
0	305	301	4	16
1	366	361	5	25
2	210	216	−6	36
3	80	87	−7	49
4	28	26	2	4
5, 6 and 7	12	10	2	4

7. **Conclusion:** Accept H_0. Poisson distribution is a good fit.

7.7 Estimation of Proportions

7.7.1 Introduction

Problems dealing with proportions, percentages or probabilities are the same. This is because a proportion multiplied by 100 is a percentage. Also a proportion with a very large number of trials is a probability.

Sample proportion = $\frac{x}{n}$ where x is the number of times an event occurs in n trials.

Sample proportion is an unbiased estimator of true proportion, the binomial parameter p.

We know that the mean and standard deviation of the number of successes are given by np and $\sqrt{np(1-p)}$ respectively, where p is the probability of success.

$$E(X) = np$$

$$\text{Var}(X) = np(1-p)$$

$$\sigma = \sqrt{np(1-p)}$$

Also,

$$E\left(\frac{X}{n}\right) = \frac{1}{n}E(X) = \frac{1}{n}np = p$$

$$\frac{\sigma}{n} = \frac{\sqrt{np(1-p)}}{n}$$

If P denotes sample proportion, then

$$E(P) = E\left(\frac{X}{n}\right) = P$$

$$\text{Var}(P) = \text{Var}\left(\frac{X}{n}\right) = \frac{1}{n^2}\text{Var}(X) = \frac{nPQ}{n^2} = \frac{PQ}{n}$$

$$\text{Standard error of } P = \sqrt{\frac{PQ}{n}}$$

If the sample is taken from a finite population of size N, then standard error of proportions

$$\text{SE}(P) = \sqrt{\frac{N-n}{N-1}\frac{PQ}{n}}$$

7.7.2 Large Sample Confidence Interval for *p*

When n is large, normal approximation is used for binomial distribution to construct confidence interval for p from the inequalities

$$-Z_{\alpha/2} < \frac{X-np}{\sqrt{np(1-p)}} < Z_{\alpha/2}$$

by replacing $\frac{x}{n}$ by p. Thus the confidence interval for p, when n is large, is given by

$$\frac{x}{n} - Z_{\alpha/2}\sqrt{\frac{\frac{x}{n}\left(1-\frac{x}{n}\right)}{n}} < p < \frac{x}{n} + Z_{\alpha/2}\sqrt{\frac{\frac{x}{n}\left(1-\frac{x}{n}\right)}{n}}$$

7.7.3 Maximum Error of Estimate

The magnitude of error committed in using sample proportion $\frac{x}{n}$ for true proportion p is given by the maximum error of estimate E, where

$$E = Z_{\alpha/2}\sqrt{\frac{p(1-p)}{n}}$$

7.7.4 Sample Size n

1. When p is known, the sample size n is given by

$$x = p(1-p)\left(\frac{Z_{\alpha/2}}{E}\right)^2$$

2. When p is unknown, then

$$n = \frac{1}{4}\left(\frac{Z_{\alpha/2}}{E}\right)^2$$

7.7.5 One-sided Confidence Interval

For $p \to 0$ and $n \to \infty$, binomial distribution is approximated by Poisson distribution with $\lambda = np$. In this case, instead of the two-sided confidence interval, one-sided confidence interval of the form

$$p < \frac{1}{2n}\chi_\alpha^2$$

is used. Here χ_α^2 is with $2(x+1)$ dof.

Example 7.13

In a random sample of 400 industrial accidents, it was found that 231 were due to unsafe working conditions. Construct a 99% confidence interval for the corresponding true proportion using large sample formula.

Solution Here $n = 400$ and $X = 231$

$$\text{Probability of success } P = \frac{x}{n} - \frac{231}{400} - 0.576$$

$$Q = 1 - P = 1 - 0.578 = 0.422$$

Confidence interval

$$P - Z_{\alpha/2}\sqrt{\frac{PQ}{n}} < p < P + Z_{\alpha/2}\sqrt{\frac{PQ}{n}}$$

$$Z_{\alpha/2} = 2.58$$

$$Z_{\alpha/2}\sqrt{\frac{PQ}{n}} = 2.58\sqrt{\frac{0.422 \times 0.578}{400}} = 0.064$$

Therefore, confidence interval is

(0.058 – 0.064, 0.578 + 0.064) = (0.514, 0.642)

Example 7.14

In Example 7.13, what can we say with 95% confidence about the maximum error if we use the sample proportion to estimate the corresponding true proportion.

Solution Here $P = 0.578$, $Q = 0.422$, $n = 400$ and $Z_{\alpha/2} = 1.96$

$$\text{Maximum error } E = Z_{\alpha/2}\sqrt{\frac{PQ}{n}}$$

$$= 1.96\sqrt{\frac{0.578 \times 0.422}{400}}$$

$$= 0.048$$

Example 7.15

In a recent study, 69 of 120 meteorites were observed to enter the earth's atmosphere with a velocity less than 26 miles/s. If we estimate the corresponding true proportion as P, what can we say with 95% confidence about the maximum error?

Solution Here sample proportion $P = 69 \div 120 = 0.58$, $Q = 1 - P = 0.42$, $n = 400$ and $Z_{\alpha/2} = 1.96$

$$\text{Maximum error } E = Z_{\alpha/2}\sqrt{\frac{PQ}{n}} = 1.96$$

$$\sqrt{\frac{0.58 \times 0.42}{120}} = 0.088$$

Example 7.16

In a study designed to investigate whether certain detonators used with explosives in coal mining meet the requirement that at least 90% will ignite the explosive when charged, it is found that 174 of 200 detonators function properly. Test the NH $P = 0.9$ against the AH $P < 0.9$ at the 0.05 LOS.

Solution Here $P = 0.9$, $Q = 1 - P = 0.1$, $n = 200$ and $p = 0.87$

1. NH H_0: $P = 0.9$
2. AH H_1: $P < 0.9$
3. LOS $\alpha = 0.05$
4. **Critical Region:** Reject H_0 if $Z < Z_{0.05} = Z_\alpha$
5. **Computations:** $Z = \dfrac{p-P}{\sqrt{\dfrac{PQ}{n}}} = \dfrac{.87 - 0.9}{\sqrt{\dfrac{0.9 \times 0.1}{200}}}$

$$= -1.41 > -1.645 = Z_\alpha$$

6. **Conclusion:** We accept H_0 since $Z > Z_\alpha$. There is no significant evidence to say that the given kind of detonator fails to meet the required standard.

EXERCISES

1. In 1950, the mean life expectancy was 50 years in India. If the life expectancies from a random sample of 11 persons are 58.2, 56.6, 54.2, 50.4, 44.2, 61.9, 57.5, 53.4, 49.7, 55.4 and 57.0, does it confirm the expected view?

   Ans: Reject H_0

   [Hint: H_0: $\mu = 50$, H_1: $\mu \neq 50$, $\bar{x} = \dfrac{598.5}{11}$,

   $s = 4.859$ and $t = 3.01$. Reject H_0 since $t = 3.01 > 2.228 = t_{0.025}$ with 10 dof.]

2. In an examination, 9 students of class A and 6 students of class B obtained the following marks:

   A: 44 71 63 59 68 46 69 54 48

   B: 52 70 41 62 36 50

   Test at 0.01 LOS whether the performance is same or not for classes A and B, assuming that the samples are drawn from normal populations having the same variance.

   Ans: Accept H_0

   $$\bar{X}_A = \frac{\Sigma X_i}{n_1} = \frac{522}{9} = 58, \bar{X}_B = \frac{311}{6} = 51.83,$$

   $$S_A^2 = \frac{\Sigma(X_i - X)}{n_i - 1} = \frac{872}{8} = 109$$

   [Hint: $S_A = 10.44$; $S_B^2 = \dfrac{804.83}{5}$, $S_B = 12.69$

   $$t = \frac{\bar{X}_A - \bar{X}_B - (\mu_1 - \mu_2)}{\sqrt{(n_1 - 1)S_A^2 + (n_2 - 1)S_B^2}} \sqrt{\frac{n_1 n_2 (n_1 + n_2 - 2)}{n_1 + n_2}} = 1.03$$

   Accept H_0 since $t = 1.03 < 3.012 = t_{0.05}$.]

3. A study shows that 16 out of 200 tractors produced on one assembly line required extensive adjustments before they could be shipped, while the same was true for 14 out of 400 tractors produced on another assembly line. At a 0.01 LOS, does this support the claim that the second production line does superior work?

   Ans: H_0 is rejected. Second production is not superior.

   [Hint: Test statistic $Z = 2.37 > Z_{0.01} = 2.33$. Hence H_0 is rejected. Second production is not superior.]

4. Use paired sample test at a 0.05 LOS to test from the following data whether the differences of the means of the weights obtained by two different weighing machines are significant.

Scale1 (wt in g)	11.23	14.36	8.33	10.50	23.42
Scale 2 (wt in g)	11.27	14.41	8.35	10.52	23.41

Scale1 (wt in g)	9.15	13.47	6.47	12.40	19.38
Scale 2 (wt in g)	9.17	13.52	6.46	12.45	19.35

   Ans: No significant difference in the two scales.

   $\bar{x} = -\dfrac{0.2}{10} = -0.02$, $S = 0.028674$, $n = 10$

   [Hint: $t = \dfrac{-0.02 - 0}{0.028/\sqrt{10}} = -2.21$; $t_{0.05} = 1.833$ with 9 dof No significant difference in the two scales.]

5. In a sample of 90 university professors, 28 own a computer. Can we conclude at 0.05 LOS that at most $\frac{1}{4}$ of the professors own a computer?

   Ans: Accept H_0. At most of $\frac{1}{4}$ of the professors do own a computer.

   [Hint: NH H_0: $P = \frac{1}{4}$ against AH H_1: $P > \frac{1}{4}$ since

   $$Z = \frac{28 - (90)\left(\frac{1}{4}\right)}{\sqrt{\left(90 \times \frac{1}{4} \times \frac{3}{4}\right)}} = 1.34 < 1.645 = Z_{0.05}$$

   Accept H_0. At most $\frac{1}{4}$ of the professors do own a computer.]

6. A study of TV viewers was conducted to find the opinion about the mega serial Ramayana. If 50% of a sample of 300 viewers from south and 48% of 200 viewers from north preferred the serial, test the claim at 0.05 LOS that (a) there is a difference of opinion between south and north and (b) Ramayana is preferred in the south.

   Ans: (a) Accept H_0. No significant difference.

   (b) Reject H_0. Ramayana is not just preferred in the south.

   [Hint: $Z = \dfrac{0.560 - 0.480}{0.0456}$

   $$\hat{P} = \frac{n_1 P_1 + n_2 P_2}{n_1 + n_2} = 0.528$$

   $$\sigma_{p_1 - p_2} = \sqrt{pq\left(\frac{1}{n_1} + \frac{1}{n_2}\right)} = 0.0456$$

   (a) Accept H_0. No significant difference between north and south viewers since $Z = 1.75$ lies in (−1.96, 1.96).

   (b) Reject H_0, since $Z = 1.75 > Z_{0.05} = 1.645$. Ramayana is not just preferred in the south.]

7. Determine whether physical handicap affects the performance of a worker in an industry with the following results.

   Test the claim that handicap has no effect on the performance at 0.05 LOS.

	Performance			
	Good	Satisfactory	Not Satisfactory	Total
Blind	21	64	17	102
Deaf	16	49	14	79
No handicap	29	93	28	150
Total	66	206	59	331

   Ans: Accept H_0. Handicap has no effect on performance.

   [Hint: $e_{11} = 20.34$, $e_{12} = 63.5$, $e_{13} = 18.18$

   $e_{21} = 15.75$, $e_{22} = 49.17$, $e_{23} = 26.74$

   $e_{31} = 29.90$, $e_{32} = 93.35$, $e_{33} = 26.74$

   $\chi^2 = 0.19472$; $0.195 < 9.488 = \chi^2_{0.05}$ with (3 − 1) (3 − 1) = 4 dof. Accept H_0. Handicap has no effect on performance.]

8. To determine the effectiveness of drugs against Aids, three types of medical treatments were tested on 50 patients with the following results:

		Drug			
		Allopathy	Homeopathy	Ayurveda	Total
Effectiveness	No relief	11	13	9	33
	Some relief	32	28	27	87
	Total relief	7	9	14	30
	Total	50	50	50	150

   Ans: Accept H_0. Three days are equally effective (homogeneous).

   [Hint: $e_{11} = 11$ $e_{12} = 11$ $e_{13} = 11$

   $e_{21} = 29$ $e_{22} = 29$ $e_{23} = 29$

   $e_{31} = 10$ $e_{32} = 10$ $e_{33} = 10$

Accept H_0 since $\chi^2 = 3.8100313 < 9.488 = \chi^2_{0.05}$ with $v = (3-1)(3-1) = 4$ dof. Three days are equally effective (homogeneous).]

9. Test for goodness-of-fit of a uniform distribution to the following data obtained when a die is tossed 120 times:

Ans: Accept H_0. Uniform distribution is good fit to the data.

$$\left[\text{Hint: } x^2 = \frac{(20-20)^2}{20} + \frac{(22-20)^2}{20} + \frac{(17-20)^2}{20} + \frac{(18-20)^2}{20} + \frac{(18-20)^2}{20} + \frac{(19-20)^2}{20} + \frac{(24-20)^2}{20} = 1.7\right]$$

Accept H_0 that die is balanced since $\chi^2 = 1.7 < 11.70 = \chi^2_{0.05}$ with $6 - 1 = 5$ dof. Uniform distribution is good fit to the data.]

10. Test for goodness-of-fit of normal distribution to the following frequency data:

Class	1.45–1.95	1.95–2.45	2.45–2.95	2.95–3.45	3.45–3.95	3.95–4.45	4.45–4.95
Frequency O_i	2	1	4	15	10	5	3

Ans: 3.05

[Hint: $e_1 = 0.5 + 2.1 + 5.9 = 8.5$
$e_2 = 10.3$ $e_4 = 7.0 + 3.5 = 10.5$; $e_3 = 10.7$]

$$\left[\text{Hint: } \chi^2 = \frac{(7-8.5)^2}{8.5} + \frac{(15-10.3)^2}{10.3} + \frac{(10-10.7)^2}{10.7} + \frac{(8-10.5)^2}{10.5} = 3.05\right]$$

FILL IN THE BLANKS

1. If $\bar{x} = 23$, $s = 6.39$, $\mu = 20$ and $n = 6$ then $t =$ ______.

Ans: 1.15

$$\left[\text{Hint: } t = \frac{\bar{x}-\mu}{s/\sqrt{n}} = \frac{23-20}{6.39/\sqrt{6}} = 1.15\right]$$

2. In Question 1, H_0: $x = 20$ min and H_1: $x > 20$ and $\alpha = 0.10$. ______ H_0.

Ans: Accept

[Hint: Accept H_0 since $t = 1.15 < 1.476 = t_{0.1}$]

3. If = 1.95, $s = 0.207$, $n = 8$ and $\mu = 1.83$, then $t =$ ______.

Ans: 1.64

$$\left[\text{Hint: } t = \frac{\bar{x}-\mu}{s/\sqrt{n}} = \frac{1.95-1.83}{0.207/\sqrt{8}} = 1.64\right]$$

4. In Question 3, H_0: $\mu = 1.83$ against H_1: $\mu > 1.83$ with 95% confidence. ______ H_0.

Ans: Accept

[Hint: Accept H_0 since $t = 1.64 < 1.895 = t_{0.05}$]

5. If sample size $n = 144$, standard deviation $\sigma = 4$ and the mean $\bar{x} = 150$, then 95% confidence interval for μ is ______.

Ans: (149.35, 150.65)

6. If 60 out of 100 students use ball pens, the maximum error for true proportion at 95% confidence level is ______.

Ans: 0.096

7. If the maximum error with 99% confidence is 0.06 and $\sigma = 1$, then the size of the sample is ______.

Ans: 1849

8. If the maximum error with 0.99 probability is 0.25 and the sample size is 400, then the standard deviation of the population is _____.

Ans: 1.93

9. A sample of 64 was taken and it was found that 15 are smokers. The standard error of proportion is _____.

Ans: 0.0275

10. The size of the sample is 16 and the standard deviation is 3. The maximum error with probability 0.95 is _____.

Ans: 1.6

11. In a random sample of 200 patients suffering from headache, 160 got relief using a particular drug. The manufacturer's claim is that his drug cures 90% of the suffers. μ and σ are _____ and _____ respectively.

Ans: $\mu = 180$ and $\sigma = 4.23$

[Hint: $\mu = np = 200 \times 0.9 = 180$ and $\sigma = \sqrt{npq} = 4.23$]

12. In Question 11, the test statistic $Z =$ _____.

Ans: –4.73

$$\left[\text{Hint: } Z = \frac{\overline{X} - \mu}{\sigma} = \frac{160 - 180}{4.23} = -4.73\right]$$

13. In Question 11, can we accept the manufacturer's claim in Question 11, if we use $\alpha = 0.01$ LOS? _____.

Ans: No

[Hint: Since $Z = -4.73 < -2.33 = Z_{0.01}$.]

14. To determine whether hypertension (HT) is dependent on smoking habits, the following table gives experimental data on 180 persons:

	Non Smokers	Moderate Smokers	Heavy Smokers
HT	21	36	30
NO HT	48	26	19

$e_{11} =$ _____

$e_{12} =$ _____

$e_{21} =$ _____

$e_{22} =$ _____

Ans: $e_{11} = 33.35$, $e_{12} = 29.97$, $e_{21} = 35.65$ and $e_{22} = 32.03$

$$\left[\text{Hint: } e_{11} = 69\left(\frac{87}{180}\right) = 33.35,\ e_{12} = 62\left(\frac{87}{180}\right) = 29.97\right.$$

$$e_{21} = 69\left(\frac{93}{180}\right) = 35.65$$

$$\left. e_{22} = 62\left(\frac{93}{180}\right) = 32.03\right]$$

15. A die is thrown 1024 times. Getting an even digit is a success. Then the standard error of true proportion is _____.

Ans: 16

16. To test the goodness-of-fit, Poisson distribution was fitted. If $\mu = 0.05$ and the expected frequency when $x = 1$ is 118, then $f(2) =$ _____.

Ans: 29.5

17. Two random samples of sizes 40 and 50 have standard deviations 10 and 15 respectively. Then variance of the difference between means of the sampling distribution is _____.

Ans: 7

8 Curve Fitting: Regression and Correlation Analysis

8.1 Introduction

The main objective of many statistical investigations is to make predictions. Such predictions require that a formula is found which relates the dependent variable to one or more independent variables.

Suppose that a dependent variable is to be predicted in terms of an independent variable. In such a case, the independent variable is observed without error or with a negligible error when compared with an error in the dependent variable. Though the independent variable is fixed at *x*, repeated measurements of the dependent variable may lead to *y* values, which differ considerably. In most situations of this kind, we are interested in the relationship between *x* and the mean of the corresponding distribution of the random variable *y*. We refer to their relationship as the regression curve of *y* on *x*.

8.2 Linear Regression

Let the mean distribution of *y*'s be given by a linear relation of the form

$$\mu(x) = \alpha + \beta x \qquad (8.1)$$

We may want the sample values as *n* points (x_i, y_i) where $i = 1, 2, \ldots, n$ in the *xy*-plane. Fit a straight line through them and use it for estimating $\mu(x)$ at values of *x* that interest us.

8.2.1 Scatter Plot

Consider the data in Table 8.1.

Table 8.1 Table of data relating *x* and *y*.

x	1	2	3	4	5	6	7	8
y	20	32	49	72	88	102	115	124

Plotting these points in the *xy* plane, we find that a straight line gives a good approximation over the range of the given data (Figure 8.1). This kind of diagram, which shows how data points are scattered, is called a *scatter plot* or *scatter diagram.*

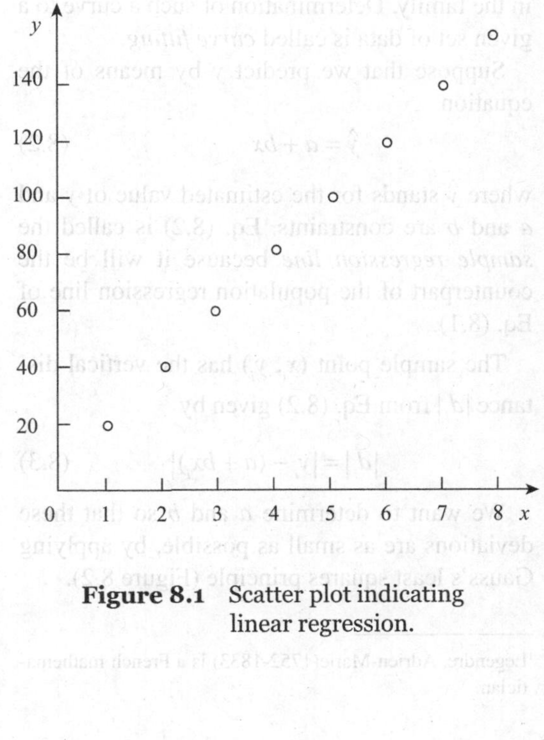

Figure 8.1 Scatter plot indicating linear regression.

8.2.2 Method of Least Squares Linear Curve

In general, more than one curve of given type will appear to fit a set of data. We will now see how we can isolate 'the best-fitting curve' by the method of least squares, which is based on Gauss's least squares principle. This principle is said to have been suggested by A.-M. Legendre[1].

Let (x_i, y_i), where $i = 1, 2, \ldots, n$, be the data points and $c: y = f(x)$ be the approximating curve. Then for a given value of x, say x_i, there will be a difference between the value y_i and the corresponding value as determined from the curve c. We denote this difference by d_i or e_i called a *deviation*, *error* or *residual* and may be positive, negative or zero. A measure of the goodness-of-fit of the curve c to the set of data is provided by the quantity $s = \sum d_i^2 \left(\text{or} \sum e_i^2\right)$. If this is small, the fit is good.

Of all curves in a given family of curves approximating a set of n data points, a curve having the property that $s = \sum d_i^2$ is a minimum is called a *best-fitting* curve (or regression curve) in the family. Determination of such a curve to a given set of data is called *curve fitting*.

Suppose that we predict y by means of the equation

$$\hat{y} = a + bx \tag{8.2}$$

where $\hat{y}$ stands for the estimated value of y and a and b are constraints. Eq. (8.2) is called the *sample regression line* because it will be the counterpart of the population regression line of Eq. (8.1).

The sample point (x_i, y_i) has the vertical distance $|d_i|$ from Eq. (8.2) given by

$$|d_i| = |y_i - (a + bx_i)| \tag{8.3}$$

We want to determine a and b so that these deviations are as small as possible, by applying Gauss's least squares principle (Figure 8.2).

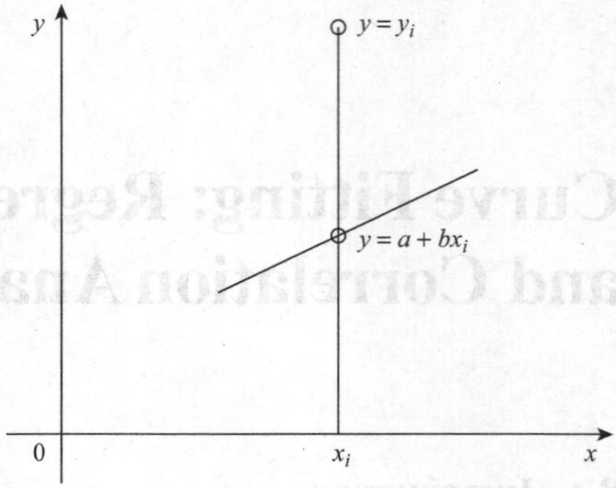

Figure 8.2 Vertical distance of a point (x_i, y_i) from straight line $y = a + bx$.

The straight line should be fitted through the given points so that the sum of the squares of the distances of those points from the straight line is minimum, where the distance is measured in the vertical direction (y-direction).

Hence the sums of the square of these distances is

$$s = \sum_{i=1}^{n} (y_i - a - bx_i)^2 \tag{8.4}$$

In the method of least squares, we now have to determine a and b such that s is minimum. Before minimizing the sum of the squared deviations to obtain the least squares estimators, we introduce the following notations:

$$S_{xx} = \sum_{i=1}^{n}(x_i - \bar{x})^2 = \sum_{i=1}^{n} x_i^2 - \frac{1}{n}\left(\sum_{i=1}^{n} x_i\right)^2$$

$$S_{yy} = \sum_{i=1}^{n}(y_i - \bar{y})^2 = \sum_{i=1}^{n} y_i^2 - \frac{1}{n}\left(\sum_{i=1}^{n} y_i\right)^2$$

$$S_{xy} = \sum_{i=1}^{n}(x_i - \bar{x})(y_i - \bar{y}) = \sum_{i=1}^{n} x_i y_i - \frac{1}{n}\left(\sum_{i=1}^{n} x_i\right)\left(\sum_{i=1}^{n} y_i\right) \tag{8.5}$$

From Calculus, we know that a necessary condition for S at Eq. (8.3) to be minimum is that the following equations hold:

$$\frac{\partial S}{\partial a} = 0 \text{ and } \frac{\partial S}{\partial b} = 0 \tag{8.6}$$

[1]Legendre, Adrien-Marie(1752-1833) is a French mathematician.

Differentiating Eq. (8.3) w.r.t. a and b, we obtain

$$\frac{\partial S}{\partial a} = -2\sum_{i=1}^{n}(y_i - a - bx_i) = 0 \text{ and}$$

$$\frac{\partial S}{\partial b} = -2\sum_{i=1}^{n}x_i(y_i - a - bx_i) = 0 \quad (8.7)$$

These equations yield the following normal equations:

1. $na + b\sum x_i = \sum y_i$ or $na + b\sum x = \sum y$

2. $a\sum x_i + b\sum x_i^2 = \sum x_i y_i$ or

$$a\sum x + b\sum x^2 = \sum xy \quad (8.8)$$

for simplicity. Its coefficient determinant is

$$\Delta = \begin{vmatrix} n & \sum x \\ \sum x & \sum x^2 \end{vmatrix} = n\sum x^2 - (\sum x)^2 = n(n-1)s_x^2$$

$$= n\sum(x_i - \bar{x})^2 \quad (8.9)$$

where $\Delta \neq 0$ since x values are not all equal. Hence the system has a unique solution.

$$\text{Also, } \Delta_a = \begin{vmatrix} \sum y & \sum x \\ \sum xy & \sum x^2 \end{vmatrix} = \sum x^2 \sum y - \sum x \sum xy \quad (8.10)$$

$$\Delta_b = \begin{vmatrix} n & \sum y \\ \sum x & \sum xy \end{vmatrix} = n\sum xy - \sum x \sum y \quad (8.11)$$

By Cramer's rule, we have

$$\therefore a = \frac{\Delta a}{\Delta} = \frac{\sum x^2 \sum y - \sum x \sum xy}{n\sum x_2 - (\sum x)^2} \text{ or } a = \bar{y} - b\bar{x} \quad (8.12)$$

$$b = \frac{\Delta b}{\Delta} = \frac{n\sum xy \sum x - \sum y}{n\sum x^2 - (\sum x)^2} \text{ or}$$

$$b = \frac{n\sum xy - \sum x \sum y}{n(n-1)s_x^2} \quad (8.13)$$

The equation for the regression curve is

$$y = a + bx = a + \left(\frac{S_{xy}}{S_{xx}}\right)x \quad (8.14)$$

The slope b is called the regression coefficient and is also given by

$$b = \frac{s_{xy}}{s_x^2} \quad (8.15)$$

8.2.3 Least Squares Quadratic (Parabolic) Curve

Let the quadratic curve

$$y = a_0 + a_1 x + a_2 x^2 \quad (8.16)$$

approximate to the given data. Using the least squares principle, we can derivate the normal equations for determining a_0, a_1 and a_2 as follows:

$$na_0 + a_1\sum x + a_2\sum x^2 = \sum y$$

$$a_0\sum x + a_1\sum x^2 + a_2\sum x^3 = \sum xy \quad (8.17)$$

$$a_0\sum x^2 + a_1\sum x^3 + a_2\sum x^4 = \sum x^2 y$$

8.2.4 Least Squares Non-linear Curves

For non-linear curve

$y = ab^x$ (exponential curve) (8.18)

$y = ax^b$ (power curve or geometric curve) (8.19)

$y = \frac{1}{a_0 + a_1 x}$ (reciprocal curve) (8.20)

A transformation of variables reduces them to linear form. Consider, e.g., Eq. (8.18).

Applying logarithms on both sides, we get

$$y = A + B_x$$

where $A = \log a$, $B = \log b$ and $y = \log y$.

8.3 Regression Analysis

In regression analysis, the nature of actual relationship between two or more variables is studied by determining the mathematical equation involving the variables.

It is mainly used to predict or estimate the dependent variable in terms of other independent variables. It is also used in optimization to determine the values of the independent variables for which the dependent variable has an extreme value.

1. **Simple Regression:** It establishes the relation between two variables.
2. **Multiple Regression:** It involves more than two variables.
3. **Linear Regression:** The relationship between the variables is linear and geometrically, it represents a straight line known as the *regression line.*
4. **Regression Line of *y* on *x*:** The linear relation

$$y = a_0 + a_1 x \tag{8.21}$$

between x and y is known as the regression line of y on x.

5. **Multiple Regression:** It is of the following form:

$$y = f(x_1, x_2, \ldots, x_n) \tag{8.22}$$

If f is linear of the form

$$f(x_1, x_2, \ldots, x_n) = a_0 + a_1 x_1 + \cdots + a_n x_n$$

then it is called a *multiple linear regressions*. Otherwise, it is called a *multiple non-linear regression.*

8.4 Inferences Based on Least Squares Estimation

Simple linear regression model is given by

$$\alpha + \beta x + \varepsilon \tag{8.23}$$

where α is the unknown intercept, β is the unknown slope parameters and ε is called the random error or disturbance. It is assumed to be normally distributed with mean $E(\varepsilon) = 0$ and variance (known as residual variance or error variance) is σ^2. In order to estimate the regression coefficients α and β, a regression line

$$\hat{y} = a_0 + a_1 x \tag{8.24}$$

is fitted based on the principle of least squares. The least squares estimates of α and β are a_0 and a_1 respectively and are given by

$$a_0 = \bar{y} - a_1 \bar{x} \tag{8.25a}$$

$$a_1 = \frac{S_{xy}}{S_{xx}} \tag{8.25b}$$

The slope of the regression line is the change in the mean of y corresponding to a unit increase in x.

8.4.1 Confidence Intervals

A $(1 - \alpha)$ 100% confidence interval for the parameter β is

$$a_1 - t_{\frac{\alpha}{2}} \frac{s_\varepsilon}{\sqrt{\frac{S_{xx}}{n}}} < \beta < a_1 + t_{\frac{\alpha}{2}} \frac{s_\varepsilon}{\sqrt{\frac{S_{xx}}{n}}} \tag{8.26}$$

where $t_{\frac{\alpha}{2}}$ is the value of t-distribution with $(n - 2)$ degrees of freedom (dof).

A $(1 - \alpha)$ 100% confidence interval for the parameter α is

$$a_0 - t_{\frac{\alpha}{2}} \frac{s_\varepsilon}{\sqrt{\frac{S_{xx}}{x}}} < \alpha < a_0 + t_{\frac{\alpha}{2}} \frac{s_\varepsilon}{\sqrt{\frac{S_{xx}}{x}}} \tag{8.27}$$

where s_ε^2 = unbiased estimate of σ^2

$$= \frac{S_{xx}S_{yy} - (S_{xy})^2}{n(n-2)S_{xx}} \tag{8.28}$$

8.4.2 Test of Hypothesis

Statistics for inferences about α and β:

1. For the slope β

$$t = \left(\frac{a_1 - \beta}{S_t}\right)\sqrt{\frac{S_{xx}}{x}} \tag{8.29}$$

2. For the intercept α

$$t = \left(\frac{a_0 - \alpha}{S_t}\right)\sqrt{\frac{nS_{xx}}{S_{xx} + (n\bar{x})^2}} \tag{8.30}$$

where t-distribution is of $(n - 2)$ dof.

Example 8.1

Fit a least squares line to the data in Table 8.2 for (a) x as independent variable. Estimate y at $x = 11$ and (b) x as dependent variable. Estimate x at $y = 2$.

Table 8.2

x	1	3	4	6	8	9	11	14
y	1	2	4	4	5	7	8	9

Solution Let (a) $y = a + bx$ and (b) $x = c + dy$ be the straight line to be fixed to the data in Table 8.2. The number of data points $n = 8$. We construct a table of sums $\sum x$, $\sum x^2$, etc. as in Table 8.3.

Table 8.3

x	y	x^2	xy	y^2
1	1	1	1	1
3	2	9	6	4
4	4	16	16	16
6	4	36	24	16
8	5	64	40	25
9	7	81	63	49
11	8	121	88	64
14	9	196	126	81
$\sum x = 56$	$\sum y = 40$	$\sum x^2 = 524$	$\sum xy = 364$	$\sum y^2 = 256$

(a) Normal equations are

$$na + b\sum x = \sum y$$

$$a\sum x + b\sum x^2 = \sum xy$$

Substituting the values, we get

$$8a + 56b = 40 \Rightarrow a + 7b = 5$$

$$56a + 524b = 364 \Rightarrow 56(5 - 76) + 524b$$

$$= 364 \Rightarrow 132b = 84 \Rightarrow b = \frac{7}{11} \quad \therefore a = \frac{6}{11}$$

The required line is $y = \frac{6}{11} + \frac{7}{11}x$

$$= 0.545 + 0636x.$$

(b) In this case, we have

$$8c + 40d = 56 \Rightarrow c + 5d = 7$$

$$40c + 256d = 364$$

$$56d = 84 \Rightarrow d = \frac{3}{2}$$

$$c = 7 - 5d = 7 - 5\left(\frac{3}{2}\right) = -\frac{1}{2}$$

The required line is $x = -\frac{1}{2} + \frac{3}{2}y$

$$= -0.5 + 1.5y.$$

Also y (at $x = 11$) = 13 and x (at $y = 2$) = $\frac{5}{2}$

Example 8.2

Fit a least squares parabola to the data in Table 8.4.

Table 8.4

x	1.2	1.8	3.1	4.9	5.7	7.1	8.6	9.8
y	4.5	5.9	7.0	7.8	7.2	6.8	4.5	2.7

Find $y_{est}(1.2)$.

Table 8.5

x	y	x^2	x^3	x^4	xy	x^2y
1.2	4.5	1.44	1.73	2.08	5.40	6.48
1.8	5.9	3.24	5.83	10.49	10.62	10.12
3.1	7.0	9.61	29.79	92.35	21.70	67.27
4.9	7.8	24.01	117.65	576.48	38.22	187.28
5.7	7.2	32.49	185.19	1055.58	41.04	233.93
7.1	6.8	50.41	357.91	2541.16	48.28	342.79
8.6	4.5	73.96	636.06	5470.12	38.70	332.82
9.8	2.7	96.04	941.19	9223.66	26.46	259.31
$\sum x =$ 42.2	$\sum y =$ 46.4	$\sum x^2 =$ 291.40	$\sum x^3 =$ 2275.35	$\sum x^4 =$ 18971.92	$\sum xy =$ 230.42	$\sum x^2y =$ 1449.00

Solution Let $y = a + bx + cx^2$ be the parabola. Then the normal equations are

$$xa + b\sum x + c\sum x^2 = \sum y$$

$$a\sum x + b\sum x^2 + c\sum x^3 = \sum xy$$

$$a\sum x^2 + b\sum x^3 + c\sum x^4 = \sum x^2 y$$

The number of data points $n = 8$. We construct the table of values $\sum x \sum x^2$, etc. as in Table 8.5.

Here $n = 8$. Substituting the values in the normal equations, we get

$$8a + 42.2b + 291.20c = 46.4$$

$$42.2a + 291.20b + 2275.35c = 230.42$$

$$291.20a + 2{,}275.35b + 18{,}971.92c = 1{,}449$$

Solving the equations, we get $a = 2.588$, $b = 2.065$ and $c = -0.211$.

Hence the required least squares parabola is

$$y = 2.588 + 2.065x - 0.211\,x^2$$

Estimated value of y at $x = 1.2$ is

$$y_{\text{est}}(1.2) = 2.588 + 2.065(1.2) - 0.211(1.2)^2$$
$$= 4.762$$

which is different from the tabulated value of $y(1.2) = 4.5$.

8.4.3 Inferences Based on Least Squares Estimates

Example 8.3

Fit a least squares straight line to the data given in Table 8.6.

Table 8.6

x	2	7	9	1	5	12
y	13	21	23	14	15	21

Solution The number of data points $n = 6$. We construct the table of values $\sum x, \sum x^2$, etc. as in Table 8.7.

Table 8.7

x	y	x^2	y^2	xy
2	13	4	169	26
7	21	49	441	147
9	23	81	529	207
1	14	1	196	14
5	15	25	225	75
12	21	144	441	252
$\sum x = 36$	$\sum y = 107$	$\sum x^2 = 304$	$\sum y^2 = 2001$	$\sum xy = 721$

Also

$$\bar{X} = \frac{36}{3} = 6$$

$$\bar{Y} = \frac{107}{6} = 17.833$$

$$S_{xx} = n\sum x_i^2 - (\sum x_i)^2 = 6(304) - (36)^2 = 528$$

$$S_{yy} = n\sum y_i^2 - (\sum y_i)^2 = 6(2001) - (107)^2 = 557$$

$$S_{xy} = n\sum x_i y_i = (\sum x_i)(\sum y_i) = 6(721) - 36(107)$$
$$= 474$$

Regression coefficient $= b = \dfrac{S_{xy}}{S_{xx}} = \dfrac{474}{528} = 0.8977$

Intercept $= a = \bar{Y} - b\bar{X} = 17.833 - (0.8977) \times 6$
$= 12.447$

The straight line of least squares is $y = a + bx = 12.45 + 0.8977x$.

Example 8.4

In Example 8.3, find the standard error of estimate s_t^2. Also, test for null hypothesis (NH) $\beta = 1.2$ against alternative hypothesis (AH) $\beta < 1.2$ at 0.05 level of significance (LOS).

Solution The standard error of estimate

$$s_\varepsilon^2 = \frac{S_{xx} \times S_{yy} - (S_{xy})^2}{n(n-2)\,S_{xx}} = \frac{(528)(557) - (474)^2}{6(6-2).528}$$

$$= 5.47822$$

$$\Rightarrow s_\varepsilon = \sqrt{5.47822} = 2.3405596 \approx 2.341$$

Test of significance

1. NH H_0: $\beta = 1.2$
2. AH H_1: $\beta < 1.2$
3. LOS $\alpha = 0.05$
4. **Critical Region:** It is a left-side one-tailed test. Reject NH H_0 if $t < -t_\alpha = -t_{0.05}$ with $n-2$ dof. From t-distribution table, the value of $t_{0.05} = 2.132$ with $n-2 = 6-2 = 4$ dof.

   Thus reject NH H_0 if $t < -2.132$.
5. **Calculations:** Test statistic t is given by

$$t = \frac{b-\beta}{s_t}\sqrt{\frac{S_{xx}}{n}}$$

   Here $n = 6$, $b = 0.8977$, $\beta = 1.2$, $s_t = 2.341$ and $S_{xx} = 528$. Substituting these values, we get

$$t = \frac{0.8977-1.2}{2.341}\sqrt{\frac{528}{6}} = 1.21137$$

6. **Conclusion:** Accept NH H_0. Since $t = -1.211 > t_\alpha = -2.132$ with 4 dof, we cannot reject NH H_0.

Example 8.5

Construct a 95% confidence interval for (a) and (b) β for Example 8.4.

Solution

(a) 95% confidence limits for α are

$$a \neq t_{\alpha/2} \times s_\varepsilon \sqrt{\frac{S_{xx} + (\overline{XX})^2}{nS_{xx}}}$$

$$= 12.45 \pm (2.776)(2.34)$$

$$\times \sqrt{\frac{528 + (6x6)^2}{6(528)}} = 12.45 \pm 4.93$$

$\therefore$ 95% confidence interval for α is (7.52, 17.38), since $t_{\alpha/2} = t_{0.05} = 2.776$ from the t-table.

(b) 95% confidence limits for β are

$$b \pm t_{\alpha/2} \times S_\varepsilon \times \sqrt{\frac{n}{S_{xx}}}$$

$$= 0.8977 \pm (2.776)(2.341)\sqrt{\frac{6}{528}}$$

$$= 0.8977 \pm 0.6925$$

$\therefore$ 95% confidence interval for β is (0.205, 1.59).

8.5 Multiple Regression

In agriculture, the crop yield (y) depends on many factors: it depends on the rainfall (x_1), the amount of fertilizers (x_2) used, the pesticides (x_3) utilized, the quality of soil (x_4) and so on. This is an example of multiple regression. In multiple regression, the dependent variable y is a function of more than one independent variable:

$$y = f(x_1, x_2, \ldots, x_n) \tag{8.31}$$

If f is a non-linear function, it is a case of non-linear regression. In multiple linear regression, f is a linear function of the form

$$y = \beta_0 + \beta_1 x_1 + \beta_2 x_2 + \cdots + \beta_k x_k \tag{8.32}$$

Response surface analysis deals with statistical methods of prediction and optimization.

8.5.1 Linear Multiple Regression

Let y depend on two variables x_1 and x_2. The relation is given by

$$y = \beta_0 + \beta_1 x_1 + \beta_2 x_2 \tag{8.33}$$

Then the linear multiple regression problem is to fit the regression plane Eq. (8.32) to a given set of ordered triples (x_{1i}, x_{2i}, y_i) (Figure 8.3).

In order to estimate the coefficients $\beta_0, \beta_1, \beta_2$, apply the method of least squares and minimize the quantity

$$\sum_{i=1}^{N}[y_i - (b_0 + b_1 x_{1i} + b_2 x_{2i})]^2$$

We obtain the following three normal equations:

$$Nb_0 + b_1\Sigma x_{1i} + b_2\Sigma x_{2i} = \Sigma y_i \tag{8.34a}$$

$$b_0\Sigma x_{1i} + b_1\Sigma x_{1i}^2 + b_2\Sigma x_{1i}x_{2i} = \Sigma x_{1i}y_i \tag{8.34b}$$

$$b_0\Sigma x_{2i} + b_1\Sigma x_{1i}x_{2i} + b_2\Sigma x_{2i}^2 = \Sigma x_{2i}y_i \tag{8.34c}$$

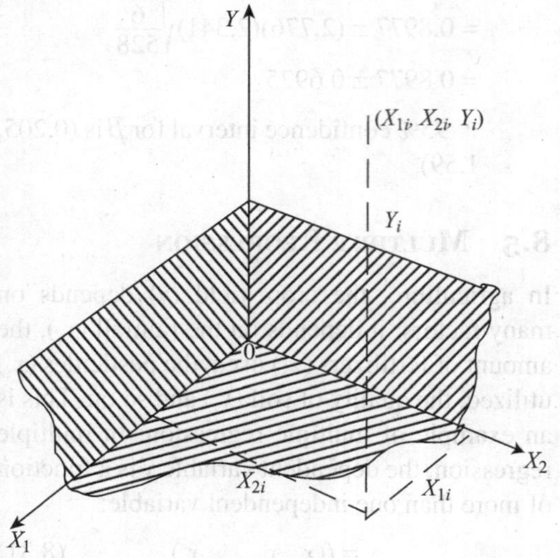

Figure 8.3 Linear multiple regression.

where b_0, b_1 and b_2 are the least squares estimates of β_0, β_1 and β_2 respectively.

If we put

$$\bar{x}_1 = \frac{\sum x_{1i}}{N}, \bar{x}_2 = \frac{\sum x_{2i}}{N} \text{ and } \bar{y} = \frac{\sum y_i}{N} \quad (8.35)$$

$$\bar{y}_i = y_i - \bar{y}, \bar{x}_{1i} = x_{1i} - \bar{x}_1 \text{ and } \bar{x}_{2i} = x_{2i} - \bar{x}_2$$

The three normal equations (Eq. 8.34) reduce to the following two equations:

$$b_1 \times \sum(\bar{x}_{1i})^2 + b_2 \sum \bar{x}_{1i}\bar{x}_{2i} = \sum \bar{x}_{1i}\bar{y}_i \quad (8.36)$$

$$b_1 \times \sum \bar{x}_{1i}\bar{x}_{2i} + b_2 \sum(\bar{x}_{2i}) = \sum \bar{x}_{2i}\bar{y}_i$$

8.5.2 Linear Multiple Regression in *K* Independent Variables

We can generate the above analysis to fit $N(k+1)$ triples $(x_{1i}, x_{2i}, \ldots, x_{ki})$, for $i = 1, 2, \ldots, N$, to the equation.

Using Eq. (8.32), we obtain the following $(K+1)$ normal equations:

$$Nb_0 + b_1 \sum x_{1i} + b_2 \sum x_{2i} + \cdots + b_K \sum x_{Ki} = \sum y_i$$

$$b_0 \sum x_{1i} + b_1 \sum x_{1i}^2 + b_2 \sum x_{1i} x_{2i} + \cdots + b_K \sum x_{1i} x_{Ki} = \sum x_{1i} y_i$$

$$b_0 \sum x_{Ki} x_{1i} + b_1 \sum x_{Ki} x_{2i} + \cdots + b_K \sum x_{Ki}^2 = \sum x_{Ki} y_i \quad (8.37)$$

Example 8.6

Find the least squares regression equation of x_1 on x_2 and x_3 from the data in Table 8.8.

Table 8.8

x_1	3	5	6	8	12	14
x_2	16	10	7	4	3	2
x_3	90	72	54	42	30	12

Solution Let $x_1 = a_0 + a_1 x_2 + a_2 x_3$

$$\bar{x}_2 = \frac{(16+10+7+4+3+2)}{6} = \frac{42}{6} = 7$$

$$\bar{x}_3 = \frac{(90+72+54+42+30+12)}{6} = \frac{300}{6} = 50$$

We shift the origin to $(\bar{x}_2, \bar{x}_3)$ by putting

$$u = x_2 - \bar{x}_2 = x_2 - 7 \text{ and } v = x_3 - \bar{x}_3 = x_3 - 50$$

Let $x_1 = a + bu + cv$. The normal equations are

$$na + b\sum u_i + c\sum v_i = \sum x_{1i}$$

$$a\sum u_i + b\sum u_i^2 + c\sum u_i v_i = \sum x_{1i} u_i$$

$$a\sum v_i + b\sum u_i v_i + c\sum v_i^2 = \sum x_i v_i$$

The number of data points $n = 6$. We construct the table of values required to feed the equations (Table 8.9).

Substituting these values into the normal equations, we get

$$6a + 0 + 0 = 48 \Rightarrow a = 8$$

$$140b + 720c = -100 \Rightarrow 7b + 36c = -20$$

$$720b + 4{,}008c = -582$$

Coefficient determinant $\Delta = \begin{vmatrix} 7 & 36 \\ 720 & 4008 \end{vmatrix}$

$= 28.056 - 25.920 = 2136$

$$\Delta_b = \begin{vmatrix} -20 & 36 \\ -582 & 4008 \end{vmatrix} = -80{,}160 + 20{,}952 = -59208$$

Table 8.9

x_1	x_2	x_3	u_i	v_i	$x_{1i}u_i$	$x_{1i}v_i$	u_iv_i	u_i^2	v_i^2
3	16	90	9	40	27	120	360	81	1,600
5	10	72	3	22	15	110	66	9	484
6	7	54	0	4	0	24	0	0	16
8	4	42	–3	–8	–24	–64	24	9	64
12	3	30	–4	–20	–48	–240	80	16	400
14	2	12	–5	–38	–70	–532	190	25	1,444
$\Sigma x_1 = 48$	$\Sigma x_2 = 42$	$\Sigma x_3 = 300$	$\Sigma u_i = 0$	$\Sigma v_i = 0$	$\Sigma x_{1i}u_i = -100$	$\Sigma x_{1i}v_i = -582$	$\Sigma u_iv_i = 720$	$\Sigma u_i^2 = 140$	$\Sigma v_i^2 = 4008$

$$\Delta_c = \begin{vmatrix} 7 & -20 \\ 720 & -582 \end{vmatrix} = -4{,}074 + 14400 = 10{,}326$$

$$b = \frac{\Delta b}{\Delta} = \frac{-59208}{2136} = -27.7191$$

$$c = \frac{\Delta c}{\Delta} = \frac{10326}{2136} = 4.8343$$

8.6 Correlation Analysis

We have so far considered problems where the independent variable is assumed to be known without error. There occur problems in which this may not happen. In such cases, both x and y values are assumed by random variables, e.g. the relationship between the tensile strength and hardness of aluminium or that between impurities in the air and the incidence of a certain disease. Problems such as these are referred to as problems of correlation analysis. It is assumed here that the data points (x_i, y_i), for i = 1, 2, …, n, are values of a pair of random variables whose joint density is given by the function $f(x, y)$.

The scatter plot provides a visual impression of the relation between the x and y values in a bivariate (two variables) data set. In many cases, the points appear to scatter about a straight line. The closeness of the scatter to a straight line can be expressed numerically in terms of the coefficient. The sample correlation coefficient can be interpreted in terms of the standardized observations

$$\frac{\text{Observations} - \text{Sample mean}}{\text{Sample standardized deviation}} = \frac{x_i - \bar{x}}{s_x} \quad (8.38)$$

where the subscript x on s distinguishes the sample variance of the x observations from the sample variance of the y observations.

$$s_x^2 = \frac{\Sigma(x_i - x)^2}{(n-1)} = \frac{S_{xx}}{(n-1)} \quad (8.39)$$

The sample correlation coefficient r is the sum of products of the standardized variable divided by $(n - 1)$, the same divisor used for variance.

$$\frac{1}{(n-1)} \sum_{i=1}^{n} \frac{(x_i - \bar{x})}{s_x} \frac{(y_i - \bar{y})}{s_y} \quad (8.40)$$

8.6.1 Types of Correlation

Let (x_i, y_i), for i = 1, 2, …, n, be the n ordered pairs of random variables x_i and y_i. Plotting these points in the xy-plane, we get a scatter diagram.

The following cases are distinguished:

1. The correlation is said to be *linear* if the points are scattered about a straight line and *non-linear* (curvilinear) if the points are scattered about a non-linear curve (parabola, exponential or power curve).
2. The correlation is said to be positive or direct if y increases as x increases; and is said to be negative or inverse if y increases as x decreases.

Examples

1. *Positive correlation*
   (a) Age and growth of a child
   (b) Distance travelled and the quantity of fuel consumed
2. *Negative correlation*
   (a) Number of workers and hours spent by them in completing a work
   (b) Number of shareholders and share of profit of each one of them.

Example 8.7

Calculate the sample correlation coefficient r using the data in Table 8.10.

Table 8.10

x	11.1	10.3	12.0	15.1	13.7	18.5	17.3	14.2	14.8	15.3
y	10.9	14.2	13.8	21.5	13.2	21.1	16.4	19.3	17.4	19.0

Solution The number of data points $n = 10$. We construct table of values $\sum x^2$, $\sum y^2$ etc. for calculating the correlation coefficient (Table 8.11).

Table 8.11

x	y	x^2	xy	y^2
11.1	10.9	123.21	120.99	118.81
10.3	14.2	106.09	146.26	201.64
12.0	13.8	144.00	165.60	190.44
15.1	21.5	228.01	324.65	462.25
13.7	13.2	187.69	180.84	174.24
18.5	21.1	342.25	390.35	445.21
17.3	16.4	299.29	283.72	268.96
14.2	19.3	201.64	274.06	372.49
14.8	17.4	219.04	257.52	302.76
15.3	19.0	234.09	290.70	361.00
$\sum x =$ 142.3	$\sum y =$ 166.8	$\sum x^2 =$ 2085.31	$\sum xy =$ 2434.69	$\sum y^2 =$ 2897.80

$$s_{xx} = \sum x_i^2 - \frac{(\sum x_i)^2}{n} = 2085.31 - \frac{1}{10}(142.3)^2$$

$$= 60.381$$

$$s_{yy} = \sum y_i^2 - \frac{(\sum y_i)^2}{n} = 2897.80 - \frac{1}{10}(166.8)^2$$

$$= 115.576$$

$$s_{xy} = \sum x_i y_i - \tfrac{1}{n}\sum x_i \sum y_i = 2434.69 - \frac{1}{10}(142.3)$$

$$(166.8) = 61.126$$

Sample correlation coefficient

$$r = \frac{S_{xy}}{\sqrt{S_{xx} \times S_{yy}}} = \frac{61.126}{\sqrt{60.381 \times 115.576}}$$

$$= 0.7317$$

r is positive. This confirms a positive association. Since $r = 0.7317$ is moderately large, the pattern of scatter is moderately narrow.

The correlation is said to be *simple* if it is between two variables. It is called *multiple* if it is between more than two variables.

The correlation coefficient r ranges between -1 and 1, i.e. $-1 \le r \le 1$ (Figure 8.4):

1. $r = 1$ if all pairs (x_i, y_i) lie exactly on a straight line having a positive slope.
2. $r = -1$ if all pairs (x_i, y_i) lie exactly on a straight line having a negative slope.
3. $r > 0$ if the scatter diagram runs from lower left up to upper right.
4. $r < 0$ if the scatter diagram runs from upper left down to lower right.

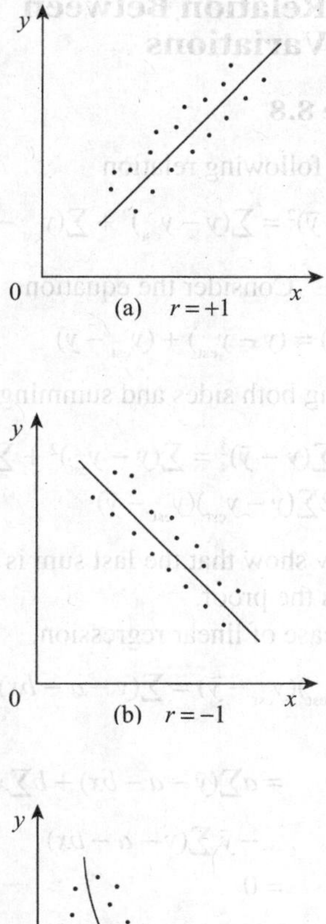

Figure 8.4 Correlation coefficient r ranges between –1 and 1.

8.7 Least Squares Line in Terms of Sample Variances and Covariance

The sample variances and covariance of x and y are given by

$$s_x^2 = \frac{\sum(x-\bar{x})^2}{n},\ s_y^2 = \frac{\sum(y-\bar{y})^2}{n},$$

$$s_{xy} = \frac{\sum(x-\bar{x})(y-\bar{y})}{n} \tag{8.41}$$

$$s_y^2 = \frac{\sum(y-\bar{y})^2}{n} \tag{8.42}$$

$$s_{xy}^2 = \frac{\sum(x-\bar{x})(y-\bar{y})}{n} \tag{8.43}$$

In terms of these, the least squares regression line of y on x is

$$y-\bar{y} = \frac{s_{xy}}{s_x^2}(x-\bar{x}) \tag{8.44}$$

and that of x on y is

$$x-\bar{x} = \frac{s_{xy}}{s_y^2}(y-\bar{y}) \tag{8.45}$$

If we formally define the sample correlation coefficient by

$$r = \frac{s_{xy}}{s_x s_y} \tag{8.46}$$

then Eq. (8.44) and (8.45) can be written as

$$\frac{y-\bar{y}}{s_y} = r\left(\frac{x-\bar{x}}{s_x}\right) \tag{8.47}$$

and

$$\frac{x-\bar{x}}{s_x} = r\left(\frac{y-\bar{y}}{s_y}\right) \tag{8.48}$$

These two lines are different unless $r = \pm 1$ in which case all sample points lie on a line. In this case, there is perfect linear correlation and regression.

If the two regression lines (Eq. (8.47) and (8.48)) are written as $y = a + bx$ and $x = c + dy$ respectively, then

$$bd = r^2 \tag{8.49}$$

8.8 Standard Error of Estimate

Let y_{est} denote the estimated value of y for a given value of x, as obtained from the regression curve of y on x. Then a measure of the scatter about the regression curve is supplied by the quantity

$$s_{yx} = \sqrt{\frac{\sum(y - y_{est})^2}{n}} \quad (8.50)$$

It is called the *standard error of estimate of y on x.* We know that $\sum(y - y_{est})^2 = d^2$. Hence out of all possible regression curves the least squares curve has the smallest standard error of estimate.

In the case of a regression line

$$y_{est} = a + bx \quad (8.51)$$

with a and b given by the normal equations

$$na + b\sum x = \sum y \quad (8.52)$$

$$a\sum x + b\sum x^2 = \sum xy \quad (8.53)$$

We have

$$s_{yx}^2 = \frac{\sum y^2 - a\sum y - b\sum xy}{n}$$

$$= \frac{\sum(y-\bar{y})^2 - b\sum(x-\bar{x})(y-\bar{y})}{n} \quad (8.54)$$

We can express s_{yx}^2 for the least squares lines in terms of variance and correlation coefficient as

$$s_{yx}^2 = s_y^2(1 - r^2) \quad (8.55)$$

From this, it follows that

$$r^2 \leq 1 \text{ for } -1 \leq r \leq 1 \quad (8.56)$$

The standard error of estimate has properties analogous to those of standard deviation.

An unbiased estimate of population variance is given by

$$\hat{s}^2 = \frac{ns^2}{(n-1)} \quad (8.57)$$

Similarly, there is an unbiased estimate of the square of the standard error of estimate. It is given by

$$\hat{s}_{yx} = \frac{ns_{yx}^2}{n-2} \quad (8.58)$$

8.8.1 Relation Between Variations

Example 8.8

Prove the following relation

$$\sum(y-\bar{y})^2 = \sum(y - y_{est})^2 + \sum(y_{est} - \bar{y})^2 \quad (8.59)$$

Solution Consider the equation

$$(y-\bar{y}) = (y - y_{est}) + (y_{est} - \bar{y}) \quad (8.60)$$

Squaring both sides and summing, we get

$$\sum(y-\bar{y})^2 = \sum(y - y_{est})^2 + \sum(y_{est} - \bar{y})^2 + 2\sum(y - y_{est})(y_{est} - \bar{y}) \quad (8.61)$$

We now show that the last sum is zero, which completes the proof.

In the case of linear regression,

$$\sum(y - y_{est})(y_{est} - \bar{y}) = \sum(y - a - bx)(a + bx - \bar{y}) \quad (8.62)$$

$$= a\sum(y - a - bx) + b\sum x(y - a - bx) - \bar{y}\sum(y - a - bx)$$

$$= 0 \quad (8.63)$$

Because of the normal equations

$$\sum(y - a - bx) = 0, \qquad \sum x(y - a - bx) = 0 \quad (8.64)$$

The result is true for non-linear regression also.

The above proof can be generalized in the case of non-linear regression using the least squares curve given by

$$y_{est} = a_0 + a_1x + a_2x^2 + \cdots + a_nx^n \quad (8.65)$$

8.8.2 Linear Correlation Coefficient

The quantity $\sum(y-\bar{y})^2$ is called the *total variation*; the sum $\sum(y - y_{est})^2$ is called the *unexplained variation* and $\sum(y_{est} - \bar{y})^2$ is called the *explained variation.* This terminology arises because the deviations $(y - y_{est})$ behave in a random or unpredictable manner while the deviations $(y_{est} - \bar{y})^2$ are explained by the least squares regression line.

In terms of s_{yx} and s_y, the correlation coefficient r is given by

$$r^2 = 1 - \frac{\sum(y - y_{est})^2}{\sum(y - \bar{y})^2} \Rightarrow r^2$$

$$= \frac{\sum(y_{est} - \bar{y})^2}{\sum(y - \bar{y})^2} = \frac{\text{Explained variation}}{\text{Total variation}} \quad (8.66)$$

$\therefore$ r^2 can be interpreted as the fraction of the total variation that is explained by the least squares regression line. The correlation coefficient can be computed from

$$r = \frac{s_{xy}}{s_x s_y} = \frac{\sum(x - \bar{x})(y - \bar{y})}{\sqrt{\sum(x - \bar{x})^2}\sqrt{\sum(y - \bar{y})^2}} \quad (8.67)$$

or $$r^2 = \frac{\text{Explained variation}}{\text{Total variation}} = \frac{\sum(y_{est} - \bar{y})^2}{\sum(y - \bar{y})^2} \quad (8.68)$$

which, for linear regression, are equivalent. Eq. (8.67) is often referred to as the *product–moment formula* for linear correlation.

Formulas equivalent to those given above, which are used in practice, are

$$r = \frac{n\sum_{xy} - (\sum x)(\sum y)}{\sqrt{[n\sum x^2 - (\sum x)^2]}\sqrt{n\sum y^2 - (\sum y)^2}} \quad (8.69)$$

and

$$r = \frac{\overline{xy} - \bar{x}.\bar{y}}{(\overline{x^2} - \bar{x}^2)(\overline{y^2} - \bar{y}^2)} \quad (8.70)$$

Example 8.9

Find the least squares regression line of y on x and that of x on y, using the data in Table 8.12.

Table 8.12

x	65	63	67	64	68	62
y	68	66	68	65	69	66

x	70	66	68	67	69	71
y	68	65	71	67	68	70

Solution The number of data points $n = 12$. We construct table of values $\sum x$, $\sum y$, $\sum x^2$ etc. (Table 8.13).

Table 8.13

x	y	x^2	xy	y^2
65	63	4,225	4,420	4,624
63	66	3,969	4,158	4,356
67	68	4,489	4,556	4,624
64	65	4,096	4,160	4,225
68	69	4,624	4,692	4,761
62	66	3,844	4,092	4,356
70	68	4,900	4,760	4,624
66	65	4,356	4,290	4,225
68	71	4,624	4,828	5,041
67	67	4,489	4,489	4,489
69	68	4,761	4,692	4,624
71	70	5,041	4,970	4,900
$\sum x =$ 800	$\sum y =$ 811	$\sum x^2 =$ 53,418	$\sum xy =$ 54,107	$\sum y^2 =$ 54,849

Let the regression line of y on x be

$$y = a + bx$$

The normal equations are

$$an + b\sum x = \sum y$$
$$a\sum x + b\sum x^2 = \sum xy$$

Substituting the computed values, we get

$$12a + 800b = 811$$
$$800a + 53{,}418b = 54{,}107$$

Solving we get

$$a = 35.82 \text{ and } b = 0.476$$

The required line is

$$y = 35.82 + 0.476x$$

Now, let the regression line of x on y be

$$x = c + dy$$

The normal equations are

$$cn + d\sum y = \sum x$$
$$c\sum y + d\sum y^2 = \sum xy$$

Substituting the computed values, we get

$$12c + 811d = 800$$
$$811c + 54{,}849d = 54{,}107$$

Solving we get

$$c = -3.38 \text{ and } d = 1.036$$

∴ The required line is

$$x = -3.38 + 1.036y$$

Example 8.10

Compute the standard error of estimate s_{yx} for the data of Example 8.9.

Solution The regression line of y on x is

$$y = 35.82 + 0.476x$$

The actual values of y are entered in Table 8.13. The estimated values of y, denoted by y_{est}, as obtained from the regression line are entered in Table 8.14.

Table 8.14

x	65	63	67	64
y	68	66	68	65
y_{est}	66.76	65.81	67.71	66.28
$y - y_{est}$	1.24	0.19	0.29	−1.28

x	68	62	70	66
y	69	66	68	65
y_{est}	68.19	65.33	69.14	67.24
$y - y_{est}$	0.81	0.67	−1.14	−2.24

x	68	67	69	71
y	71	67	68	70
y_{est}	68.19	67.71	68.66	69.02
$y - y_{est}$	2.81	−0.71	−0.66	0.38

Now

$$s_{yx}^2 = \frac{\Sigma(y - y_{est})^2}{n}$$

$$= \frac{(1.24)^2 + (0.19)^2 + \cdots + (0.38)^2}{12} = 1.642$$

$$\therefore s_{yx} = \sqrt{1.642} = 1.28$$

Example 8.11

Compute the following for the data of Example 8.9:

(a) Explained variation
(b) Unexplained variation
(c) Total variation

Solution We have $\bar{y} = \frac{\Sigma y}{n} = \frac{811}{12} = 67.58$.

Using the values y_{est} from Table 8.14, we construct Table 8.15.

Table 8.15

$y_{est} - \bar{y}$	−0.82	−1.77	0.13	−1.30
$y_{est} - \bar{y}$	0.61	−2.25	1.56	−0.34
$y_{est} - \bar{y}$	0.61	0.13	1.08	2.04

(a) Explained variation

$$\Sigma(y_{est} - \bar{y})^2 = (-0.82)^2 + \cdots + (2.04)^2 = 19.22$$

(b) Unexplained variation

$$\Sigma(y - y_{est})^2 = ns_{yx}^2 = 19.70$$

(c) Total variation

$$\Sigma(y - \bar{y})^2 = 19.22 + 19.70 = 38.92$$

Example 8.12

Find the following for the data of Example 8.9:

(a) Coefficient of determination
(b) Coefficient of correlation

Solution

(a) Coefficient of determination

$$r^2 = \frac{\text{Explained variation}}{\text{Total variation}} = \frac{19.22}{38.92} = 0.4938$$

(b) Coefficient of correlation

$$r = \pm\sqrt{0.4938} = \pm 0.7027$$

Since the variable y_{est} increases as x increases, the correlation is positive, i.e. $r = 0.7027$.

8.8.3 Test of Hypothesis for Correlation Coefficient

We often have to estimate the population correlation coefficient ρ from the sampling correlation coefficient r or to test hypothesis concerning ρ. For this, we must know the sampling distribution of r. In case $\rho = 0$, this distribution is symmetric and a statistic having Student's t-distribution can be used. For $\rho \neq 0$, the distribution is skewed. In such a case, a transformation due to Fisher produces a statistic which is approximately normally distributed.

1. **Test of Hypothesis when $\rho = 0$** Here we use the fact that the statistic $t = \dfrac{r\sqrt{n-2}}{\sqrt{1-r^2}}$ has Student's t-distribution with $(n-2)$ dof.
2. **Test of Hypothesis when $\rho \neq 0$** Here we use the fact that the statistic (Fisher's Z-transformation)

$$Z = \frac{1}{2} ln\left(\frac{1+r}{1-r}\right) = 1.1513 \times \log_{10}\left(\frac{1+r}{1-r}\right)$$

is approximately normally distributed with mean and standard deviation given by

$$\mu_z = \frac{1}{2} ln\left(\frac{1+\rho}{1-\rho}\right) = 1.1513 \log_{10}\left(\frac{1+\rho}{1-\rho}\right)$$

$$\sigma_Z = \frac{1}{\sqrt{n-3}} \qquad (8.71)$$

Example 8.13

A correlation coefficient based on a sample of size 18 was computed to be 0.32. Can we conclude at an LOS (a) 0.05 and (b) 0.01 that the corresponding population correlation coefficient is significantly greater than zero?

Solution

1. NH H_0: $\rho = 0$
2. AH H_1: $\rho > 0$
3. LOS: (a) $\alpha = 0.05$ and (b) $\alpha = 0.01$
4. **Critical Regions:** Reject H_0 if

   (a) $t > t_{0.95} = 1.75$

   (b) $t > t_{0.99} = 2.58$

   for $v = n - 2 = 18 - 2 = 16$ dof
5. **Computation:** $t = \dfrac{r\sqrt{n-2}}{\sqrt{1-r^2}}$

$$= \frac{0.32\sqrt{18-2}}{\sqrt{1-(0.32)^2}} = 1.35$$

6. **Conclusions:**

   (a) On the basis of a one-tailed test of Student's t-distribution at a 0.05 level, we would reject H_0 if $t > t_{0.95} = 1.75$ for $v = 16$ dof. Since $t = 1.35 < 1.75$, we cannot reject H_0 at $\alpha = 0.05$ level.

   (b) Since we cannot reject H_0 at 0.05 level, we certainly cannot reject it at 0.01 level.

Example 8.14

A correlation coefficient based on a sample of size 24 was computed to be $r = 0.75$. Can we reject the hypothesis that the population correlation coefficient is as small as (a) $\rho = 0.60$ and (b) $\rho = 0.50$ at a 0.05 significance level?

Solution

1. NH H_0: (a) $\rho = 60$ and (b) $\rho = 50$
2. AH H_1: (a) $\rho > 60$ and (b) $\rho > 50$
3. LOS $\alpha = 0.05$
4. **Computation:** $Z = 1.1513 \log\left(\dfrac{1+0.75}{1-0.75}\right)$

$$= 0.9750$$

(a) $\mu_Z = 1.1513 \times \log\left(\dfrac{1+0.60}{1-0.60}\right) = 0.6932$

$$\sigma_Z = \frac{1}{\sqrt{n-3}} = \frac{1}{\sqrt{21}} = 0.2182$$

The standardized variable is then

$$Z = \frac{z - \mu_z}{\sigma_z} = \frac{0.9730 - 0.6932}{0.2182} = 1.28$$

(b) $\mu_Z = 1.1513 \times \log\left(\frac{1 + 0.50}{1 - 0.50}\right)$

$= 1.1513 \times \log 3 = 0.5493$

$$Z = \frac{0.9730 - 0.5493}{0.2182} = 1.94$$

5. **Conclusions:**

(a) Using a one-tailed test of normal distribution, we would reject H_0 only if $Z > 1.64$. Since $Z = 1.28 < 1.64$, we cannot reject H_0, i.e. the population correlation coefficient ρ is as small as 0.60 at a 0.05 LOS.

(b) In this case, $Z = 1.94 > 1.28$. Therefore, we can reject H_0, i.e. the population correlation coefficient is as small as $\rho = 0.50$ at a 0.05 LOS.

8.9 Spearman's Rank Correlation

Often there arise problems in which quantitative measurement of data is not possible and only qualitative assessment has to be made. In such cases, the usual Pearson correlation cannot be calculated. To overcome this difficulty, C. E. Spearman[2] in 1906 developed a non-parametric correlation coefficient as follows.

Let a group of n individuals be arranged in order of merit in respect of two characteristics A and B. The ranks in the two characteristics are, in general, different. Suppose A stands for intelligence and B for beauty. An intelligent person need not be beautiful. Let (x_i, y_i), for $i = 1, 2, \ldots, n$, be the ranks of n individuals, in respect of characteristics A and B respectively. Pearsonian coefficient correlation between the ranks x_i and y_i is called the *rank correlation coefficient* between the characteristics A and B for that group of individuals. Since each of the variables assumes values 1, 2, 3, …, n in some order, we have

$$\bar{x} = \bar{y} = \frac{1}{n}(1 + 2 + \cdots + n) = \frac{1}{2}(n + 1)$$

If x and y be the deviations of x and y from their means, then

$$\sum x_i = \sum (x_i - \bar{x})^2 = \sum x_i^2 + n(\bar{x})^2$$
$$- 2\bar{x}\sum x_i = \sum n^2 + \frac{n}{4}(n + 1)^2 - 2\frac{n + 1}{2}\sum n$$
$$= \frac{n(n + 1)(2n + 1)}{6} + \frac{n}{4}(n + 1)^2 - \frac{n}{4}(n + 1)^2$$
$$= \frac{1}{12}(n^3 - n)$$

Similarly, $\sum y_i = \frac{1}{12}(n^3 - n)$

Now, let $d_i = x_i - y_i = (x_i - \bar{x}) - (y_i - \bar{y}) = x_i - y_i$

$$\Rightarrow \sum d_i^2 = \sum x_i^2 + \sum y_i^2 - 2\sum x_i y_i$$
$$\Rightarrow \sum x_i y_i = \frac{1}{2}\left(\sum x_i^2 + \sum y_i^2 - d_i^2\right)$$
$$= \frac{1}{12}(x^3 - n) - \frac{1}{2}\sum d_i^2$$

Hence the rank correlation coefficient between these variables is

$$r_s = \frac{\sum x_i y_i}{\sqrt{\sum x_i^2 \sum y_i^2}} = \frac{\frac{1}{12}(x^3 - n) - \frac{1}{2}\sum d_i^2}{\frac{1}{12}(x^3 - n)}$$

$$= 1 - \frac{6\sum d_i^2}{n^3 - n}$$

Note If there occurs a tie between two values of x (or y), assign the mean of the ranks to both these places.

Example 8.15

Ten participants in a contest are ranked by two judges as shown in Table 8.16.

Table 8.16

x	1	6	5	10	3	2	4	9	7	8
y	6	4	9	8	1	2	3	10	5	7

Calculate the rank correlation coefficient r_s.

Solution We construct the table of values for calculating the rank correlation coefficient (Table 8.17).

[2]Spearman, Charles Edward (1863-1945) is a psychologist.

Table 8.17

$d_i = x_i - y_i$	−5	2	−4	2	2
d_i^2	25	4	16	4	4

$d_i = x_i - y_i$	0	1	−1	2	1	
d_i^2	0	1	1	4	1	$\sum d_i^2 = 60$

Hence

$$r_s = 1 - \frac{6\sum d_i^2}{n^3 - n} = 1 - \frac{6 \times 60}{990} = 0.6$$

Example 8.16

The marks secured by applicants in the selection test (x) and in the proficiency test (y) are given in Table 8.18.

Table 8.18 Marks secured by applicants in selection and proficiency tests.

Serial No.	1	2	3	4	5	6	7	8	9
x	10	15	12	17	13	16	24	14	22
y	30	42	45	46	33	34	40	35	39

Calculate the rank correlation coefficient.

Solution Assigning ranks 1–9 according to the decreasing order of marks in x and in y, we construct the table of values for calculating the rank correlation coefficient (Table 8.19).

Table 8.19

x	10	15	12
y	30	42	45
x_i = rank in x	9	5	8
y_i = rank in y	9	3	2
$d_i = x_i - y_i$	0	2	6
d_i^2	0	4	36

x	17	13	16
y	46	33	34
x_i = rank in x	3	7	4
y_i = rank in y	1	8	7
$d_i = x_i - y_i$	2	−1	−3
d_i^2	4	1	9

x	24	14	22	
y	40	35	39	
x_i = rank in x	1	6	2	
y_i = rank in y	4	6	5	
$d_i = x_i - y_i$	−3	0	−3	$\sum d_1 = 0$
d_i^2	9	0	9	$\sum d_i^2$

$$\therefore r_s = 1 - \frac{6\sum d_i^2}{n^3 - n} = 1 - \frac{6 \times 72}{9 \times 80} = 1 - 0.6 = 0.4$$

Example 8.17

Ten students were ranked in laboratory (x) and lecture (y) portions (Table 8.20). Find the rank correlation coefficient.

Table 8.20

Laboratory (x)	8	3	9	2	7	10	4	6	1	5
Lecture (y)	9	5	10	1	8	7	3	4	2	6

Solution We construct the table of values for calculating the rank correlation coefficient (Table 8.21).

Table 8.21

$d_i = x_i - y_i$	−1	−2	−1	1	−1	3
d_i^2	1	4	1	1	1	9

$d_i = x_i - y_i$	1	2	−1	−1	$\sum d^2 = 24$
d_i^2	1	4	1	1	$\sum d_i^2 = 24$

$$r_s = 1 - \frac{6\sum d_i^2}{n(n^2-1)} = 1 - \frac{6(24)}{10(10^2-1)} = 0.8545$$

Example 8.18

Calculate the rank correlation coefficient for the data in Table 8.13 of Example 8.9.

Solution Arranging x values in ascending order of magnitude, we get

x: 62 63 64 65 66 67 67 68 68 69 70 71 (1)

Since the 6th and 7th places represent the same value 67, we assign a mean rank 6.5 to both these places. Similarly, the 8th and 9th places are assigned the rank 8.5. We have

1 2 3 4 5 6.5 6.5 8.5 8.5 10 11 12 (2)

Again, arranging y values in ascending order of magnitude, we get

y: 65 65 66 66 67 68 68 68 68 69 70 71 (3)

Since the 6–9th places represent the same value 68, we assign the mean rank (6 + 7 + 8 + 9)/4 = 7.5 to these places. We have

1.5 1.5 3.5 3.5 5 7.5 7.5 7.5 7.5 10 11 12 (4)

Using the correspondences (1) and (2), (3) and (4), we get Table 8.22.

Table 8.22

x	65	63	67	64	68	62	70	66	68	67	69	71
y	68	66	68	65	69	66	68	65	71	67	68	70
Rank in x (x_i)	4	2	6.5	3	8.5	1	11	5	8.5	6.5	10	12
Rank in y (y_i)	7.5	3.5	7.5	1.5	10	3.5	7.5	1.5	12	5	7.5	11
d_i	−3.5	−1.5	−1.0	1.5	−1.5	−2.5	3.5	3.5	−3.58	1.5	2.5	1.0
d_i^2	12.25	2.25	1.00	2.25	2.25	6.25	12.25	12.25	12.25	2.25	6.25	1.00

$$r_s = 1 - \frac{6\sum d_i^2}{n(n^2-1)} = 1 - \frac{6(72.50)}{12(12^2-1)} = 0.7465$$

8.10 Correlation for Bivariate Frequency Distribution

When the set of data is very large, it is arranged in the form of a bivariate frequency table or bivariate frequency distribution, as shown in Table 8.23.

Table 8.23 Class intervals for x.

Y \ X	$X_{Li} - X_{Ui}$	…	$X_{Lk} - X_{Uk}$	Total
$Y_{Li} - Y_{Ui}$	f_{11}	…	f_{1k}	
⋮	⋮	f_{ij}	⋮	
$Y_{Lm} - Y_{Um}$ $Y_{Lm} - Y_{Um}$	f_{m1}	…	f_{mk}	
Total				N

Assume that

X is grouped into k classes.

Y is grouped into m classes.

f_{ij}, or f simply, is the cell frequency of the ith X class interval and jth Y class interval.

X_{Li} and X_{Ui} X_{Li}, X_{Ui} denote the lower and upper limits of the ith class.

Y_{L_i} and Y_{U_i} denote the lower and upper limits of the jth class.

A blank cell represents zero cell frequency.

$$\text{Total frequency} = N = \sum_{j=1}^{m}\sum_{i=1}^{k} f_{ij}$$

Let X_i be the mid-value (class mark) of the ith X class and Y_j be the mid-value (class mark) of the jth Y class

Put $U_x = \dfrac{X - A}{C_1}$ and $U_y = \dfrac{\tau Y - B}{C_2}$

where

C_1 = class size of X intervals
C_2 = class size of Y intervals
A = assumed class mark for X classes
B = assumed class mark for Y classes
f_X = marginal frequencies of X (column sums of f_{ij}'s)
f_Y = marginal frequencies of Y (row sums of f_{ij}'s)

Note that $\sum f_x = \sum f_y = N$

Note also that $fU_X U_Y$ is denoted by a number in the box at right top corner of each cell (Table 8.24).

Now the correlation coefficient is given by

$$r = \frac{N\sum fU_X U_Y - (\sum f_X U_X)(\sum f_Y U_Y)}{\sqrt{[N\sum f_X U_X^2 - (\sum f_X U_X)^2]}\sqrt{[N\sum f_Y U_Y^2 - (\sum f_Y U_Y)^2]}}$$

Notes

1. U_X and U_Y turn out to be 0, ±1, ±2, …
2. In general $C_1 = C_2$
3. Check Total frequency $N = \sum f_X = \sum f_Y$
4. Check $\sum fU_X U_Y = \sum fU_Y U_X$ obtained from row sums and column sums

Example 8.19

Find the correlation coefficient for the bivariate distribution in Table 8.25.

Table 8.24 Correlation table.

	Mid-value X	X_1	X_2	…	X_k					
Mid-value Y	U_X / U_Y					f_Y		$f_Y U_Y$	$f_Y U_Y^2$	Sum of boxed numbers in each row
Y_1										
Y_2										
Y_3										
⋮										
Y_m										
	f_X					$N = \sum f_X$	$= \sum f_Y$	$\sum f_Y U_Y$	$\sum f_Y U_Y^2$	$\sum fU_X U_Y$
$f_X U_X$						$\sum f_X U_X$		CHECK		
$f_X U_X^2$						$\sum f_X U_X^2$				
Sum of boxed numbers in each column						$\sum fU_X U_Y$				

Table 8.25

Y \ X	15-25	25-35	35-45	45-55	55-65	65-75
15-25	1	1				
25-35	2	12	1			
35-45		4	10	1		
45-55			3	6	1	
55-65	1			2	4	2
65-75					1	2

Test NH H_0 $\rho = 0$ against AH H_1 $\rho \neq 0$ at 0.05 LOS. Determine whether there is a relationship between X and Y.

Solution We construct the bivariate correlation table (Table 8.26).

We have $N = 53$, $\sum fU_XU_Y = 86$, $\sum fU_X = 10$,

$\sum fU_Y = 16$, $\sum fU_X^2 = 98$, $\sum fU_Y^2 = 92$,

$$U_X = \frac{X - 40}{10} \text{ and } U_Y = \frac{Y - 40}{10}$$

The coefficient of correlation r for the bivariate distribution is

$$r = \frac{N\sum fU_XU_Y}{\sqrt{[N\sum f_XU_X^2 - (\sum f_XU_X)^2]}}$$

Table 8.26 Bivariate correlation table.

Mid-value Y		Mid-value X	15-25	25-35	35-45	45-55	55-65	65-75				
		X	20	30	40	50	60	70				
		U_X	–2	–1	0	1	2	3	Total f	fU_Y	fU_Y^2	fU_XU_Y
		U_Y										
15-25	20	–2	1 [4]	1 [2]					2	–4	8	6
25-35	30	–1	2 [4]	12 [12]	1 [0]				15	–15	15	16
35-45	40	0		4 [0]	10 [0]	1 [0]			15	0	0	0
45-55	50	1			3 [0]	6 [6]	1 [2]		10	10	10	8
55-65	60	2				2 [4]	4 [16]	2 [12]	8	16	32	32
65-75	70	3					1 [6]	2 [18]	3	9	27	24
		Total f	3	17	14	9	6	4	$N =$ 53	16	92	86
		fU_X	–6	–17	0	9	12	12	10			
		fU_X^2	12	17	0	9	24	36	98	CHECK		
		fU_XU_Y	8	14	0	10	24	30	86			

$$- \frac{(\sum f_X U_X)(\sum f_Y U_Y)}{\sqrt{[N\sum f_Y U_Y^2 - (\sum f_Y U_Y)^2]}}$$

$$= \frac{53 \times 86 - 10 \times 16}{\sqrt{53 \times 98 - 100}\sqrt{53 \times 92 - 256}} = \frac{4398}{4850}$$

$$= 0.91$$

Test of significance

1. NH H_0: $\rho = 0$
2. AH H_1: $\rho \neq 0$
3. LOS $\alpha = 0.05$
4. **Critical Region:** Reject H_0 if $Z < -1.96$ or $Z > 1.96$, where $Z = (\sqrt{N-3})\,Z^*$.
5. **Computation:** The value of Z^* corresponding to $r = 0.91$ is

$$Z^* = \frac{1}{2}\ln\left(\frac{1+r}{1-r}\right) = \frac{1}{2}\ln\left(\frac{1+0.91}{1-0.91}\right) = 1.5275$$

$$\therefore Z = (\sqrt{N-3})\,Z^* = (\sqrt{53-3})\,Z^*$$

$$= 7.071 \times 1.5275 = 10.801$$

$$\therefore z = \sqrt{n-3}\; z^* = \sqrt{53-3}\; z^*$$

$$= 7.071 \times 1.5275 = 10.801$$

6. **Conclusion:** Reject NH H_0 of non-linear association because $Z = 10.801 > 1.96$. So, we conclude that there is a relationship between the two observations.

EXERCISES

1. Fit a least squares straight line to the following data:

x	2	7	9	1	5	12
y	13	21	23	14	15	21

Ans: $y = 12.45 + 0.8977x$

$\sum x = 36$, $\sum y = 107$, $\sum x^2 = 3.04$,

$\sum y^2 = 2001$, $\sum xy = 721$

$\bar{x} = \frac{36}{6} = 6$, $\bar{y} = \frac{107}{6} = 17.833$

[Hint: $n = 6$, $s_{xx} = n\sum x^2 - \left(\sum x\right)^2$

$= 6(304) - (36)^2$

$= 528$

$s_{yy} = x\sum y^2 - \left(\sum y\right)^2 = 6(2001) - (107)^2$

$= 557$

$s_{xy} = n\sum$

$y = a + bx = 12.45 + 0.8977x$]

2. Fit a second-degree parabola to the following data using method of least squares:

x	0	1	2	3	4
y	1	1.8	1.3	2.5	6.3

Ans: $Y = 1.42 - 1.07x + 0.55x^2$

[Hint: Normal equations are

$5a + 10b + 30c = 12.9;$

$10a + 30b + 100c = 37.1$

$30a + 100b + 354c = 130.3$

Solving we get $y = 1.42 - 1.07x + 0.55x^2$]

3. Fit a parabola $y = a + bx + cx^2$ to the following data:

x	1	2	3	4	5	6	7	8	9
y	2	6	7	8	10	11	11	10	9

Ans: $a = -1$, $b = 3.55$ and $c = -0.27$

4. (a) Predict y when $x = 210$ by fitting a least squares straight line to the following data:

   (b) Determine 95% confidence interval for α and β.

   (c) Test NH H_0 $\beta = 0$ against AH H_1 $\beta \neq 0$ at 0.05 LOS.

x	20	60	100	140	180
y	0.18	0.37	0.35	0.78	0.56

x	220	260	300	340	380
y	0.75	1.18	1.36	1.17	1.65

Ans: (a) $y = 0.867$

(d) $-0.164 < \alpha < 0.302$ and $0.00348 < \beta < 0.004119$

(e) Reject NH H_0 $\beta = 0$, since $t = 8.36 > 2.306$

$$N = 10, \sum X = 2{,}000, \sum X^2 = 5{,}32{,}000,$$

$$\sum Y = 8.35, \sum XY = 21{,}754$$

[Hint: s_{xx}=13,20,000, s_{yy}=21.3745, s_{xy}=5054, $s_E^2 = 0.0253$

(a) $y = a + bx = 0.069 + (0.0038 \times 210) = 0.867$]

5. Find y when $x_1 = 10$ and $x_2 = 6$ from the least squares regression equation of y on x_1 and x_2 for the following data:

y	90	72	54	42	30	12
x_1	3	5	6	8	12	14
x_2	16	10	7	4	3	2

Ans: $y = 40$

[Hint: $n = 6, \sum y_i = 300, \sum x_{1i} = 48, \sum x_{2i} = 42,$

$$\sum x_{1i} \sum x_{2i} = 236$$

$$\sum x_{1i}^2 = 474, \sum x_{2i}^2 = 434, \sum x_{1i} y_i = 1818,$$

$$\sum x_{2i} y_i = 2820$$

$$y = 61.40 - 3.65\,x_1 + 2.54\,x_2$$

$$y(10.6) = 40.14 \cong 40]$$

6. (a) Estimate the blood pressure (BP) of a woman of age 45 years from the following data which shows the ages (x) and systolic BP (y) of 12 women.

Age (x)	56	42	72	36	63	47
BP (y)	147	125	160	118	149	128

Age (x)	55	49	38	42	68	60
BP (y)	150	145	115	140	152	155

(b) Are the two variable ages (x) and BP (y) correlated?

Ans: (a) $y = 132$

(c) Age (x) and BP (y) are strongly positively correlated.

[Hint: $\sum x = 628$;

$\sum y = 1{,}684; \sum x^2 = 34{,}416; \sum y^2 = 2{,}38{,}822;$

$\sum xy = 89{,}894; n = 12.$

Substituting in $y = a + bx$

$= 80.777 + 1.138(45) = 132$

$r = 0.8961$. Therefore, age (x) and BP (y) are strongly positively correlated.]

7. Find the curve of best fit of the type $y = ae^{bx}$ to the following data by the method of least squares:

x	1	5	7	9	12
y	10	15	12	15	21

Ans: $y = 9.4754e^{0.059}$

8. Find the least squares regression equation of x_1 on x_2 and x_3 from the following data:

x_1	3	5	6	8	12	14
x_2	16	10	7	4	3	2
x_3	90	72	54	42	30	12

Ans: $x_1 = 16.1 + 0.417x_2 - 0.22x_3$

9. Calculate the correlation coefficient for the heights of fathers and their sons.

x	65	66	67	67	68	69	70	72
y	67	68	65	68	72	72	69	71

Ans: $r = 0.603$

10. Obtain the rank correlation coefficient for the following data:

x	68	64	75	50	64	80	75	40	55	64
y	62	58	68	45	81	60	68	48	50	70

Ans: $r = 0.545$

11. For $n = 12$ and $r = 0.75$, test NH H_0 $P = 0.6$ against AH H_1 $P > 0.6$ at an $\alpha = 0.05$ LOS.

Ans: H_0 cannot be rejected.

[Hint: Reject H_0 if $Z > 1.65 = Z_{0.05}$. Z_f corresponding to $r = 0.75$ is 0.973 and $\mu_Z = 0.693$. So,

$$Z = \frac{Z_f - \text{ì}_Z}{\left(\frac{1}{\sqrt{x-3}}\right)} = 0.840$$

Hence, H_0 cannot be rejected.]

12. Find the correlation coefficient for the following bivariate frequency distribution:

	20-24	25-29	30-34	35-39	40-44
20-24	20	10	3	2	
25-29	4	28	6	4	
30-34		5	11		
35-39			2		
40-44					

Ans: $r = 0.613$

[Hint: $N = 100$, $\sum U_X U_Y = 138$

$\sum f_X U_X = -80$, $\sum f_Y U_Y = -100$

$\sum f_X U_X^2 = 150$, $\sum f_Y U_Y^2 = 204$

$$r = \frac{N\sum f U_X U_Y - (\sum f_X U_X)(\sum f_Y U_Y)}{\sqrt{[N\sum f_X U_X^2 - (\sum f_X U_X)^2]}\sqrt{[N\sum f_Y U_Y^2 - (\sum f_Y U_Y)^2]}}$$

$= 0.613$]

FILL IN THE BLANKS

1. The purpose of _____ is to estimate the dependent variable from the independent variable.

Ans: curve fitting

2. The process of estimation of the dependent variable from the independent variable is called _____.

Ans: regression

3. Of all the curves in a given family of curves approximating a set of n data points, a curve having the property that the sum of squares of deviations, viz. $\sum d_i^2$, is a minimum is called a _____.

Ans: best-fitting curve

4. (Explained variation)/(total variation) = _____.

Ans: $\sum(y_{\text{est}} - \bar{y})^2/\sum(y - \bar{y})^2$

5. Coefficient of rank correlation r = _____.

Ans: $1 - 6\sum d^2/n(n^2 - 1)$

6. In Fisher's Z-transformation, the test statistic is _____.

Ans: $Z = \frac{1}{2}\ln\left(\frac{1+r}{1-r}\right)$

7. If the equation of the regression line is $y = 0.545 + 0.636x$, then the estimated value of y when $x = 12$ is _____.

Ans: 8.2

8. If the slope of the regression line of y on x is $b = 0.476$ and that of the regression line of x on y is $d = 1.036$, then the correlation coefficient r is _____.

Ans: 0.7022

9. The two lines of regression $y = a + bx$ and $x = c + dy$ always pass through the point _____.

Ans: $(\bar{x}, \bar{y})$

10. If $s_{xy} = 72$ and $S_x^2 = 50$ then the slope b of the regression line $y = a + bx$ is b = _____.

Ans: 1.44

$$\left[\text{Hint: } b = \frac{S_{xy}}{S_x^2} = \frac{72}{50} = 1.44\right]$$

11. If $\bar{x} = 3.2$ and $\bar{y} = 4.5$, then the intercept a of the regression line $y = a + 1.2x$ is $a =$ _____.

Ans: 0.66

[Hint: $a = \bar{y} - b\bar{x} = 4.5 - 1.2 \times 3.2 = 0.66$]

12. If the coefficient of rank correlation between x and y is $P = 0.8$ and $\sum d_i^2 = 33$, then $n =$ _____.

Ans: 10

$$\left[\text{Hint: } n(n^2 - 1) = \frac{6\sum d_i^2}{1 - P} = \frac{6 \times 33}{1 - 0.8} = 10 \times 99 = 10\,(10^2 - 1) \Rightarrow n = 10\right]$$

13. If rank correlation coefficient is $r = 0.4$, $\sum d_i = 72$ and no rank is repeated, then $n =$ _____.

Ans: 9

$$\left[\text{Hint: } n(n^2 - 1) = (n - 1)n(n + 1) = \frac{6\sum d_i^2}{1 - r} = \frac{6 \times 72}{1 - 0.4} = \frac{6 \times 72}{0.6} = 720 = 8 \times 9 \times 10 \Rightarrow n = 9\right]$$

14. If $r = 0.4$ and $n = 16$, then the standard error SE(r) = _____.

Ans: 0.21

$$\left[\text{Hint: SE}(r) = \frac{1 - r^2}{\sqrt{n}} = \frac{1 - (0.4)^2}{\sqrt{16}} = 0.21\right]$$

15. In Question 14, the probability error of the correlation coefficient is _____.

Ans: 0.1416

[Hint: PE(r) $= 0.6745\ SE = 0.6745 \times 0.21 = 0.1416$]

16. If $r = -0.32$ and $n = 64$, then the standard error SE(r) = _____.

Ans: 0.1122

$$\left[\text{Hint: SE}(r) = \frac{1-r^2}{\sqrt{n}} = \frac{1-(-0.32)^2}{\sqrt{64}} = \frac{0.8976}{8} = 0.1122\right]$$

17. In Question 16, the probability error of the correlation coefficient is _____.

Ans: 0.0757

[Hint: PE(r) $= 0.6745 \times 0.1122 = 0.757$]

18. If $\sum d_i^2 = 104$ for the following pair of observations

x	10	15	12	17	13	16	24	14	22	20
y	30	42	45	46	33	34	40	35	39	38

then the rank correlation is _____.

Ans: 0.3697

$$\left[\text{Hint: } r = 1 - \frac{6\sum d_i^2}{n(n^2 - 1)} = 1 - \frac{6 \times 104}{10(10^2 - 1)} = 1 - \frac{624}{990} = 0.3697\right]$$

19. If $s_{xx} = 8$, $s_{yy} = 42$, $s_{xy} = 16$ and the standard error SE(r) = 0.0476, then $n =$ _____.

Ans: 25

$$\left[\text{Hint: } r = \frac{s_{xy}}{\sqrt{s_{xx}s_{yy}}} = \frac{16}{\sqrt{8 \times 42}}\right.$$

$$1 - r^2 = 1 - 0.7619 = 0.238$$

$$\left.\sqrt{n} = \frac{1 - r^2}{s_E} = \frac{0.238}{0.0476} = 5\right]$$

20. If variables x and y are independent, then the angle between the two regression lines is _____.

Ans: $\frac{\pi}{2}$

21. The lines of regression are coincident if the correlation coefficient is _____.

Ans: −1

22. The lines fitted by the least squares method is $y = a + 3.25x$ and the means of x and y are 20 and 72 respectively, then $a =$ _____.

Ans: 7

[Hint: $a = \bar{y} - 3.25\,\bar{x} = 72 - 3.25 \times 20 = 7$]

23. The arithmetic mean of the coefficient of regressions is _____.

Ans: the correlation coefficient

24. If $\sigma_x \sigma_y = \sigma$ and the angle between the regression lines is $\tan^{-1}\frac{4}{3}$, then the coefficient of correlation is _____.

Ans: $\frac{1}{3}$

$$\left[\text{Hint: } K = \frac{\sigma_x \sigma_y}{\sigma_x^2 \sigma_y^2} = \frac{\sigma^2}{2\sigma^2}\right.$$

$$= \frac{1}{2} \Rightarrow \tan\theta = \left(\frac{1-r^2}{r}\right)$$

$$K = \frac{1-r^2}{2r} = \frac{4}{3}$$

$$\Rightarrow 3r^2 + 8r - 3 = (3r-1)(r+3)$$

$$\left. = 0 \Rightarrow r = \frac{1}{3}\ (|rK|)\right]$$

25. If θ is the angle between two regression lines, $r = 0.25$ and $\sigma_y = 2\sigma_x$, then $\tan\theta =$ _____.

Ans: $\frac{3}{2}$

$$\left[\text{Hint: } K = \frac{\sigma_x\sigma_y}{\sigma_x^2\sigma_y^2} = \frac{2\sigma_x^2}{\sigma_x^2 + 4\sigma_x^2} = \frac{2}{5}\right.$$

$$\tan\theta = K\left(\frac{1-r^2}{r}\right) = \frac{2}{5}\,\frac{1-\left(\frac{1}{4}\right)^2}{\left(\frac{1}{4}\right)}$$

$$\left. = \frac{2}{5}\,\frac{15}{4} = \frac{3}{2}\right]$$

26. If $\sum x_i = 30$, $\sum x_i y_i = 25$, $\sum y_i = 50$, $\sum x_i^2 = 95$, then the slope of the regression line is _____.

Ans: 4

$$\left[\text{Hint: } b = \frac{\sum x_i y_i - \frac{\sum y_i}{x}}{\sum x_i^2 - 2x_i\left(\frac{\sum x_i}{n}\right)} = \frac{25 - \frac{50}{10}}{95\text{-}30 \times \frac{30}{10}} = 4\right]$$

27. The coefficient of correlation is $\frac{2}{3}$. Also $\sigma_y = 3\sigma_x$ Then the angle between the regression lines is _____.

Ans: $\tan^{-1}\frac{1}{4}$

$$\left[\text{Hint: } \tan\theta = \frac{1-r^2}{r}\,\frac{\sigma_y\sigma_x}{\sigma_y^2\sigma_x^2} = \frac{1-\left(\frac{2}{3}\right)^2}{\frac{2}{3}}\right.$$

$$\left.\frac{3\sigma_x^2}{3^2\sigma_x^2\sigma_x^2} = \frac{5}{9}\,\frac{3}{2}\,\frac{3}{10} = \frac{1}{4}\right]$$

28. If $\left(\frac{\sigma_y}{\sigma_x}\right) = m$ the correlation coefficient $r = \frac{2}{3}$ and the regression lines is $\tan^{-1}\frac{1}{4}$ then $m =$ _____.

Ans: 3 or $\frac{1}{3}$

$$\left[\text{Hint: } \tan\theta = \frac{1-r^2}{r}\,\frac{\sigma_y\sigma_x}{\sigma_y^2 + \sigma_x^2} \Rightarrow \frac{1}{4}\right.$$

$$= \frac{1-\left(\frac{2}{3}\right)^2}{\frac{2}{3}}\,\frac{m\sigma_x^2}{(m^2+1)\sigma_x^2} \Rightarrow 3(m^2+1)$$

$$= 10m \Rightarrow (3m-1)(m-3) = 0$$

$$\left.\Rightarrow m = 3 \text{ or } \frac{1}{3}\right]$$

29. If $\sum x = 146$, $\sum y = 102$, $\sum xy = 1334$ and $s_{xy} = 93$, then $n =$ _____.

Ans: 12

$$\left[\text{Hint: } xy = \sum xy - \frac{\sum x \sum y}{n} \Rightarrow n = \frac{\sum x \sum y}{\sum xy - s_{xy}}\right.$$

$$\left. = \frac{146 \times 102}{1334 - 93} = \frac{73 \times 17 \times 12}{1241} = 12\right]$$

30. If $s_{xx} = 126$, $s_{yy} = 123$ and $s_{xy} = 93$, then sample correlation coefficient $r =$ _____.

Ans: 0.747

$$\left[\text{Hint: } r = \frac{s_{xy}}{\sqrt{s_{xx}s_{yy}}} = \frac{93}{\sqrt{126 \times 123}} = 0.747\right]$$

Queueing Theory

9

9.1 Introduction

It is an important area of applied probability theory. A queue is a waiting line, an orderly line of one or more persons or jobs, e.g. a waiting line at a ticket booking counter at a cinema hall, railway or bus station. Queues are also common in computer systems. There are queues of enquiries waiting to be processed by an interactive computer system, e.g. queues of database records and queues of input–output requests.

9.2 Queues or Waiting Lines

They are not only of human beings or animals but also of the aeroplanes seeking to land at a busy airport, trains waiting outside the railway station for clearance of a platform and signal permitting entry of a train at a busy railway station, ships to be unloaded, machine parts to be assembled, vehicles waiting for a green light at a traffic signal point, calls arriving at a telephone switch board, jobs waiting for processing by a computer and so on.

9.3 Elements of a Basic Queueing System

It consists of the following elements:

1. Customer population or sources
2. Input traffic or arrival pattern
3. Service time distribution
4. Service facility: Number of servers
5. Queue discipline (service discipline)
6. Customers' behaviour

We will deal with these one by one.

9.3.1 Customer Population or Sources

Customers from a population or source enter a queueing system to receive some type of service. The word 'customer' is used here in the generic sense. It does not necessarily refer to animate things. It may mean an enquiry message requiring transmission and processing.

If the population is finite, it is difficult to model the problem. In infinite population systems, the number of customers has no effect on the arrival pattern and it is therefore easy to model.

9.3.2 Input Traffic or Arrival Pattern

It is the arrival of people or things from a customer population and entering the waiting line in anticipation of their turn to receive service.

The ability of a queueing system to provide service for an arriving system of customers depends not only on the mean arrival rate λ, but also on the pattern of their arrival. If customer arrivals are evenly spaced, the service facility can provide better service than if they arrive in clusters:

$$0 \le t_0 < t_1 < t_2 \dots$$

The random variables $\tau_k = t_k - t_{k-1}$ $(k = 1, 2, \dots)$ are called inter-arrival times. It is usual to describe this by $A(t) = 1 - e^{-\lambda t}$ (exponential distribution).

9.3.3 Service Time Distribution

It is the customers or things forming into an orderly waiting line waiting to receive service when their turn comes and get served at the service point.

The exponential distribution is used to describe the service time of a server. The distribution function is

$$\omega(t) = P(s \le t) = 1 - e^{-\mu t} \qquad (9.1)$$

where μ stands for the average service rate.

9.3.4 Service Facility: Number of Servers

It is the point at which service is provided and the people or jobs in the queue system get the service they require. There may be one or more servers (channels) attending the waiting line or lines.

A server is an entity capable of performing the required service for a customer.

The simplest queueing system is the one with single server system. A multi server system has 'C' identical servers and provides service to 'C' customers at a time.

9.3.5 Queue Discipline (Service Discipline)

The rule for selecting the next customer to receive is given in detail as follows:

1. **FCFS:** First come, first served
   or **FIFO:** First in, first out
2. **LCFS:** Last come, first served
   or **LIFO:** Last in, first out
3. **RSS:** Random selection for service
   or **SIRO:** Service in random order
4. **PRI:** Priority service

Some customers get preferred treatment.

The first one is the most commonly prevalent queue discipline.

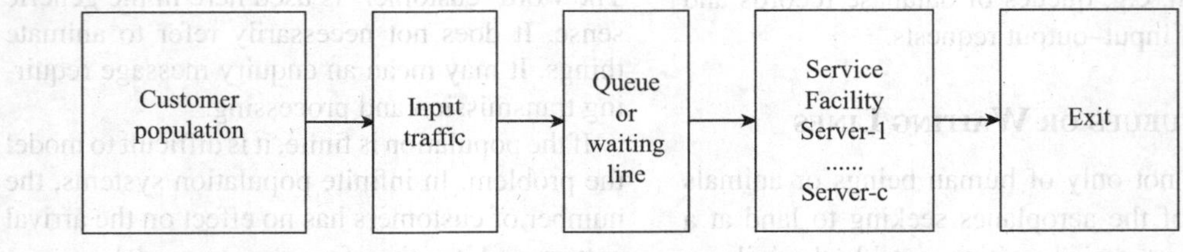

Figure 9.1 Elements of a basic queueing system.

Table 9.1 Typical queueing systems.

S. No.	Queueing System	Customer	Server(s)
1	Railway reservation System	Traveller wanting reservation or information	Railway employee with his computer system.
2	Cash with drawal from bank account	Customer of a bank	ATM (automated teller machine)
3	Airline reservation	Traveller wanting reservation or information	Agent plus terminal to a computer reservation system
4	Taxis at a taxi stand	Traveller	Owner of taxis with a communication system
5	Interactive enquiry system	Inquiry from a terminal	Communication line plus a computer

9.3.6 Customer's Behaviour

In general, some customers behave in one of the following ways.

1. **Balking:** A customer who refuses to enter a queuing system because the queue is too long is said to be balking.
2. **Reneging:** A customer who leaves the queue without receiving service because of excessive queueing time is said to have reneged.
3. **Jockeying:** Customers may leave one queue and join another which is shorter expecting quicker service. This kind of behaviour on the part of a customer is called jockeying.
4. **Priority:** In some applications, some customers are given preferential service before others regardless of their time of arrival. This is a case of priority in giving service in a queueing system.

A block diagram describing the various elements of a basic queueing system is shown in Figure 9.1.

Some typical queueing systems are presented in Table 9.1.

9.4 Description of a Queueing System

The primary random variables in a queueing system are illustrated in Figure 9.2.

9.5 Classification of Queueing Systems

The different queueing systems may be classified as follows:

1. Queueing system with single queue and single service station (Figure 9.3a)
2. Queueing system with single queue and several service stations (Figure 9.3b)
3. Queueing system with several queue and several service stations (Figure 9.3c)
4. Complex queueing system (Figure 9.4)

9.6 Queueing Problem

The queueing theory is concerned with the statistical description of the behaviour of the queueing system, modeling and solving the queueing problem.

In a specified queueing system, the problem is to determine the following statistical parameters:

1. **Probability Distribution of Queue Length** When the nature of probability distributions of the arrivals and service patterns is given, the probability distribution of queue length can be obtained. Further, we can also estimate the probability that there is no queue.

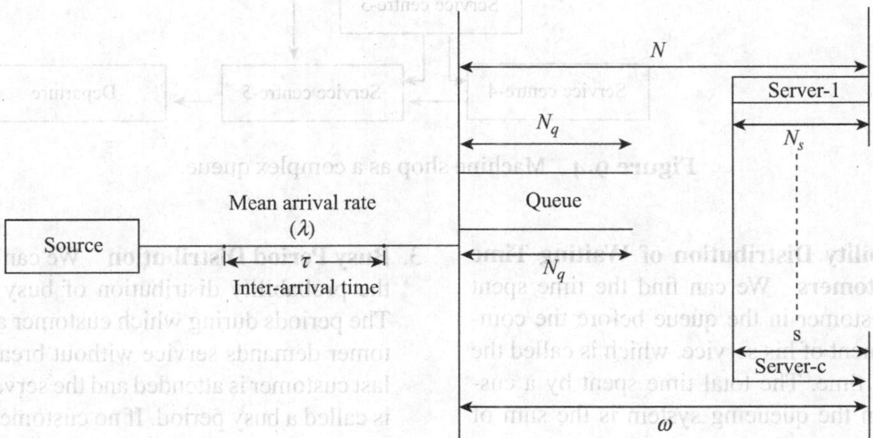

Figure 9.2 Queueing theory random variables.

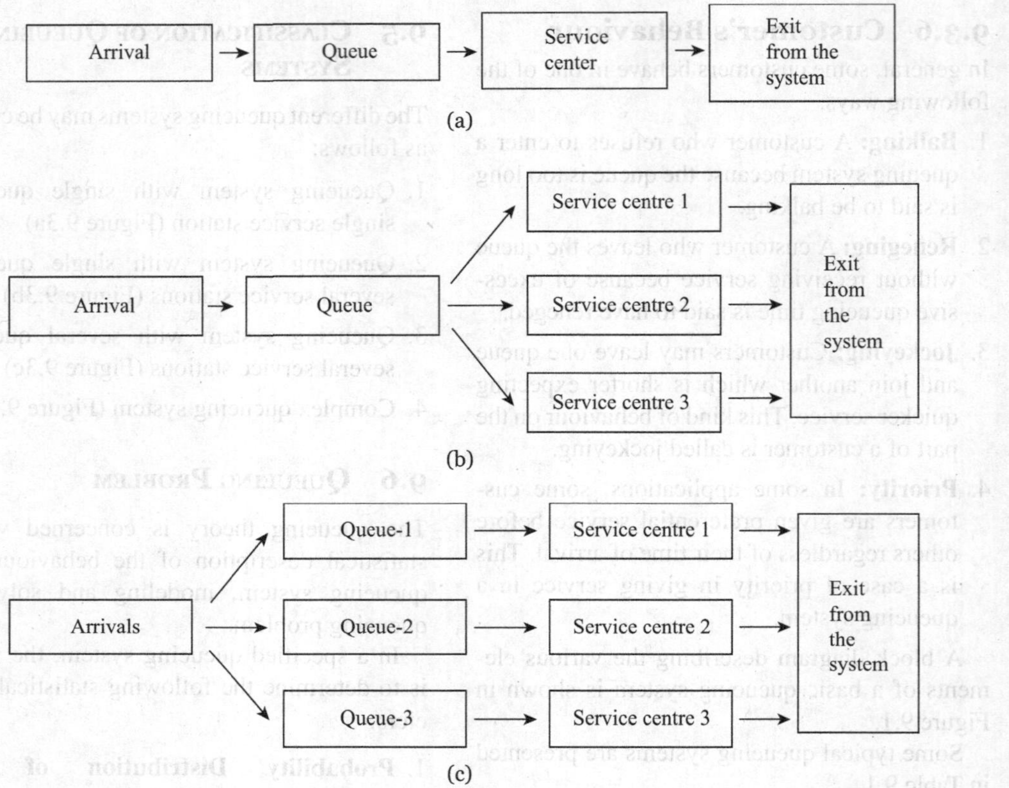

Figure 9.3 Queueing systems.

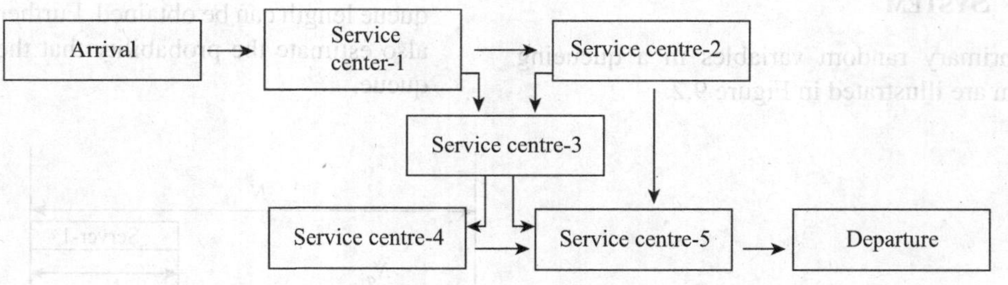

Figure 9.4 Machine shop as a complex queue.

2. **Probability Distribution of Waiting Time of Customers** We can find the time spent by a customer in the queue before the commencement of his service, which is called the waiting time. The total time spent by a customer in the queueing system is the sum of waiting time and service time.

Total time spent = waiting time + service time.

3. **Busy Period Distribution** We can estimate the probability distribution of busy periods. The periods during which customer after customer demands service without break till the last customer is attended and the server is free, is called a busy period. If no customer is present and the server is free, it is called an idle period. The study of busy periods is of great

importance in cases where technical features of the server and his capacity for continuous operation must be taken into account.

9.7 States of Queueing Theory

In a queueing theory analysis, we have to distinguish between the following two states:

1. Transient state
2. Steady state

9.7.1 Transient State

A queueing system is said to be in a transient state if its operating characteristics are dependent on time.

This occurs usually in the early stages of the operation of the queueing system.

9.7.2 Steady State

A queueing system is said to be in a steady state if its operating characteristics are independent of time.

Let $p_n(t)$ be the probability that there are n units in the system at time t.

In the steady state, rate of change of $p_n(t)$ is zero, i.e.

$$\frac{dp_n(t)}{dt} = 0 \text{ as } t \to \infty \tag{9.2}$$

The basic queueing theory definitions and notations are given in Table 9.2.

Table 9.2 Basic queueing theory notations and definitions.

Notation	Definition
n	Number of units in the system
$p_n(t)$	Transient state probability that exactly n calling units are in the queueing system at time t
E_n	State of having n calling units in the system
p_n	Study state probability of having n units in the system
λ_n	Mean arrival rate of customers (expected numbers of arrivals per unit time)
μ_n	Mean service rate (expected number of customers served per unit time)
λ	Mean arrival rate when λ_n is constant for all n.
μ	Mean service rate when μ_n is constant for all $n \geq 1$
S	Number parallel service stations
$P = \lambda/\mu_s$	Traffic intensity (or utilization factor) for server facility, i.e. the expected fraction of time the servers are busy
$\phi_T(n)$	Probability of n services in time T_1 given that servicing is going on throughout T
Line length or queue size	Number of services in the queueing system
Queue length	Line length (queue size) – (number of units being served)
$\psi(\omega)$	Probability density function (pdf) of waiting time in the system
L_s	Expected line length, i.e. expected number of customers in the system
L_q	Expected queue length, i.e. expected number of customers in the queue
W_s	Expected waiting time per customer in the system
W_q	Expected waiting time per customer in the queue
$(W / W > 0)$	Expected waiting time of a customer who has to wait
$(L / L > 0)$	Expected length of non-empty queues, i.e. expected number of customers in the queue when there is a queue
$P(W > 0)$	Probability of a customer having to wait for service

9.8 Probability Distribution in Queueing Systems

The arrival pattern of customers in a queueing system changes from one system to another. But one pattern of common occurrence which is relatively easy to model is that of 'completely random arrivals', which means that the number of arrivals in unit time has a Poisson distribution. In this, the intervals between successive arrivals are distributed negative exponentially, i.e. in terms of $e^{-\lambda t}$.

9.8.1 Distribution of Arrivals: Poisson Process (Pure Birth Process)

The model in which only arrivals are considered is called a pure birth model. In this, departures are not taken into account. Here, 'birth' refers to a new calling unit in the system and 'death' refers to the departure of a served unit. Thus, pure birth models are not of practical importance study of pure birth models is important for understanding completely random arrival problems.

We establish in the following theorem that in a completely random arrival queueing system, the probability distribution is that of Poisson distribution.

Theorem (Probability Distribution of Arrivals) If in a queueing system, the arrivals are completely random, then the probability distribution of the number of arrivals in a fixed time-interval follows a Poisson distribution.

Proof We prove the theorem based on the following axioms:

1. Let there be n units in the system at time t. Let the probability that exactly one arrival will occur during a small time interval δt be given by $\lambda\delta t + 0(\delta t)$. Here λ is the arrival rate which is independent of t and $0(\delta t)$ contains terms of order higher than the first.
2. Let δt be so small that the probability of more than one arrival in time interval δt is $0[(\delta t)^2]$ $0((\delta t)^2)$ which is negligibly small.
3. The number of arrivals in non-overlapping intervals is statistically independent, i.e. the process has independent increments.

Let $p_n t$ be the probability of n arrivals in a time intervals of length t. Clearly $n \geq 0$, we now construct differential difference equations governing the process in two different situations, viz. $n > 0$ and $n = 0$.

Case 1: $n > 0$ There may be two mutually exclusive ways of having n units of time:

1. There are n units in the system at a time t and no arrival takes place. So, there are n units at time $t + \delta t$ (Figure 9.5).

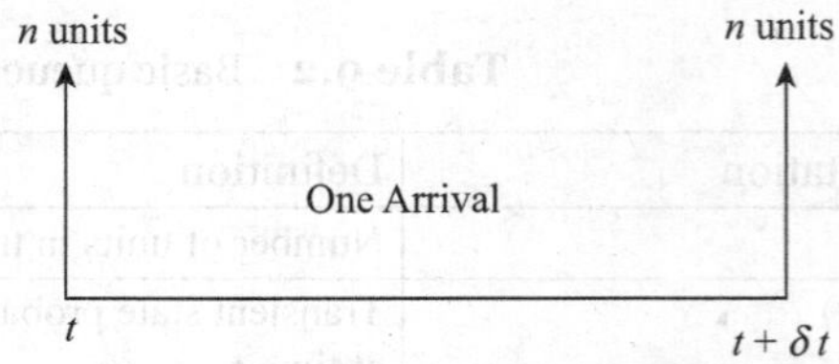

Figure 9.5

We compute the following probability:

Probability of exactly one arrival in time interval

$$\delta t = \lambda\delta t \tag{9.3}$$

Probability of no arrival (complement of this case) $= 1 - \lambda\delta t$ (9.4)

Now, probability of these two combined events described in (1)

= (probability of n units at time t) × (probability of no arrivals)

$$= p_n(t)\,(1 - \lambda\delta t) \tag{9.5}$$

2. There are $(n - 1)$ units in the system at time t and one arrival takes place in time interval δt (Figure 9.6).

Probability of these two combined events described in (2)

= (probability of $n - 1$ units at time t) × (probability of one arrival during time interval δt

$$= p_{n-1}(t)\lambda\delta t \tag{9.6}$$

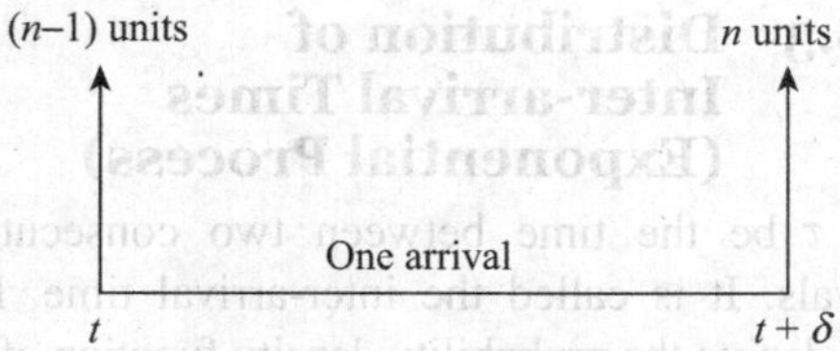

Figure 9.6

Adding Eqs. (9.5) and (9.6), we get the probability of n arrivals at time $t + \delta t$:

$$p_n(t+\delta t) = p_n(t)(1 - \lambda\delta t) + p_{n-1}(t)\lambda\delta t \quad (9.7)$$

Case 2: $n = 0$ In this case, we obtain

$p_n(t + \delta t) = p(\text{no unit at time } t) + p(\text{no arrival during interval } \delta t)$

$$p_0(t)(1 - \lambda\delta t) \quad (9.8)$$

Adding Eqs. (9.7) and (9.8), we get

$$p_n(t+\delta t) - p_n(t) = p_n(t)(-\lambda\delta t) + p_{n-1}(t)\,\lambda\delta t \quad \text{for } n > 0 \quad (9.9)$$

$$p_0(t+\delta t) - p_0(t) = p_0(t)(-\lambda\delta t) \quad \text{for } n = 0 \quad (9.10)$$

On dividing equations by δt and taking limits as $\delta t \to 0$, we obtain

$$\lim_{\delta t \to 0} \frac{p_n(t+\delta t) - p_n(t)}{\delta t}$$

$$= p_n(t) = -\lambda p_n(t) + \lambda p_{n-1}(t) \quad \text{for } n > 0 \quad (9.11)$$

$$\lim_{\delta t \to 0} \frac{p_n(t+\delta t) - p_n(t)}{\delta t} = p_0'(t) = -\lambda p_0(t) \quad \text{for } n = 0 \quad (9.12)$$

We have to solve the differential difference equations under the initial conditions stated as follows:

$$p_0'(t) = -\lambda p_0(t),\ p_0(0) = 1 \quad \text{for } n = 0 \quad (9.13)$$

$$p_n(t) + \lambda p_n(t) + \lambda p_{n-1}(t),\ p_n(0) = 0 \quad \text{for } n > 0 \quad (9.14)$$

To Solve the Differential Equation Eq. (9.13) This is a first order and first degree differential equation. Separating the variables, we have

$$\frac{dp_0}{p_0} = -\lambda dt \quad (9.15)$$

On integration, we get

$\log p_0 = -\lambda t + \text{constant}$ or $p_0(t) = Ae^{-\lambda t}$

When $t = 0$, $p_0 = 1$. $\therefore A = 1$

Hence the solution is

$$p_0(t) = e^{-\lambda t} \quad (9.16)$$

Iterative Method of Solution for Eq. (9.14) It is a differential-cum-difference equation. We apply iterative method and solve the system for each n.

For $n = 1$, we have from Eq. (9.14)

$$p_1'(t) + \lambda p_1(t) = \lambda p_0(t) = \lambda e^{-\lambda t} \text{ by Eq. (9.16)} \quad (9.17)$$

This is a linear equation Integrating Factor $= e^{-\lambda t}$, Now multiplying Eq. (9.17) by $e^{-\lambda t}$, we write

$$d[p_1(t)e^{-\lambda t}] = \lambda dt \quad (9.18)$$

Integrating, we get

$$p_1(t)e^{-\lambda t} = \lambda t + B$$

When $t = 0$, $p_1 = 0 \Rightarrow B = 0$. $\therefore$ Therefore, we have

$$p_1(t) = \lambda t e^{-\lambda t} \text{ or } \frac{\lambda t}{1!} e^{-\lambda t} \quad (9.19)$$

For $n = 2$, we have from Eq. (9.14)

$$p_2(t) + \lambda p_2(t) = \lambda p_1(t)\ \lambda \frac{(\lambda t)}{1!} e^{-\lambda t} \text{ by Eq. (9.19)}$$

Multiplying Eq. (9.19) by $e^{-\lambda t}$, we can write

$$d[p_2(t)\, e^{-\lambda t}] = \lambda \frac{(\lambda t)}{1!} dt \quad (9.20)$$

Integrating we get $p_2(t)\, e^{\lambda t} = \dfrac{(\lambda t)^2}{2!} + C$

When $t = 0$, $p_2 = 0 \Rightarrow C = 0$. Therefore, we have

$$p_2(t) = \frac{(\lambda t)^2}{2!} e^{-\lambda t} \quad (9.21)$$

Proceeding like this we finally obtain

$$p_n(t) = \frac{(\lambda t)^n}{n!} e^{-\lambda t} \quad (9.22)$$

which is the formula for Poisson distribution.

This completes the proof of the theorem.

9.8.2 Properties of Poisson Process of Arrivals

We have known that if n is the number of arrivals during the time interval t, then the law of probability in Poisson process is given by

$$p_n(t) = \frac{(\lambda t)^2 e^{-\lambda t}}{n!} \qquad n = 0, 1, 2,..$$

$$p_n(t) = (\lambda t)^n e^{-\lambda t}/n! \qquad n = 0, 1, 2,.. \quad (9.23)$$

where λt is the parameter.

1. We know that if the mean $E(n) = \lambda t$ and $\text{Var}(n) = \lambda t$, the average (expected) number of arrivals in unit time is
$$\frac{E(n)}{t} = \lambda \quad (9.24)$$
mean arrival rate (or input rate)
2. Let us consider the time interval $(t_1, t + \delta t)$, where δt is sufficiently small. Then
$$p_0(\delta t) = \text{prob (no arrival in time } \delta t) \quad (9.25)$$
Putting $n = 0$ and $t = \delta t$ in Eq. (9.24), we get
$$p_0(\delta t) = e^{-\lambda \delta t} = 1 - \lambda \delta t + 0(\delta t) \quad (9.26)$$
When δt is small
$$p_0(\delta t) = 1 - \lambda \delta t \quad (9.27)$$
This means that the probability of no arrival in time δt is $1 - \lambda \delta t$. Similarly, $P_1(\delta t)$ can be calculated.
$$p_1(\delta t) = \frac{\lambda \delta t}{1!} e^{-\lambda \delta t} = \lambda \delta t\left(1 - \lambda \delta t + \frac{(\lambda \delta t)^2}{2!} + \cdots\right)$$
$$= \lambda \delta t + 0(\delta t) \quad (9.28)$$
Neglecting $0(\delta t)$
$$p_1(\delta t) = \lambda \delta t$$
$$p_1(\delta t) = \frac{(\lambda \delta t)^2}{2!} e^{-\lambda \delta t} = \lambda (\delta t)^2 \left(1 - \delta t + \frac{(\lambda \delta t)^2}{2!} \cdots\right) \quad (9.29)$$
Neglecting $0(\delta t)$
$$p_1(\delta t) = 0$$

Similarly $p_3(\delta t) = P_4(\delta t) = \cdots = 0 \quad (9.30)$

Generally, $p_n \delta t$ = negligibly small quantity for all $n > 1$.

9.8.3 Distribution of Inter-arrival Times (Exponential Process)

Let τ be the time between two consecutive arrivals. It is called the inter-arrival time. Let $a(\tau)$ denote the probability density function of τ.

The distribution function $g(x)$ and the probability density function are related by $\frac{dg(x)}{dx} = f(x)$. If $p_0(\tau)$ denotes the probability distribution function for no arrival in time τ and $a(\tau)$ denotes the corresponding probability density function of t then they are related by $\frac{dp_0(\tau)}{d\tau} = a(\tau)$. We now prove an important theorem.

Theorem Let n, the number of arrivals in time t, follow the Poisson distribution:

$$p_n(t) = \frac{(\lambda t)}{n!} e^{-\lambda t} \quad (9.31)$$

Then the inter-arrival time obeys the negative exponential law

$$a(\tau) = \lambda e^{-\lambda t} \quad (9.32)$$

and vice versa.

Proof Let t_0 be the instant at which arrival occurs initially. Suppose there is no arrival in the intervals $(t_0, t_0 + \tau)$ and $(t_0 + \tau, t_0 + \tau + \delta t)$ so that $t_0 + \tau + \delta t$ will be instant of subsequent arrival (Figure 9.7).

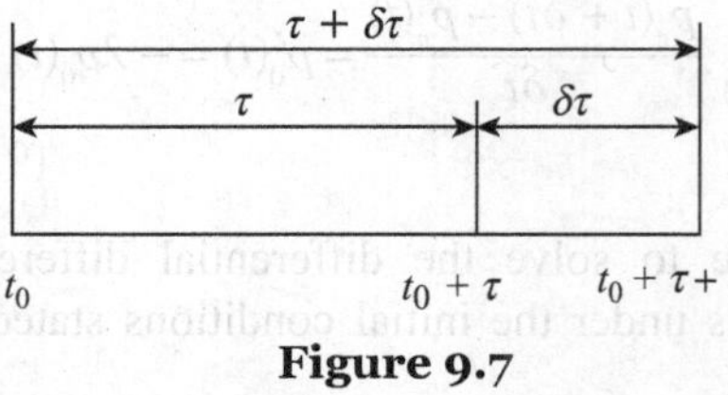

Figure 9.7

Putting $t = \tau + \delta\tau$ and $n = 0$ in Eq. (9.31)

$$p_0(\tau + \delta\tau) = \frac{[\lambda(\tau + \delta\tau)^{\circ} e^{-\lambda(\tau+\delta\tau)}]}{0!}$$

$= e^{-\lambda\tau}[1 - \lambda \delta t + 0(\delta\tau)]$, on expanding $e^{-\lambda\tau}$ (9.33)

We know that $p_0(\tau) = e^{-\lambda\tau}$. Therefore, we have from Eq. (9.33)

$$p_0(\tau+\delta\tau) - p_0(\tau) = p_0(\tau)[-\lambda\delta\tau + 0(\delta\tau)]$$

Dividing by $\delta\tau$ and taking limits as $\delta\tau \to 0$, we get

$$\frac{dp_0(\tau)}{d\tau} = -\lambda\, p_0(\tau)\left[\because \lim_{\partial\tau\to 0}\frac{0(\delta t)}{\delta\tau} = 0\right] \quad (9.34)$$

The LHS of Eq. (9.34) denotes the probability density function of τ. If we denote it by $a(\tau)$, we can write Eq. (9.34) as

$$a(\tau) = \lambda p_0(\tau) \quad (9.35)$$

We know that $p_0(\tau) = e^{-\lambda\tau}$. Putting this value of $p_0(\tau)$ in Eq. (9.35), we have

$$a(\tau) = \lambda e^{-\lambda\tau} \quad (9.36)$$

Eq. (9.36) gives the exponential law of probability for the inter-arrival time τ with mean $1/\lambda$ and variance $1/\lambda^2$, i.e.

$$E(\tau) = \frac{1}{\lambda} \text{ and } \mathrm{Var}(\tau) = \frac{1}{\lambda^2} \quad (9.37)$$

where λ is the mean arrival rate.

In a similar way, we can prove the converse of the theorem.

9.8.4 Markovian Property of Inter-arrival Times

Theorem The Markovian property of inter-arrival times states that at any instant of the time until the next arrival occurs is independent of the time that has elapsed since the occurrence of the last arrival (Figure 9.8), i.e.

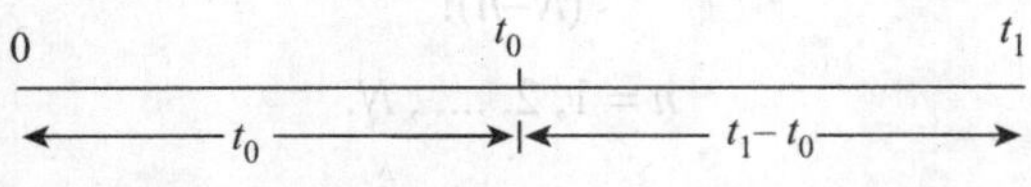

Figure 9.8 Markovian property of inter-arrivaltimes.

$$P(\tau \geq t_1 \mid \tau \geq t_0) = P(0 \geq t \leq (t_1 - t_0)) \quad (9.38)$$

Proof We have

$$P(\tau \geq t_1 \mid \tau \geq t_0) = \frac{P(\tau \geq t_1 \mid \tau \geq t_0)}{P(\tau \geq t_0)} \quad (9.39)$$

Conditional Probability Since the inter-arrival times are exponentially distributed

$$= \frac{\int_{t_1}^{\infty} \lambda e^{-\lambda t}dt}{\int_{t_0}^{\infty} \lambda e^{-\lambda t}dt} = \frac{e^{-\lambda t_1} e^{-\lambda t_0}}{e^{-\lambda t_0}} \Rightarrow P(\tau \geq t_1 \mid \tau \geq t_0)$$

$$= 1 - e^{-\lambda(t_1 - t_0)} \quad (9.40)$$

Since $P(0 \leq \tau \leq t_1 - t_0) = \int_0^{t_1 - t_0} \lambda e^{-\lambda t}dt = 1 - e^{-\lambda(t_1 - t_0)}$

$$P(\tau \geq t_1 \mid \tau \geq t_0) = P(0 \geq t \leq (t_1 - t_0)) \quad (9.41)$$

This proves the Markovian property of inter-arrival times.

9.8.5 Probability Distribution of Departures (Pure Death Process)

Let there be N customers in the system at time $t = 0$. Assume that no arrivals (births) can occur in the system; departures occur at the rate of μ per unit time, i.e. the output rate is μ. We derive the distribution of departures from the system on the basis of the following three axioms (Figure 9.9):

1. P(one departure during δt) $= \mu\delta t + 0((\delta t)^2)$ $= \mu\delta t$

   $\because 0((\delta t)^2)$ is negligible.

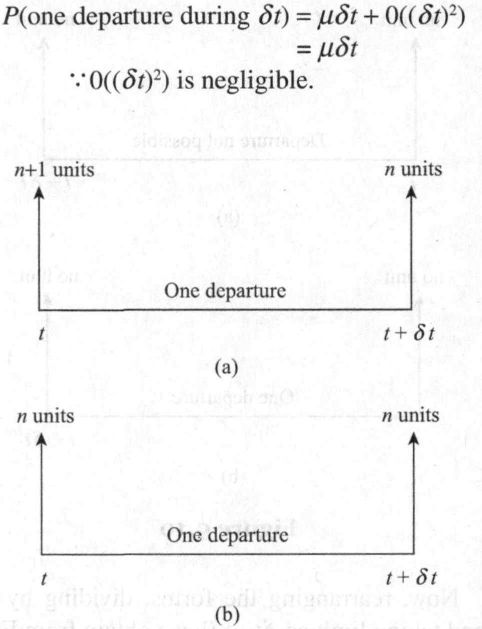

Figure 9.9

2. P(more than one departure during δt) = $0[(\delta t)^2] = 0$
3. The number of departures in non-overlapping intervals is statistically independent and are identically distributed random variables, i.e. the process $N(t)$ has independent increments.

First we obtain the differential difference equation in three mutually exclusive ways:

Case 1: $0 < n < N$ Proceeding exactly as in the pure birth process, we obtain

$$P_n(t+\delta t) = P_n(t)(1+\delta t) + P_{n+1}(t)\,\mu\delta t \quad (9.42)$$

Case 2: $x = N$ Since there are exactly N units in the system $P_{n+1}(t) = 0$ for $n = N$

$$P_N(t+\delta t) = P_N(t)(1-\mu\delta t) \quad (9.43)$$

Case 3: $n = 0$ Since there is no unit in the system at time t, the question of departure during the interval δt does not arise. So probability of no departure is unity in this case (Figure 9.10).

$$P_0(t+\delta t) = P_0(t) + P_1(t)\,\mu\delta t \quad (9.44)$$

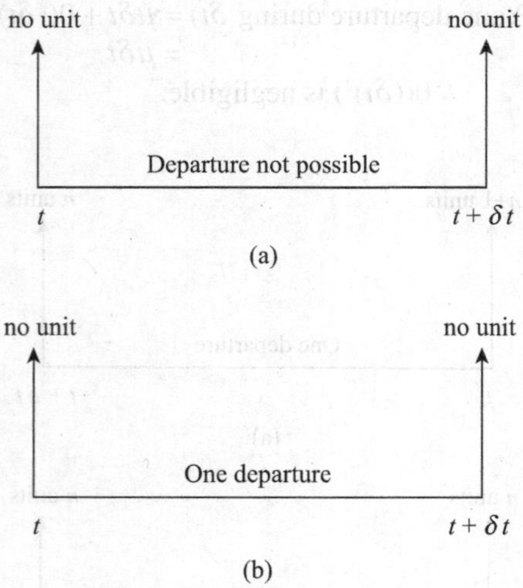

Figure 9.10

Now, rearranging the forms, dividing by δt and taking limit as $\delta t \to 0$, we obtain from Eqs. (9.42–9.44)

$$P'_N(t) = -\mu P_N(t) \quad \text{for } n = N \quad (9.45)$$

$$P'_N(t) = -\mu P_n(t) - \mu P_{n+1}(t) \quad \text{for } 0 < n < N \quad (9.46)$$

$$P'_N(t) = \mu P_n(t) \quad \text{for } n = 0 \quad (9.47)$$

To solve Eq. (9.45) This is a first-order and first-degree differential equation. Separating the variables and integrating, we get

$$\frac{P_N(t)}{P_N(t)} = -\mu \Rightarrow P_N(t) = Ae^{-\mu t}$$

When $t = 0$, P_N and so $A = 0$, we have

$$P_N(t) = e^{-\mu t} \quad (9.48)$$

To solve Eq. (9.46). Putting $n = N - 1$ in Eq. (9.46)

$$P'_{N-1}(t) + \mu P_{N-1}(t) = \mu P_N(t) = \mu e^{-\mu t} \text{ by } (9.48)$$

Multiplying this by $e^{\mu t}$, we can write

$$d[P_{N-1}(t)e^{\mu t}] = \mu dt \Rightarrow P_{N-1}(t)e^{\mu t} = \mu t + B$$

When $t = 0$, $P_{N-1} = 0$ which implies that $B = 0$

$$P_{N-1}(t) = \frac{\mu t}{1!}e^{-\mu t}$$

Putting $x = N - 2$ in Eq. (9.46) and proceeding as above, we obtain

$$P_{N-2}(t) = \frac{(\mu t)^2}{2!}e^{-\mu t}$$

By induction, we obtain

$$P_n(t) = \frac{(\mu t)^{N-n}}{(N-n)!}\,e^{-\mu t} \quad (9.49)$$

$$n = 1, 2, \ldots, N.$$

To find $P_0(t)$ Since $1 = \sum_{n=1}^{N} P_n(t) = P_0(t) + \sum_{n=1}^{N} P_n t$, we have

$$P_0(t) = 1 - \sum_{n=1}^{N} P_n(t) = 1 - \sum_{n=1}^{N} \frac{(\mu t)^{N-n}}{(N-n)!}\,e^{-\mu t} \quad (9.50)$$

Combining the results at Eqs. (9.49) and (9.50), we get

$$P_n(t) = \begin{cases} \dfrac{(\mu t)^{N-n}}{(N-n)!} e^{-\mu t} \text{ for } n = 1, 2, \dots, N \\ 1 - \sum_{x=1}^{N} \dfrac{(\mu t)^{N-n}}{(N-n)!} e^{-\mu t} \text{ for } n = 0 \end{cases} \quad (9.51)$$

Thus, the number of departures in time t follows the truncated Poisson distribution.

9.8.6 Derivation of Service Time Distribution

Let τ be the random variable denoting the service time and t be the possible value of τ.

Suppose $s(t)$ and $s(\tau)$ are the cumulative density function and probability density function of τ respectively.

To find $s(t)$ for the Poisson departure cases, it has been observed that the probability of no service during time 0 to t is equivalent to the probability of having no departure during this period. Thus,

$$P(\text{service time } \tau \geq t) = P(\text{no departure during } t) = P_N(t)$$

where these are N units in the system and no arrival takes place after N.

$$P_N(t) = e^{-\mu t} \quad (9.52)$$

Now

$$s(t) = P(\tau \leq t) = 1 - P(\tau \geq t) \text{ or } s(t) = 1 - e^{-\mu t}$$

Differentiating both sides w.r.t. 't', we get

$$\frac{d}{dt} s(t) = s(t) = \begin{cases} \mu e^{-\mu t} & t \geq 0 \\ 0 & t < 0 \end{cases} \quad (9.53)$$

The service time distribution is 'exponential' with mean = $1/\mu$ and variance = $1/\mu^2$, Mean service time = $1/\mu$. (9.54)

9.9 Kendall's Notation for Representing Queueing Models

A queueing model may be completely specified in the following symbolic form $(a/b/c):(d/e)$, where

a = probability law for the arrival (or inter-arrival) time

b = probability law according to which customers are being served

c = number of service stations (or channels)

d = capacity of the system

e = queue discipline

9.10 Basic Probabilistic Queueing Models

We consider the following probabilistic queueing models:

Model 1 (Erlang Model): Birth and Death Model This model is symbolically represented as $(M|M|1):(\infty|FCFS)$.

Here the first M denotes Poisson arrival (exponential inter-arrival), the second M Poisson departure (exponential service time) and 1 denotes single server and $(\infty|FCFS)$ stands for first come, first served service discipline.

Model 2 (General Erlang Model) This model is also represented by the symbol $(M|M|1):(\infty|FCFS)$ as model 1. But this is a general model in which the rate of arrival and service depend on the length n of the time.

Model 3 This model is also represented by the symbol $(M|M|1):(N|FCFS)$. In this model, capacity of the system is limited. n takes values from 0 to a finite number n, say. Clearly the number of arrivals will not exceed the number N in any case.

We first derive equations of model 1, from which steady-state equations are obtained using the following conditions:

$$\lim_{t\to\infty} P'_n(t) = 0 \text{ and } (n \geq 0)$$

$$\lim_{t\to\infty} P_n(t) = P_n \quad (9.55)$$

which is independent of t.

9.10.1 Governing Equations of Model 1 [(*M*|*M*|1):(∞|FCFS)]

The probability that there will be *n* units ($n > 0$) in the system at time $t + \delta t$ may be expressed as the sum of the following three independent compound probabilities: by application of the fundamental properties of probability, Poisson arrivals and exponential service times.

1. **Product of the following three probabilities (Figure 9.11):**

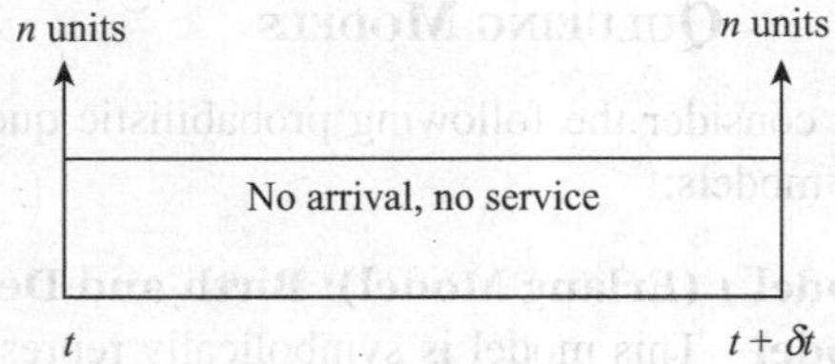

Figure 9.11

(a) There are *n* units in the system at time $t = P_n(t)$.

(b) There is no arrival during the time interval $\delta t = P_0(\delta t) = \lambda \delta t$.

(c) There is no service during the time interval $\delta t = \theta_{\delta t}(0) = 1 - \mu \delta t$.

The product of these three probabilities is given by

$$P_{n-1}(t)(1-\lambda\delta t)(1-\mu\delta t) = P_n(t)$$
$$[1-(\lambda+\mu)\delta t] + 0(\delta t) \qquad (9.56)$$

2. **Product of the following three probabilities (Figure 9.12):**

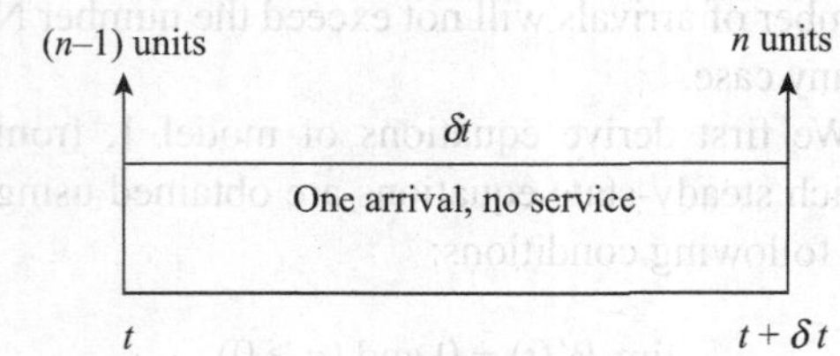

Figure 9.12

(a) There are $(n - 1)$ units in the system at time $t = P_{n-1}(t)$.

(b) There is one arrival during the time interval $\delta t = P_1(\delta t) = \lambda \delta t$.

(c) There is no servicing during the time interval $\delta t = \theta_{\delta t}(0) = 1 - \mu \delta t$.

The product of these probabilities is given by

$$P_{n-1}(t)(\lambda\delta t)(1-\mu\delta t) = \lambda P_{n-1}(t)\delta t + 0(\delta t) \quad (9.57)$$

3. **Product of the following three probabilities (Figure 9.13):**

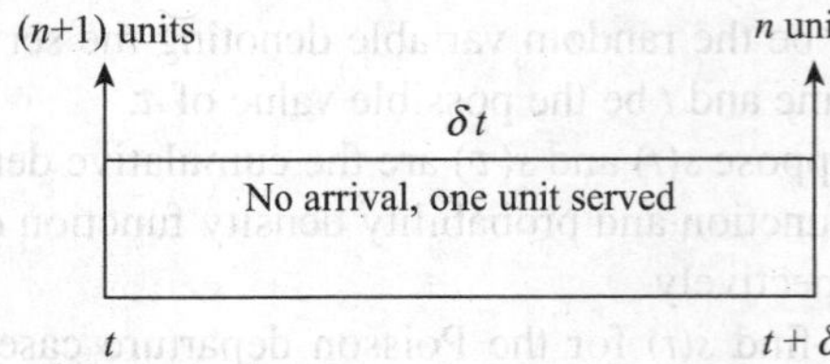

Figure 9.13

Probability that

1. There are $(n + 1)$ units in the system at time $t = P_{n+1}(t)$.
2. There is no arrival during the time interval $\partial t = P_0(\delta t) = \lambda \delta t$
3. There is one service during the time interval $\delta t = \phi_{\delta t}(1)\mu \delta t$

The product of these probabilities is given by

$$P_{n+1}(t) = (1-\lambda\delta t)(\mu\delta t) = P_{n+1}(t)\mu\delta t + 0(\delta t) \qquad (9.58)$$

By adding the above three independent compound probabilities Eqs. (9.56–9.58), we obtain the probability of *n* units in the system at time $t + \delta t$:

$$P_n(t+\delta t) = P_n(t)[1-(\lambda+\mu)\,\delta t] + P_{n-1}(t)\lambda\delta t + P_{n+1}(t)\mu\delta t + 0(\delta t)$$

Transposing, dividing both sides by δt and taking limits as $\delta t \to 0$

$$\lim_{\delta t\to 0}\left[\frac{P_n(t+\delta t) - P_n(t)}{\delta t}\right] = \lim_{\delta t\to 0} -(\lambda+\mu)\,P_n(t) + \lambda P_{n-1}(t) + P_{n+1}(t) + \frac{0(\delta t)}{\delta t}\Bigg] \qquad (9.59)$$

$$\Rightarrow \frac{dP_n(t)}{dt} = -(\lambda + \mu)\, P_n(t) + \lambda P_{n-1}(t) + P_{n+1}(t) \tag{9.60}$$

$$\because \lim_{\delta t \to 0} \frac{0(\delta t)}{\delta t} = 0$$

In the same way, the probability that there will be no unit, $n = 0$, in the system at time $t + \delta t$ will be the sum of the following two independent probabilities.

Probabilities that

1. There is no unit in the system at time t and no arrival in time $\delta t = P_0(t) + (1 - \lambda\delta t)$
2. There is no unit in the system at time t, one unit served in δt and no arrival in δt

$$= P_1(t) + (1 - \lambda\delta t)(\mu\delta t)) = P_1(t)\mu\delta t + 0(\delta t) \tag{9.61}$$

Adding these two probabilities, we have

$$P_0(t + \delta t) = P_0(t)(1 - \lambda\delta t) + P_1(t)\mu\delta t + 0(\delta t) \tag{9.62}$$

Now

$$\lim_{\delta t \to 0}\left[\frac{P_0(t+\delta t)P_0(t)}{\delta t} = \lim_{\delta t \to 0}[-\lambda P_0(t) + \mu P_0(t)] + \frac{0(\delta t)}{\delta t}\right] \text{ for } n = 0 \tag{9.63}$$

$$\frac{dP_0(t)}{dt} = -\lambda P_0(t) + \mu P_0(t)$$

When we consider steady-state imposing the conditions

$$\lim_{\delta t \to 0} \text{ for } n > 0 \text{ and } \lim_{\delta t \to 0} P_n(t) = P_n \tag{9.64}$$

where P_n is independent of time.

Consequently, the steady-state difference equations obtained are

$$-(\lambda + \mu)P_n + \lambda P_{n-1} + \mu P_{n+1} = 0 \quad \text{for } n > 0 \tag{9.65}$$

$$\lambda P_0 + \mu P_1 = 0 \quad \text{for } n = 0 \tag{9.66}$$

9.10.2 Solution of the System of Difference Equations of Model 1

We have to solve the system of difference equations

$$-\lambda P_0 + \mu P_1 = 0 \quad \text{for } n = 0 \tag{9.67}$$

$$-(\lambda + \mu)P_n + \lambda P_{n-1} + \mu P_{n+1} = 0 \quad \text{for } n > 0 \tag{9.68}$$

These equations can be written as

$$P_1 = \frac{\lambda}{\mu} P_0 \tag{9.69}$$

$$P_{n+1} = \left(\frac{\lambda}{\mu} + 1\right)P_n - \frac{\lambda}{\mu} P_{n-1} \tag{9.70}$$

We will prove by induction that

$$P_n = \left(\frac{\lambda}{\mu}\right)^n P_0 \quad \text{for } n = 0, 1, 2, \ldots \tag{9.71}$$

For $n = 0$, it is obvious. For $n = 1$, it is true by Eq. (9.69). Assume that Eq. (9.71) holds for $n = 0, 1, 2, \ldots, (m - 1)$ and m.

So that we have among other equations

$$P_{m+1} = \left(\frac{\lambda}{\mu}\right)^{m-1} P_0 \tag{9.72}$$

$$P_m = \left(\frac{\lambda}{\mu}\right)^m P_0 \tag{9.73}$$

Replacing n by m in Eq. (9.71), we have

$$P_{m+1} = \left(\frac{\lambda}{\mu} + 1\right) P_m - \frac{\lambda}{\mu} P_{m-1}$$

$$= \left(\frac{\lambda}{\mu} + 1\right)\left(\frac{\lambda}{\mu}\right)^m P_0 - \frac{\lambda}{\mu} \cdot \left(\frac{\lambda}{\mu}\right)^{m-1} P_0 \text{ by}$$

Eqs. (9.72) and (9.73)

$$= \left(\frac{\lambda}{\mu}\right)^{m+1} P_0$$

This proves by induction that Eq. (9.71) holds for $n = 0, 1, 2, \ldots$

Now, using the fact that $\sum_{n=0}^{\infty} P_n = 1$, we obtain

$$P_0\left[1+\frac{\lambda}{\mu}+\left(\frac{\lambda}{\mu}\right)^2+...\right]=1 \text{ or } P_0\left(\frac{1}{1-\lambda/\mu}\right)=1 \text{ or}$$

$$P_0 = 1-\frac{\lambda}{\mu} \quad (9.74)$$

since arrival rate (λ) < service rate (μ) and the infinite geometric progression.

Substituting the value of P_0 at Eq. (9.74) into Eq. (9.71), we get

$$P_n = \left(\frac{\lambda}{\mu}\right)^n\left(1-\frac{\lambda}{\mu}\right) = \rho^n(1-\rho) \quad (9.75)$$

where $\left(\rho = \frac{\lambda}{\mu} < 1, n \geq 0\right)$

Equations (9.74) and (9.75) give the required probability distribution of queue length.

9.10.3 Measures of Model 1

1. **To find the expected (average) number of units in the system L_s** By definition of the expected value:

$$L_s = \sum_{n=1}^{\infty} nP_n = \sum_{n=1}^{\infty} n\left(\frac{\lambda}{\mu}\right)^n\left(1-\frac{\lambda}{\mu}\right) = \left(1-\frac{\lambda}{\mu}\right)\left(\frac{\lambda}{\mu}\right)$$

$$\sum_{n=1}^{\infty} n\left(\frac{\lambda}{\mu}\right)^{n-1}$$

$$= \left[\left(1-\frac{\lambda}{\mu}\right)\frac{\lambda}{\mu}\, 1+2\left(\frac{\lambda}{\mu}\right)+3\left(\frac{\lambda}{\mu}\right)^2+...+...\right]$$

$$= \frac{\lambda}{\mu}\left[1+\frac{\lambda}{\mu}+\left(\frac{\lambda}{\mu}\right)^2+...\right] \quad (9.76)$$

$$= \frac{\lambda}{\mu}\cdot\frac{1}{1-\frac{\lambda}{\mu}} = \frac{\rho}{1-\rho}...$$

where $\frac{1}{1-\rho}$ is the sum of infinite GP with CR $\rho = \frac{\lambda}{\mu} < 1.$

2. **To find the expected (average) queue length L_q** There are $(n - 1)$ units in the queue excluding the one being serviced:

$$L_q = \sum_{n-1}^{\infty}(n-1)P_n = \sum_{n=1}^{\infty} nP_n = \sum_{n=1}^{\infty} P_n$$

$$= \sum_{n=1}^{\infty} nP_n - \left(\sum_{n=1}^{\infty} P_n - P_0\right) \quad (9.77)$$

$$= L_s - 1 + \left(1-\frac{\lambda}{\mu}\right) = L_s - \frac{\lambda}{\mu}\left[\sum_{n=1}^{\infty} P_n = 1, P_0 = 1\frac{\lambda}{\mu}, = \rho\right]$$

$$= \frac{\lambda}{\mu}\cdot\frac{1}{1-\frac{\lambda}{\mu}} - \frac{\lambda}{\mu} = \rho\left(\frac{1}{1-\rho}-1\right) = \frac{\rho^2}{1-\rho}$$

$$L_s = \cdot\frac{1}{1-\frac{\lambda}{\mu}}$$

3. **To find the mean (or expected) waiting time in the queue (excluding the service time) W_q** The expected time spent in queue is given by

$$W_q = \int_0^{\infty} \omega\psi(\omega)d\omega = \int_0^{\infty} \omega\lambda\left(1-\frac{\lambda}{\mu}\right)e^{(\mu-\lambda)\omega}\, d\omega$$

$$= \lambda\left(1-\frac{\lambda}{\mu}\right)\left[\frac{\omega.e^{-(\mu-\lambda)\omega}}{-(\mu-\lambda)} - \frac{1}{(\mu-\lambda)^2}e^{-(\mu-\lambda)\omega}\right]_0^{\infty}$$

$$= \lambda\cdot\frac{\mu-\lambda}{\mu}\frac{1}{(\mu-\lambda)^2} \quad (9.78)$$

$$= \frac{\lambda}{\mu(\mu-\lambda)}$$

4. **To find the expected waiting time in the system (including service time W_s**

W_s = Expected waiting time in the system

= Expected waiting time in queue + expected service time

$$= W_q + \frac{1}{\mu} = \frac{\lambda}{\mu(\mu-\lambda)} + \frac{1}{\mu} = \frac{1}{\mu-\lambda}$$

expected service time or mean service time $= \frac{1}{\mu}$

5. **To find the expected waiting time of a customer who has to wait $(W \mid W > 0)$** The expected length of the busy period is given by

$$(W|W>0) = \int_0^{\infty} \omega\psi(\omega>0)d\omega$$

$$= \int_0^{\infty} \omega(\mu\lambda)e^{-(\mu-\lambda)\omega}d\omega$$

$$(\mu-\lambda)\left[\omega\cdot\frac{e^{-(\mu-\lambda)\omega}}{-(\mu-\lambda)}-1\frac{e^{-(\mu-\lambda)\omega}}{(\mu-\lambda)^2}\right]_0^\infty \tag{9.79}$$

$$(\mu-\lambda)\frac{1}{(\mu-\lambda)^2}-\frac{1}{\mu-\lambda}=\frac{1}{\mu(1-\rho)}$$

where $\rho=\dfrac{\lambda}{\mu}$

6. **To find the expected length of non-empty queue ($L \mid L>0$)** By the conditional probability,

$(L \mid L>0) = L_s/P$ (an arrival has to wait; $L>0$)

$= L_s/(1-P_0)$ ($\because$ probability of an arrival not to wait is P_0)

$$=\frac{\left(\frac{\lambda}{\mu}\right)\Big/\left(1-\frac{\lambda}{\mu}\right)}{\left(\frac{\lambda}{\mu}\right)}=\frac{\mu}{\mu-\lambda}=\frac{1}{1-\rho} \tag{9.80}$$

7. **To find the variance of queue length**

$$\sum_{x=1}^{\infty} n^2P_n-\left(\sum_{x=1}^{\infty} nP_n\right)^2=\sum_{x=1}^{\infty} n^2P_n-L_s^2$$

$$\sum_{x=1}^{\infty} n^2(1-\rho)\rho^n-\left(\frac{\rho}{1-\rho}\right)^2 \quad \because P_n=\rho^n(1-\rho)$$

$$\text{Var}(n)=L_s=\frac{\rho}{1-\rho} \tag{9.81}$$

$$(1-\rho)\rho\,[1+2^2\rho+3^2\rho^2+\ldots]-\frac{\rho^2}{(1-\rho)^2}$$

$$(1-\rho)\rho\frac{1+\rho}{(1-\rho)^3}-\frac{\rho^2}{(1-\rho)^2}=\frac{\rho}{(1-\rho)^2}$$

Let $s=1+2^2\rho+3^2\rho^2+\ldots$

Integrating both sides w.r.t. ρ from 0 to ρ

$$\int_0^\rho s\,d\rho=\rho+2\rho^2+\ldots=\rho(1-\rho)^{-2}$$

Now differentiating w.r.t ρ

$$s=\frac{1}{(1-\rho)^2}+\frac{2\rho}{(1-\rho)^3}=\frac{1+\rho}{(1-\rho)^3}$$

8. **To find the probability of arrivals during the service time of any given customer**
The arrivals are Poisson and service times are exponential. So, the probability of r arrivals during the service time of any specified customer is given by

$$k_r=\int_0^\infty P_r(t)s(t)dt=\int_0^\infty \frac{e^{-\lambda t}(\lambda t)^r}{r!}\mu e^{-\mu t}dt$$

$$=\frac{\lambda^r\mu}{r!}\int_0^\infty e^{-(\lambda+\mu)t}t^r dt=\frac{\lambda^r\mu\Gamma(r+1)}{r!(\lambda+\mu)^{r+1}}$$

$$=\left(\frac{\lambda}{\lambda+\mu}\right)^r\frac{\mu}{(\lambda+\mu)} \tag{9.82}$$

$$\left[\because\int_0^\infty p^re^{-at}\,t^m\,dt=\frac{\Gamma(n+1)}{a^{n+1}}\right]=\frac{r!}{a^{n+1}}$$

9.10.4 Inter-relations between L_s, L_q, W_s, W_q

It can be established under general conditions of arrival, departure and service discipline that

$$L_s=\lambda W_s \tag{9.83}$$

$$L_q=\lambda W_q \tag{9.84}$$

hold. By definition,

$$W_q=W_s-\frac{1}{\mu} \tag{9.85}$$

Multiplying both sides of this equation by λ and using the above general relations, we get

$$L_q=L_s-\frac{\lambda}{\mu}$$

Example 9.1

In a railway marshalling yard, goods trains arrive at the rate of 30 trains/day. Assuming that the inter-arrival time follows an exponential distribution and the service time (the time taken to hump a train) distribution is also exponential with an average 30 min. Calculate

(a) Average number of trains in the queue.

(b) Probability that the queue size exceeds 10

If the input of trains increases to an average 33 trains/day, what will be change in (a) and (b)? Establish the formula you use in your calculations.

Solution Here

$$\lambda = \frac{30}{60 \times 24} = \frac{1}{48} \text{ trains/min}$$

$$\mu = \frac{1}{36} \text{ trains/min}$$

$$\rho = \frac{\lambda}{\mu} = \frac{36}{48} = 0.75$$

(a) $L_s = \rho/(1-\rho) = \dfrac{0.75}{1-0.75} = 3$ trains

(b) $P(\text{queue size} \ge 10) = \rho^{10} = (0.75)^{10} = 0.06$

When the input increases to 33 trains/day,

$$\lambda = 1/43,\ \mu = 1/36$$

$$\therefore \rho = \lambda/\mu = 36/43 = 0.84$$

Then,

(a) $L_s = \dfrac{0.84}{0.16} = 5$ trains

(b) $P(\text{queue size} \ge 10) = (0.84)^{10} = 0.2$

Example 9.2

In Example 9.1, calculate

(a) Expected waiting time in the queue

(b) Probability that the number of trains in the system exceeds 10

(c) Average number of trains in the queue

Solution Here $\lambda = \dfrac{1}{48}$, $\mu = \dfrac{1}{36}$ and $\rho = 0.75$

(a) Expected waiting time in the queue

$$W_q = \frac{\lambda}{\mu(\mu-\lambda)} = \frac{\frac{1}{48}}{\frac{1}{36}\left(\frac{1}{36}-\frac{1}{48}\right)} = 108 \text{ min}$$

$= 1\text{h } 48\text{min}$

(b) $P(\text{queue size} \ge 10) = \rho^{10} = (0.75)^{10} = 0.06$

(c) $L_q = \dfrac{\lambda}{\mu(\mu-\lambda)} = \dfrac{\left(\frac{1}{48}\right)^2}{\frac{1}{36}\left(\frac{1}{36}-\frac{1}{48}\right)} = \dfrac{108}{48}$

$= 2.25$ or 2 trains

Example 9.3

A TV repairman finds that the time spent on his jobs has an exponential distribution with mean 30 min. If he repairs sets in the order in which they come in and if the arrival of sets is nearly Poisson with an average rate of 10 per an 8-h/day, what is the repairman's expected idle time each day? How many jobs are ahead of the average set just brought in?

Solution Here $\mu = \dfrac{1}{30}$, $\lambda = \dfrac{10}{8 \times 60} = \dfrac{1}{48}$

$\therefore$ Expected numbers of jobs are

$$L_s = \frac{\lambda/\mu}{1-\lambda/\mu} = \frac{\lambda}{\mu-\lambda} = \frac{\frac{1}{48}}{\frac{1}{30}-\frac{1}{48}}$$

$= 1\frac{2}{3}$ jobs

The fraction of hours for which the repairman is busy (i.e. traffic intensity) is λ/μ.

The number of hours for which the repairman remains busy in an 8h/day

$$8\left(\frac{\lambda}{\mu}\right) = 8 \times \frac{30}{48} = 5\text{h}$$

The time for which the repairman remains idle in an 8-h/day = 8 – 5 = 3 h.

Example 9.4

At what average rate must a clerk at a super market work in order to ensure a probability of 0.90 that the customer will not have to wait longer than 12 min. It is assumed that there is only one customer to which customers arrive in a Poisson fashion at an average rate of 15 per hour. The length of service by the clerk has an exponential distribution.

Solution Here

$$\frac{\lambda}{\mu}=\frac{15}{60}=\frac{1}{4} \text{ customer/min } \mu=?$$

$$P(\text{waiting time} \geq 12) = 1 - 0.90 = 0.10$$

$$\therefore \int_{12}^{\infty} \lambda\left(1-\frac{\lambda}{\mu}\right)e^{-(\mu-\lambda)\omega}\,d\omega = 0.10 \quad \text{or}$$

$$\lambda\left(1-\frac{\lambda}{\mu}\right)\frac{e^{-(\mu-\lambda)\omega}}{(\mu-\lambda)}\Bigg|_{12}^{\infty} = 0.10 \Rightarrow e^{3-12\mu} = 0.4\mu$$

or $\frac{1}{\mu} = 2.48$ min/service

Example 9.5

Arrivals at a telephone booth are considered to be Poisson with an average time of 10 min between one arrival and the next. The length of a phone call assumed to be distributed exponentially with mean 3 min.

(a) What is the probability that a person arriving at the booth will have to wait?

(b) What is the average length of the queues that form from time to time?

(c) The telephone department will install a second booth when convinced that an arrival would expect to have to wait at least 3 min for the phone. By how much the flow of arrival be increased in order to justify a second booth?

Solution Here $\lambda = \frac{1}{10}$ and $\mu = \frac{1}{3}$

(a) $P(w>0) = 1 - P_0 = \frac{\lambda}{\mu} = \frac{1}{10} \times \frac{3}{1} = \frac{3}{10} = 0.3$

(b) $(L \mid L>0) = \frac{\mu}{\mu-\lambda} = \frac{\frac{1}{3}}{\frac{1}{3}-\frac{1}{10}} = 1.43$ persons

(c) $W_q = \frac{\lambda}{\mu(\mu-\lambda)}$

Since $W_q = 3$, $\mu = \frac{1}{3}$ and $\lambda = \lambda'$ (for second booth), we have

$$3 = \frac{\lambda'}{\frac{1}{3}\left(\frac{1}{3}-\lambda'\right)} \Rightarrow \lambda' = 0.16$$

∴ Increase in the arrival rate = 0.16 – 0.10 = 0.06 arrivals/min.

Example 9.6

In Example 9.5, a telephone booth with Poisson arrivals spaced 10 min apart on the average and exponential call lengths averaging 3 min.

(a) What is the probability that an arrival will have to wait more than 10 min before the phone is free?

(b) What is the probability that it will take him more than 30 min altogether to wait for phone and complete his call?

(c) Estimate the fraction of a day that the phone will be in use.

(d) Find the average number of units in the system.

Solution Here $\lambda = 0.1$ arrival/min and $\mu = 0.33$ service/min

$$\int_{10}^{\infty} \psi(\omega)d\omega = \int_{10}^{\infty} 1 - \frac{\lambda}{\mu}\lambda e^{-(\mu-\lambda)\omega}d\omega$$

(a) $P(\text{waiting time} \geq 10) =$

$$= -\frac{\lambda}{\mu}e^{-(\mu-\lambda)\omega}\Bigg|_{10}^{\infty} = 0.3e^{-2.3} = 0.03$$

(b) $P(\text{waiting time in the system} \geq 10)$

$$= \int_{10}^{\infty} (\mu-\lambda)e^{-(\mu-\lambda)\omega}d\omega = e^{-10(\mu-\lambda)} = e^{-2.3}$$

(c) The fraction of a day that the phone will be busy (i.e. traffic intensity) is

$$\rho = \frac{\lambda}{\mu} = 0.3$$

(d) Average number of units in the system is

$$L_s = \frac{\lambda}{\mu-\lambda} = \frac{\frac{1}{10}}{\frac{1}{3}-\frac{1}{10}} = \frac{3}{7} = 0.43 \text{ customer}$$

9.10.5 Model 2(A)—general Erlang Model $(M|M|1)$:$(|$FCFS) (Birth and Death Process)

To Obtain the System of Steady-state Equations Let $\lambda = \lambda_n$ be the arrival rate and $\mu = \mu_n$ be the service rate depending upon n.

Then proceeding as in the case of model 1, we get

$P_n(t + \delta t)$

$$= P_n(t)[1 - (\lambda_n + \mu_n)\delta t] + P_{n-1}(t)\lambda_{n-1}\delta t + P_{n+1}(t)\,\mu_{n+1}\delta t + 0(\delta t) \quad \text{for } n > 0 \quad (9.86)$$

$$P_0 = (t + \delta t) = P_0(t)(1 - \lambda_0\delta t) + P_1(t)\mu_1\delta t + 0(\delta t) \quad \text{for } n = 0 \quad (9.87)$$

Dividing by δt, transposing and taking the limit as $\delta t \to 0$, we obtain

$$\frac{dP_n(t)}{dt} = -(\lambda_n + \mu_n)P_n(t) + \lambda_{n-1}P_{n-1}(t) + \mu_{n+1}P_{n+1}(t) \quad (9.88)$$

$$\frac{dP_0}{dt} = -\lambda_0 P_0(t) + \mu_1 P_1(t) \quad (9.89)$$

Taking $P_n'(t) = 0$ and $P_0' = 0$ in the steady state, we obtain

$$-(\lambda_n + \mu_n)P_n + \lambda_{n-1}P_{n-1} + \mu_{n+1}P_{n+1} = 0 \text{ for } n > 0 \quad (9.90)$$

$$-\lambda_0 P_0(t) + \mu_1 P_1(t) = 0 \quad \text{for } n = 0 \quad (9.91)$$

9.10.6 Solution of the System of the Difference Equations of Model 2(A)

From Eq. (9.91), we get

$$P_1 = \frac{\lambda_0}{\mu_1} P_0 \quad (9.92)$$

From Eq. (9.90), we get

$$P_{n+1} = \frac{\lambda_n + \mu_n}{\mu_{n+1}} P_n - \frac{\lambda_{n-1}}{\mu_{n+1}} P_{n-1} \quad (9.93)$$

For $n = 1$, $P_2 = \frac{\lambda_1 \mu_1}{\mu_2} P_1 - \frac{\lambda_0}{\mu_2} P_0$

$$= \left(\frac{\lambda_1 + \mu_1}{\mu_2} \cdot \frac{\lambda_0}{\mu_1} - \frac{\lambda_0}{\mu_2}\right) P_0 = \frac{\lambda_0 \lambda_1}{\mu_1 \mu_2} P_0 \quad (9.94)$$

Generally, we obtain

$$P_n = \frac{\lambda_0 \lambda_1 \cdots \lambda_{n-2} \lambda_{n-1}}{\mu_1 \mu_2 \cdots \mu_{n-1} \mu_n} P_0 \quad \text{for } n \geq 1 \quad (9.95)$$

Since $\sum_{n=0}^{\infty} P_n = P_0 + P_1 + P_2 + \cdots$

$$= P_0 \left(1 + \frac{\lambda_0}{\mu_1} + \frac{\lambda_0 \lambda_1}{\mu_1 \mu_2} + \cdots\right) = 1$$

$$\Rightarrow P_0 = \frac{1}{s} \quad (9.96)$$

where $s = 1 + \frac{\lambda_0}{\mu_1} + \frac{\lambda_0}{\mu_1}\frac{\lambda_0}{\mu_2} + \cdots$

The series is meaningful only if it is convergent.

Case 1: $(\lambda_n = \lambda, \mu_n = \mu)$ In this case,

$$s = 1 + \frac{\lambda}{\mu} + \left(\frac{\lambda}{\mu}\right)^2 + \cdots \frac{1}{1 - \frac{\lambda}{\mu}}$$

$$= \frac{\mu}{\mu - \lambda} \left(\text{for } \lambda/\mu < 1\right)$$

$$\therefore P_0 = \frac{1}{s} = 1 - \frac{\lambda}{\mu} \text{ and } P_n = \left(\frac{\lambda}{\mu}\right)^n \left(1 - \frac{\lambda}{\mu}\right) \quad (9.97)$$

which is exactly the result we have obtained for model 1.

Case 2: $\left(\lambda_n = \frac{\lambda}{\mu + 1}, \mu_n = \mu\right)$ In this case, the arrival rate λ_n depends on n inversely and the rate of service μ_n is independent of n. It is called the case of *queue with discouragement*.

Now

$$s = 1 + \frac{\lambda}{\mu} + \frac{1}{2}\left(\frac{\lambda}{\mu}\right)^2 + \frac{1}{2.3}\left(\frac{\lambda}{\mu}\right)^3 + \cdots = e^{\lambda/\mu} = e^{\rho}$$

$$\therefore P_0 = \frac{1}{s} = e^{\lambda/\mu} = e^{-\rho}$$

$$P_1 = \frac{\lambda}{\mu} P_0 = \rho e^{-\rho}, \; P_2 = \frac{1}{2}\left(\frac{\lambda}{\mu}\right)^2 P_0 = \frac{\rho^2}{2!} e^{-\rho}, \cdots$$

$$P_n = \frac{1}{n!}\left(\frac{\lambda}{\mu}\right)^n P_0 = \frac{\rho^n e^{-\rho}}{n!} \text{ for } n = 0, 1, 2, \ldots \quad (9.98)$$

In this case, P_n follows the Poisson distribution where $\frac{\lambda}{\mu} = \rho$ is constant: $\rho > 1$ but finite.

Case 3: $(\lambda_n = \lambda, \mu_n = \mu)$ This is a case of infinite stations. In this case,

$$s = 1 + \frac{\lambda}{\mu} + \frac{1}{2.1}\left(\frac{\lambda}{\mu}\right)^2 + \frac{1}{3.2.1}\left(\frac{\lambda}{\mu}\right)^3 + \cdots$$

$$= 1 + \rho + \frac{1}{2!}\rho^2 + \cdots = e^{\rho} \qquad (9.99)$$

$$\therefore P_0 = e^{-\rho} \text{ and } P_n = e^{-\rho}\frac{\rho_n}{n!}$$

which follows the Poisson distribution law.

Example 9.7

Problems arrive at a computing centre in a Poisson fashion at an average rate of 5 per day. The rules of the computing centre are that any man waiting to get his problem solved must aid the man whose problem is being solved. If the time to solve a problem with one man has an exponential distribution with mean time of $\frac{1}{3}$ day, and if the average solving time is inversely proportional to the number of people working on the problem, approximate the expected time in the centre for a person entering the line.

Solution Here $\lambda = 5$ problems/day and $\mu = 3$ problems/day (mean service rate with one unsolved problem).

The expected number of persons working at any specified instant is

$$L_s = \int_{n=0}^{\infty} nP_n$$

$$\text{where } P_n = \frac{1}{n!}\left(\frac{\lambda}{\mu}\right)^n e^{-\lambda/\mu} \qquad \text{(Case 2)}$$

$$\sum_{n=0}^{\infty} n.\frac{1}{n!}\left(\frac{\lambda}{\mu}\right)^n e^{-\lambda/\mu} = e^{-\lambda/\mu}\frac{\lambda}{\mu}\left[\sum_{n=0}^{\infty}\frac{1}{n!}\left(\frac{\lambda}{\mu}\right)^n\right]$$

$$= e^{-\lambda/\mu}\frac{\lambda}{\mu}e^{-\lambda/\mu} = \frac{\lambda}{\mu}$$

Substituting the value of λ and μ, we get

$$L_s = \frac{5}{3} \text{ persons}$$

The average solving time which is inversely proportional to the number of people working on the problem is $\frac{1}{5}$ day/problem.

$\therefore$ Expected time for a person entering the line is

$$\frac{1}{5} \times L_s \text{ days} = \frac{1}{5} \times \frac{5}{3} \times 24 \text{ h} = 8 \text{ h}$$

9.10.7 Model 3 (M|M|1):(N|FCFS)

Consider the case where the capacity of the system is limited, say N.

The physical interpretation for this model is that

1. There is room only for N units in the system (as in a parking lot) or
2. Arriving customers will go for their service elsewhere permanently, if the waiting time is too long ($\leq N$)

To Obtain Steady State Difference Equations The simplest way of deriving the equation is to treat the present model as a special case of Model 2 where

$$\lambda_n = \begin{cases} \lambda, & \{n = 0, 1, 2, ..., (N-1)\} \\ 0, & n \geq N \end{cases}$$

$$\mu_n = \mu \text{ for } n = 1, 2, 3, \ldots$$

Now, following the derivation of equation in Model 1, we get

$$P_0(t + \delta t) = P_0(t)(1 - \lambda\delta t) + P_1(t)\mu\delta t + 0(\delta t)$$

$$\text{for } n = 0 \qquad (9.100)$$

$$P_n(t + \delta t) = P_n(t)\,[1 - (\lambda + \mu)\delta t] + P_{n-1}(t)\lambda\delta t$$
$$+ P_{n-1}(t)\mu\delta t + 0(\delta t) \text{ for } n = 1, 2, \ldots, (N-1) \qquad (9.101)$$

$$P_N(t + \delta t) = P_n(t)[1 - (0 + \mu)\delta t] + P_{n-1}(t)\,\lambda\delta t +$$
$$0\lambda\mu\delta t + 0(\delta t) \text{ for } n = N, P_{N+1} = 0, \lambda = 0 \qquad (9.102)$$

Dividing the equations by δt and taking the limit as $\delta t \to 0$, we get

$$P_0'(t) = -\lambda P_0(t) + \mu P_1(t) \text{ for } n = 0 \qquad (9.103)$$

$$P'_n(t) = -(\lambda + \mu) P_n(t) + \lambda P_{n-1}(t) + \mu P_{n+1}(t)$$

for $n = 1, 2, \ldots, (N-1)$ (9.104)

$$P'_N(t) = -\mu P_N(t) + \lambda P_{N-1}(t) \quad \text{for } n = N \tag{9.105}$$

In the case of steady state, as $t \to \infty$ and $P_n(t) \to P_n$ (independent of t) and hence $P'_n(t) \to 0$, we obtain

$$-\lambda P_0 + \mu P_1 = 0 \Rightarrow P_1 = \frac{\lambda}{\mu} P_0(t) \text{ for } n = 0 \tag{9.106}$$

$$-(\lambda + \mu) P_n + \lambda P_{n-1} + \mu P_{n+1} = 0 \Rightarrow P_{n+1}$$

$$= \frac{\lambda + \mu}{\mu} P_n - \frac{\lambda}{\mu} P_{n-1} \text{ for } n = 1, 2, \ldots, (N-1) \tag{9.107}$$

$$-\mu P_N + \lambda P_{N-1} = 0 \Rightarrow P_N = \frac{\lambda}{\mu} P_{N-1} \text{ for } n = N \tag{9.108}$$

9.10.8 Solution of the System of Difference Equations of Model 3

$$P_1 = \frac{\lambda}{\mu} P_0 \quad \text{for } n = 0 \tag{9.109}$$

$$P_2 = \left(1 + \frac{\lambda}{\mu}\right) P_1 - \frac{\lambda}{\mu} P_0 = \left(\frac{\lambda}{\mu}\right)^2 P_0 \text{ for } n = 1 \tag{9.110}$$

Generally,

$$P_n = \left(\frac{\lambda}{\mu}\right)^n P_0 \quad \text{for } n < N \tag{9.111}$$

$$n = N, P_{N+1} = 0 \tag{9.112}$$

We have

$$1 = \sum_{x=0}^{\infty} P_n = \sum_{x=0}^{\infty} \left(\frac{\lambda}{\mu}\right)^n P_0$$

$$= \frac{1-\rho^{N+1}}{1-\rho} \Rightarrow P_0 = \frac{1-\rho}{1-\rho^{N+1}} \tag{9.113}$$

Substituting the values of P_0 into

$$P_n = \left(\frac{\lambda}{\mu}\right)^n P_0 = \rho^n P_0 = \rho^n \left(\frac{1-\rho}{1-\rho NH}\right)$$

for $x = 0, 1, 2, \ldots, n$ (9.114)

Measures of Model 3

$$L_s = \sum_{x=0}^{\infty} n P_n = \sum_{x=0}^{N} \left(\frac{1-\rho}{1-\rho^{N+1}}\right) P^n = \frac{1-\rho}{1-\rho^{N+1}}$$

$$\Rightarrow \sum_{n=0}^{N} n P^n = P_0 \sum_{n=0}^{N} n \rho^n \tag{9.115}$$

$$L_q = L_s - \frac{\lambda}{\mu} \tag{9.116}$$

$$W_s = \frac{L_s}{\lambda} \tag{9.117}$$

$$W_q = W_s - \frac{1}{\mu} = \frac{L_q}{\lambda} \tag{9.118}$$

Example 9.8

If for a period of 2 h in a day (8–10 am) trains arrive at the yard every 20 min, but the service time continues to remain 36 min, then calculate for this period

(a) Probability that the yard is empty

(b) Average queue length

on the assumption that the time capacity of the yard is limited to 4 trains only.

Solution Here $\rho = \dfrac{\lambda}{\mu} = \dfrac{36}{20} = 1.8 > 1$ and $N = 4$

(a) $P_0 = \dfrac{\rho - 1}{\rho^{5-1}} = \dfrac{1.8 - 1}{(1.8)^{5-1}} = 0.04$

(b) Average queue size $= P_0 \sum_{x=0}^{4} n\rho^n$

$$= (0.04)(\rho + 2\rho^2 + 3\rho^3 + 4\rho^4)$$

$$= 0.04 \times 72.0 = 2.9 \cong 3 \text{ trains}$$

EXERCISES

1. What is a queueing theory problem?
2. Explain a queueing system. What is meant by transient and steady of a queueing system?
3. Describe the basic elements of a queueing system.
4. Show that the number of arrivals n in a queue in time t follows the Poisson distribution, stating the assumptions clearly.
5. Show that the distribution of the number of births up to time t in a simple birth process follows the Poisson law.
6. Explain what do you mean by Poisson process. Derive the Poisson distribution, given that the probability of single arrival during a small time interval s_t is λs_t and that of more than one arrival is negligible.
7. Show that if the inter-arrival times are negative exponentially disturbed, the number of arrivals in a time period is a Poisson process and conversely.
8. State that in the three axioms underlying the exponential assumptions, can two events occur during a very small interval.
9. Establish the probability distribution formula for pure death process.
10. Explain Kendall's notations for representing queueing models.
11. Explain $(M|M|1)$:$(\infty|FCFS)$ queueing model. Derive and solve the difference equations in steady state of the model.
12. Explain $(M|M|1)$ queue model in the transient state. Derive steady-state difference equations and their solutions.
13. Show that average number of units in an $(M|M|1)$ model is .
14. Discuss $(M|M|1)$:$(\infty|FCFS)$ queueing model and find the expected line length $E(L_s)$ in the system.
15. For the $(M|M|1)$ queueing system, find
    (a) Expected value of queue length n
    (b) Probability distribution of waiting time w.
16. A telephone booth functions with Poisson arrivals spaced 10 min apart on the average, and exponential call length averaging 3 min.
    (a) What is the probability that an arrival will have to wait more than 10 min before the phone is free?
    (b) What is the probability that it will take a customer more than one 10 min altogether to wait for phone and complete his call?
    (c) Estimate the fraction of a day that the phone will be in use.
    (d) Find the average number of units in the system.

Ans: (a) 0.03, (b) 0.10, (c) 0.3 and (d) 0.43 customer

[Hint: (a) $\lambda = 0.1$ and $\mu = 0.33$

$$P(\text{waiting time} \geq 10) = \int_{10}^{\infty} \psi(w)dw$$

$$= \int_{10}^{\alpha} \left(1 - \frac{\lambda}{\mu}\right) \lambda e^{-(\mu-\lambda)w}\, dw = 0.03$$

(b) $P(\text{waiting time} \geq 10) = \int_{10}^{\infty} (\mu - \lambda) \times e^{-(\mu-\lambda)w}\, dw = e^{-2.3} = 0.10$

(c) Traffic intensity $\rho = \dfrac{\lambda}{\mu} = 0.3$

(d) $L_s = \dfrac{\lambda}{\mu - \lambda} \dfrac{\frac{1}{10}}{\frac{1}{3} - \frac{1}{10}} = \dfrac{3}{7}$

$= 0.43$ customer]

17. The inter-arrival times at a service centre are exponential with an average time of 10 min. The length of the service time is assumed

to be exponentially distributed with mean 6 min. Find

(a) Probability that a person arriving at the booth will have to wait.

(b) Average length for the queue that forms and the average time that an operator spends in the queueing system.

(c) Manager of the centre will install a second booth when an arrival would have to wait 10 min or more for the service. By how much must the rate of arrival be increased in order to justify a second booth?

(d) Probability that an arrival will have to wait for more than 12 min for service and to obtain tools.

(e) Probability that there will be 6 or more operators waiting for service.

Ans: (a) 0.6, (b) 1/4 h, (c) $\lambda = 6.25$ and $\mu = 60$, (d) 0.27 and (e) $(0.6)^6$

[Hint:

(a) Probability of waiting $= 1 - p_0$

$$= 1 - (1 - \rho) = \rho = \frac{\lambda}{\mu} = 0.6$$

(b) $L_q = \dfrac{\rho^2}{(1-\rho)} = 0.9;\ L_s = L_q + \dfrac{\lambda}{\mu}$

$$= 0.9 + 0.6 = 1.5 \text{ hence } w_s = \frac{L_s}{\lambda} = \frac{1.5}{6} = \frac{1}{4}\text{h}$$

(c) $w_q = \dfrac{L_q}{\mu} = \dfrac{0.9}{6} = 9\text{min};\ w_q \geq 10 \text{ min}$

$$= \frac{\lambda}{\mu(\mu - \lambda)} = \frac{1}{6} \text{ hence } \lambda = 6.25 \text{ and } \mu = 60$$

(d) $P(w \geq 12) = \displaystyle\int_{\frac{12}{60}}^{\infty} \rho(\mu - \lambda)e^{-(\mu-\lambda)w}\, dw$

$$= -\rho.e^{-(\mu-\lambda)w}\Big|_{\frac{12}{60}}^{\infty} = 0.6e^{\frac{-4}{5}} = 0.27$$

(e) Probability of six or more operators waiting for service $= \rho^6 = (0.6)^6$]

18. A company distributes its products by trucks loaded at its only loading station. Both company's trucks and constructor's trucks are used for this purpose. It was found out that on an average every 5 min, one truck arrived and the average loading time was 3 min. 50% of the trucks belong to the contractor. Find

(a) Probability that a truck has to wait

(b) Waiting time of a truck that waits

(c) Expected waiting time of constructor's trucks per day, assuming a 24-h shift

Ans: (a) 0.6, (b) 7.5 min and (c) 10.8 h

[Hint: Here

(a) Probability that a truck has to wait

(b) $w_s = \dfrac{1}{\mu - \lambda} = \dfrac{1}{20 - 12} = \dfrac{1}{8}\text{h} = 7.5 \text{ min}$

(c) Expected time of contractor's truck per day

= (Number of trucks/day) × (Contractor's percentage) × (Expected waiting time of a truck)

$$= 12 \times 24 \times \frac{50}{100} \times \frac{\lambda}{\mu(\mu - \lambda)}$$

$$= 144 \times \frac{12}{20} \times 8 = \frac{54}{5} = 10.8\text{h}$$

19. A road transport company has bus reservation clerk on duty at a time. He handles information of bus schedules and makes reservations. Customers arrive at a rate of 8 per hour and the clerk can service 12 customers on an average per hour. After starting your assumptions, answer the following:

(a) What is the average number of customers waiting for the service of the clerk?

(b) What is the average time a customer has to wait before getting the service?

Ans: (a) $L_s = 2$ and $L_q = 1.33$, and (b) 15 min

[Hint: Here $\lambda = 8$ and $\mu = 12$

(a) $L_s = \dfrac{\lambda}{\mu - \lambda} = \dfrac{8}{12 - 8} = 2$

$L_q = \dfrac{\lambda^2}{\mu(\mu - \lambda)} = \dfrac{82}{12(12 - 8)} = 1.33$

(b) $w_s = \dfrac{1}{\mu - \lambda} = \dfrac{1}{12 - 8} = \dfrac{1}{4}$ h = 15min]

FILL IN THE BLANKS

1. FCFS means _____.

Ans: first come, first served

2. FILO means _____.

Ans: first in, last out

3. FIFO means _____.

Ans: first in first out

4. LIFO means _____.

Ans: last in first out

5. SIRO means _____.

Ans: service in random order

6. A customer leaves the queue because the queue is too long and he has no time to wait. This type of customer's behaviour is called _____.

Ans: balking

7. The customer leaves the queue due to impatience. This type of customer's behaviour is called _____.

Ans: reneging

8. In certain applications, some customers are served before others regardless of their order of arrival. Thus, some customers are shown _____ over others.

Ans: priority

9. Customers change from one waiting line to another shorter line. This behaviour is called _____.

Ans: jockeying

10. The term _____ refers to the arrival of a new calling unit in the queuing system.

Ans: birth

11. The term _____ refers to the departure of a severed unit.

Ans: death

12. If x is the number of arrivals during time interval t, then the law of probability in Poisson process is given by $p_x(b)$ = _____.

Ans: $\dfrac{(\lambda t)^n e^{\frac{t}{-1}}}{n!}$ *for* $n = 0, 1, 2, \ldots$

13. In Question 12, the term λt is the _____.

Ans: parameter

14. $\sum_{n=20}^{\infty} P_n = P_1 + P_2 + \ldots =$ _____.

Ans: 1

15. $P_0 =$ _____.

Ans: $\left(1 - \dfrac{\lambda}{\mu}\right)$

16. $P_n =$ _____.

Ans: $\left(\dfrac{\lambda}{\mu}\right)^n \left(1 - \dfrac{\lambda}{\mu}\right)$ or $\rho^n (1 - \rho)$

17. $L_s =$ _____.

Ans: $\dfrac{\frac{\lambda}{\mu}}{1 - \frac{\lambda}{\mu}}$ or $\dfrac{\rho}{1 - \rho}$

18. $L_q =$ ______.

Ans: $\dfrac{\left(\dfrac{\lambda}{\mu}\right)^2}{1-\dfrac{\lambda}{\mu}}$ or $\dfrac{\rho^2}{1-\rho}$

19. $w_s =$ ______.

Ans: $\dfrac{1}{\mu-\lambda}$

20. $w_q =$ ______.

Ans: $\dfrac{\lambda}{\mu(\mu-\lambda)}$

Appendix A: Statistical Tables

A.1 Binomial Distribution Function

Table A.1 Binomial probability sums: $\sum_{x=0}^{r} b\,(x;\,n,\,p) = \sum_{x=0}^{r} \binom{n}{k} p^{x}\,(1-p)^{n-x}$

		p									
n	*r*	0.10	0.20	0.25	0.30	0.40	0.50	0.60	0.70	0.80	0.90
1	0	0.9000	0.8000	0.7500	0.7000	0.6000	0.5000	0.4000	0.3000	0.2000	0.1000
	1	1.0000	1.0000	1.0000	1.0000	1.0000	1.0000	1.0000	1.0000	1.0000	1.0000
2	0	0.8100	0.6400	0.5625	0.4900	0.3600	0.2500	0.1600	0.0900	0.0400	0.0100
	1	0.9900	0.9600	0.9375	0.9100	0.8400	0.7500	0.6400	0.5100	0.3600	0.1900
	2	1.0000	1.0000	1.0000	1.0000	1.0000	1.0000	1.0000	1.0000	1.0000	1.0000
3	0	0.7290	0.5120	0.4219	0.3430	0.2160	0.1250	0.0640	0.0270	0.0080	0.0010
	1	0.9720	0.8960	0.8438	0.7840	0.6480	0.5000	0.3520	0.2160	0.1040	0.0280
	2	0.9990	0.9920	0.9844	0.9730	0.9360	0.8750	0.7840	0.6570	0.4880	0.2710
	3	1.0000	1.0000	1.0000	1.0000	1.0000	1.0000	1.0000	1.0000	1.0000	1.0000
4	0	0.6561	0.4096	0.3164	0.2401	0.1296	0.0625	0.0256	0.0081	0.0016	0.0001
	1	0.9477	0.8192	0.7383	0.6517	0.4752	0.3125	0.1792	0.0837	0.0272	0.0037
	2	0.9963	0.9728	0.9492	0.9163	0.8208	0.6875	0.5248	0.3483	0.1808	0.0523
	3	0.9999	0.9984	0.9961	0.9919	0.9744	0.9375	0.8704	0.7599	0.5904	0.3439
	4	1.0000	1.0000	1.0000	1.0000	1.0000	1.0000	1.0000	1.0000	1.0000	1.0000
5	0	0.5905	0.3277	0.2373	0.1681	0.0778	0.0312	0.0102	0.0024	0.0003	0.0000
	1	0.9185	0.7373	0.6328	0.5282	0.3370	0.1875	0.0870	0.0308	0.0067	0.0005
	2	0.9914	0.9421	0.8965	0.8369	0.6826	0.5000	0.3174	0.1631	0.0579	0.0086
	3	0.9995	0.9933	0.9844	0.9692	0.9130	0.8125	0.6630	0.4718	0.2627	0.0815
	4	1.0000	0.9997	0.9990	0.9976	0.9898	0.9688	0.9222	0.8319	0.6723	0.4095
	5		1.0000	1.0000	1.0000	1.0000	1.0000	1.0000	1.0000	1.0000	1.0000

(Contd.)

		p									
n	*r*	0.10	0.20	0.25	0.30	0.40	0.50	0.60	0.70	0.80	0.90
6	0	0.5314	0.2621	0.1780	0.1176	0.0467	0.0156	0.0041	0.0007	0.0001	0.0000
	1	0.8857	0.6554	0.5339	0.4202	0.2333	0.1094	0.0410	0.0109	0.0016	0.0001
	2	0.9841	0.9011	0.8306	0.7443	0.5443	0.3438	0.1792	0.0705	0.0170	0.0013
	3	0.9987	0.9830	0.9624	0.9295	0.8208	0.6563	0.4557	0.2557	0.0989	0.0158
	4	0.9999	0.9984	0.9954	0.9891	0.9590	0.8906	0.7667	0.5798	0.3447	0.1143
	5	1.0000	0.9999	0.9998	0.9993	0.9959	0.9844	0.9533	0.8824	0.7379	0.4686
	6		1.0000	1.0000	1.0000	1.0000	1.0000	1.0000	1.0000	1.0000	1.0000
7	0	0.4783	0.2097	0.1335	0.0824	0.0280	0.0078	0.0016	0.0002	0.0000	
	1	0.8503	0.5767	0.4449	0.3294	0.1586	0.0625	0.0188	0.0038	0.0004	0.0000
	2	0.9743	0.8520	0.7564	0.6471	0.4199	0.2266	0.0963	0.0288	0.0047	0.0002
	3	0.9973	0.9667	0.9294	0.8740	0.7102	0.5000	0.2898	0.1260	0.0333	0.0027
	4	0.9998	0.9953	0.9871	0.9712	0.9037	0.7734	0.5801	0.3529	0.1480	0.0257
	5	1.0000	0.9996	0.9987	0.9962	0.9812	0.9375	0.8414	0.6706	0.4233	0.1497
	6		1.0000	0.9999	0.9998	0.9984	0.9922	0.9720	0.9176	0.7903	0.5217
	7			1.0000	1.0000	1.0000	1.0000	1.0000	1.0000	1.0000	1.0000
8	0	0.4305	0.1678	0.1001	0.0576	0.0168	0.0039	0.0007	0.0001	0.0000	
	1	0.8131	0.5033	0.3671	0.2553	0.1064	0.0352	0.0085	0.0013	0.0001	
	2	0.9619	0.7969	0.6785	0.5518	0.3154	0.1445	0.0498	0.0113	0.0012	0.0000
	3	0.9950	0.9437	0.8862	0.8059	0.5941	0.3633	0.1737	0.0580	0.0104	0.0004
	4	0.9996	0.9896	0.9727	0.9420	0.8263	0.6367	0.4059	0.1941	0.0563	0.0050
	5	1.0000	0.9988	0.9958	0.9887	0.9502	0.8555	0.6846	0.4482	0.2031	0.0381
	6		0.9991	0.9996	0.9987	0.9915	0.9648	0.8936	0.7447	0.4967	0.1869
	7		1.0000	1.0000	0.9999	0.9993	0.9961	0.9832	0.9424	0.8322	0.5695
	8				1.0000	1.0000	1.0000	1.0000	1.0000	1.0000	1.0000
9	0	0.3874	0.1342	0.0751	0.0404	0.0101	0.0020	0.0003	0.0000		
	1	0.7748	0.4362	0.3003	0.1960	0.0705	0.0195	0.0038	0.0004	0.0000	
	2	0.9470	0.7382	0.6007	0.4628	0.2318	0.0898	0.0250	0.0043	0.0003	0.0000
	3	0.9917	0.9144	0.8343	0.7297	0.4826	0.2539	0.0994	0.0253	0.0031	0.0001
	4	0.9991	0.9804	0.9511	0.9012	0.7334	0.5000	0.2666	0.0988	0.0196	0.0009
	5	0.9999	0.9969	0.9900	0.9747	0.9006	0.7461	0.5174	0.2703	0.0856	0.0083
	6	1.0000	0.9997	0.9987	0.9957	0.9750	0.9102	0.7682	0.5372	0.2618	0.0530
	7		1.0000	0.9999	0.9996	0.9962	0.9805	0.9295	0.8040	0.5638	0.2252
	8			1.0000	1.0000	0.9997	0.9980	0.9899	0.9596	0.8658	0.6126
	9					1.0000	1.0000	1.0000	1.0000	1.0000	1.0000

n	r	p = 0.10	0.20	0.25	0.30	0.40	0.50	0.60	0.70	0.80	0.90
10	0	0.3487	0.1074	0.0563	0.0282	0.0060	0.0010	0.0001	0.0000		
	1	0.7361	0.3758	0.2440	0.1493	0.0464	0.0107	0.0017	0.0001	0.0000	
	2	0.9298	0.6778	0.5256	0.3828	0.1673	0.0547	0.0123	0.0016	0.0001	
	3	0.9872	0.8791	0.7759	0.6496	0.3823	0.1719	0.0548	0.0106	0.0009	0.0000
	4	0.9984	0.9672	0.9219	0.8497	0.6331	0.3770	0.1662	0.0474	0.0064	0.0002
	5	0.9999	0.9936	0.9803	0.9527	0.8338	0.6230	0.3669	0.1503	0.0328	0.0016
	6	1.0000	0.9991	0.9965	0.9894	0.9452	0.8281	0.6177	0.3504	0.1209	0.0128
	7		0.9999	0.9996	0.9984	0.9877	0.9453	0.8327	0.6172	0.3222	0.0702
	8		1.0000	1.0000	0.9999	0.9983	0.9893	0.9536	0.8507	0.6242	0.2639
	9				1.0000	0.9999	0.9990	0.9940	0.9718	0.8926	0.6513
	10					1.0000	1.0000	1.0000	1.0000	1.0000	1.0000
11	0	0.3138	0.0859	0.0422	0.0198	0.0036	0.0005	0.0000			
	1	0.6974	0.3221	0.1971	0.1130	0.0302	0.0059	0.0007	0.0000		
	2	0.9104	0.6174	0.4552	0.3127	0.1189	0.0327	0.0059	0.0006	0.0000	
	3	0.9815	0.8369	0.7133	0.5696	0.2963	0.1133	0.0293	0.0043	0.0002	
	4	0.9972	0.9496	0.8854	0.7897	0.5328	0.2744	0.0994	0.0216	0.0020	0.0000
	5	0.9997	0.9883	0.9657	0.9218	0.7535	0.5000	0.2465	0.0782	0.0117	0.0003
	6	1.0000	0.9980	0.9924	0.9784	0.9006	0.7256	0.4672	0.2103	0.0504	0.0028
	7		0.9998	0.9988	0.9957	0.9707	0.8867	0.7037	0.4304	0.1611	0.0185
	8		1.0000	0.9999	0.9994	0.9941	0.9673	0.8811	0.6873	0.3826	0.0896
	9			1.0000	1.0000	0.9993	0.9941	0.9698	0.8870	0.6779	0.3026
	10					1.0000	0.9995	0.9964	0.9802	0.9141	0.6862
	11						1.0000	1.0000	1.0000	1.0000	1.0000
12	0	0.2824	0.0687	0.0317	0.0138	0.0022	0.0002	0.0000			
	1	0.6590	0.2749	0.1584	0.0850	0.0196	0.0032	0.0003	0.0000		
	2	0.8891	0.5583	0.3907	0.2528	0.0834	0.0193	0.0028	0.0002	0.0000	
	3	0.9744	0.7946	0.6488	0.4925	0.2253	0.0730	0.0153	0.0017	0.0001	
	4	0.9957	0.9274	0.8424	0.7237	0.4382	0.1938	0.0573	0.0095	0.0006	0.0000
	5	0.9995	0.9806	0.9456	0.8821	0.6652	0.3872	0.1582	0.0386	0.0039	0.0001
	6	0.9999	0.9961	0.9857	0.9614	0.8418	0.6128	0.3348	0.1178	0.0194	0.0005
	7	1.0000	0.9994	0.9972	0.9905	0.9427	0.8062	0.5618	0.2763	0.0726	0.0043
	8		0.9999	0.9996	0.9983	0.9847	0.9270	0.7747	0.5075	0.2054	0.0256
	9		1.0000	1.0000	0.9998	0.9972	0.9807	0.9166	0.7472	0.4417	0.1109
	10				1.0000	0.9997	0.9968	0.9804	0.9150	0.7251	0.3410

(*Contd.*)

		p									
n	*r*	0.10	0.20	0.25	0.30	0.40	0.50	0.60	0.70	0.80	0.90
	11					1.0000	0.9998	0.9978	0.9862	0.9313	0.7176
	12						1.0000	1.0000	1.0000	1.0000	1.0000
13	0	0.2542	0.0550	0.0238	0.0097	0.0013	0.0001	0.0000			
	1	0.6213	0.2336	0.1267	0.0637	0.0126	0.0017	0.0001	0.0000		
	2	0.8661	0.5017	0.3326	0.2025	0.0579	0.0112	0.0013	0.0001		
	3	0.9658	0.7473	0.5843	0.4206	0.1686	0.0461	0.0078	0.0007	0.0000	
	4	0.9935	0.9009	0.7940	0.6543	0.3530	0.1334	0.0321	0.0040	0.0002	
	5	0.9991	0.9700	0.9198	0.8346	0.5744	0.2905	0.0977	0.0182	0.0012	0.0000
	6	0.9999	0.9930	0.9757	0.9376	0.7712	0.5000	0.2288	0.0624	0.0070	0.0001
	7	1.0000	0.9980	0.9944	0.9818	0.9023	0.7095	0.4256	0.1654	0.0300	0.0009
	8		0.9998	0.9990	0.9960	0.9679	0.8666	0.6470	0.3457	0.0991	0.0065
	9		1.0000	0.9999	0.9993	0.9922	0.9539	0.8314	0.5794	0.2527	0.0342
	10			1.0000	0.9999	0.9987	0.9888	0.9421	0.7975	0.4983	0.1339
	11				1.0000	0.9999	0.9983	0.9874	0.9363	0.7664	0.3787
	12					1.0000	0.9999	0.9987	0.9903	0.9450	0.7458
	13						1.0000	1.0000	1.0000	1.0000	1.0000
14	0	0.2288	0.0440	0.0178	0.0068	0.0008	0.0001	0.0000			
	1	0.5846	0.1979	0.1010	0.0475	0.0081	0.0009	0.0001			
	2	0.8416	0.4481	0.2811	0.1608	0.0398	0.0065	0.0006	0.0000		
	3	0.9559	0.6982	0.5213	0.3552	0.1243	0.0287	0.0039	0.0002		
	4	0.9908	0.8702	0.7415	0.5842	0.2793	0.0898	0.0175	0.0017	0.0000	
	5	0.9985	0.9561	0.8883	0.7805	0.4859	0.2120	0.0583	0.0083	0.0004	
	6	0.9998	0.9884	0.9617	0.9067	0.6925	0.3953	0.1501	0.0315	0.0024	0.0000
	7	1.0000	0.9976	0.9897	0.9685	0.8499	0.6047	0.3075	0.0933	0.0116	0.0002
	8		0.9996	0.9978	0.9917	0.9417	0.7880	0.5141	0.2195	0.0439	0.0015
	9		1.0000	0.9997	0.9983	0.9825	0.9102	0.7207	0.4158	0.1298	0.0092
	10			1.0000	0.9998	0.9961	0.9713	0.8757	0.6448	0.3018	0.0441
	11				1.0000	0.9994	0.9935	0.9602	0.8392	0.5519	0.1584
	12					0.9999	0.9991	0.9919	0.9525	0.8021	0.4154
	13					1.0000	0.9999	0.9992	0.9932	0.9560	0.7712
	14						1.0000	1.0000	1.0000	1.0000	1.0000
15	0	0.2059	0.0352	0.0134	0.0047	0.0005	0.0000				
	1	0.5490	0.1671	0.0802	0.0353	0.0052	0.0005	0.0000			
	2	0.8159	0.3980	0.2361	0.1268	0.0271	0.0037	0.0003	0.0000		

n	*r*	0.10	0.20	0.25	0.30	0.40	0.50	0.60	0.70	0.80	0.90
	3	0.9444	0.6482	0.4613	0.2969	0.0905	0.0176	0.0019	0.0001		
	4	0.9873	0.8358	0.6865	0.5155	0.2173	0.0592	0.0094	0.0007	0.0000	
	5	0.9978	0.9389	0.8516	0.7216	0.4032	0.1509	0.0338	0.0037	0.0001	
	6	0.9997	0.9819	0.9434	0.8689	0.6098	0.3036	0.0951	0.0152	0.0008	
	7	1.0000	0.9958	0.9827	0.9500	0.7869	0.5000	0.2131	0.0500	0.0042	0.0000
	8		0.9992	0.9958	0.9848	0.9050	0.6964	0.3902	0.1311	0.0181	0.0003
	9		0.9999	0.9992	0.9963	0.9662	0.8491	0.5968	0.2784	0.0611	0.0023
	10		1.0000	0.9999	0.9993	0.9907	0.9408	0.7827	0.4845	0.1642	0.0127
	11			1.0000	0.9999	0.9981	0.9824	0.9095	0.7031	0.3518	0.0556
	12				1.0000	0.9997	0.9963	0.9729	0.8732	0.6020	0.1841
	13					1.0000	0.9995	0.9948	0.9647	0.8329	0.4510
	14						1.0000	0.9995	0.9953	0.9648	0.7941
	15							1.0000	1.0000	1.0000	1.0000
16	0	0.1853	0.0281	0.0100	0.0033	0.0003	0.0000				
	1	0.5147	0.1407	0.0635	0.0261	0.0033	0.0003	0.0000			
	2	0.7892	0.3518	0.1971	0.0994	0.0183	0.0021	0.0001			
	3	0.9316	0.5981	0.4050	0.2459	0.0651	0.0106	0.0009	0.0000		
	4	0.9830	0.7982	0.6302	0.4499	0.1666	0.0384	0.0049	0.0003		
	5	0.9967	0.9183	0.8103	0.6598	0.3288	0.1051	0.0191	0.0016	0.0000	
	6	0.9995	0.9733	0.9204	0.8247	0.5272	0.2272	0.0583	0.0071	0.0002	
	7	0.9999	0.9930	0.9729	0.9256	0.7161	0.4018	0.1423	0.0257	0.0015	0.0000
	8	1.0000	0.9985	0.9925	0.9743	0.8577	0.5982	0.2839	0.0744	0.0070	0.0001
	9		0.9998	0.9984	0.9929	0.9417	0.7728	0.4728	0.1753	0.0267	0.0005
	10		1.0000	0.9997	0.9984	0.9809	0.8949	0.6712	0.3402	0.0817	0.0033
	11			1.0000	0.9997	0.9951	0.9616	0.8334	0.5501	0.2018	0.0170
	12				1.0000	0.9991	0.9894	0.9349	0.7541	0.4019	0.0684
	13					0.9999	0.9979	0.9817	0.9006	0.6482	0.2108
	14					1.0000	0.9997	0.9967	0.9739	0.8593	0.4853
	15						1.0000	0.9997	0.9967	0.9719	0.8147
	16							1.0000	1.0000	1.0000	1.0000
17	0	0.1668	0.0225	0.0075	0.0023	0.0002	0.0000				
	1	0.4818	0.1182	0.0501	0.0193	0.0021	0.0001	0.0000			
	2	0.7618	0.3096	0.1637	0.0774	0.0123	0.0012	0.0001			
	3	0.9174	0.5489	0.3530	0.2019	0.0464	0.0064	0.0005	0.0000		
	4	0.9779	0.7582	0.5739	0.3887	0.1260	0.0245	0.0025	0.0001		

(*Contd.*)

		p									
n	r	0.10	0.20	0.25	0.30	0.40	0.50	0.60	0.70	0.80	0.90
	5	0.9953	0.8943	0.7653	0.5968	0.2639	0.0717	0.0106	0.0007	0.0000	
	6	0.9992	0.9623	0.8929	0.7752	0.4478	0.1662	0.0348	0.0032	0.0001	
	7	0.9999	0.9891	0.9598	0.8954	0.6405	0.3145	0.0919	0.0127	0.0005	
	8	1.0000	0.9974	0.9876	0.9597	0.8011	0.5000	0.1989	0.0403	0.0026	0.0000
	9		0.9995	0.9969	0.9873	0.9081	0.6855	0.3595	0.1046	0.0109	0.0001
	10		0.9999	0.994	0.9968	0.9652	0.8338	0.5522	0.2248	0.0377	0.0008
	11		1.0000	0.9999	0.9993	0.9894	0.9283	0.7361	0.4032	0.1057	0.0047
	12			1.0000	0.9999	0.9975	0.9755	0.8740	0.6113	0.2418	0.0221
	13				1.0000	0.9995	0.9936	0.9536	0.7981	0.4511	0.0826
	14					0.9999	0.9988	0.9877	0.9226	0.6904	0.2382
	15					1.0000	0.9999	0.9979	0.9807	0.8818	0.5182
	16						1.0000	0.9998	0.9977	0.9775	0.8332
	17						1.0000	1.0000	1.0000	1.0000	1.0000
18	0	0.1501	0.0180	0.0056	0.0016	0.0001	0.0000				
	1	0.4503	0,0991	0.0395	0.0142	0.0013	0.0001				
	2	0.7338	0.2713	0.1353	0.0600	0.0082	0.0007	0.0000			
	3	0.9018	0.5010	0.3057	0.1646	0.0328	0.0038	0.0002			
	4	0.9718	0.7164	0.5787	0.3327	0.0942	0.0154	0.0013	0.0000		
	5	0.9936	0.8671	0.7175	0.5344	0.2088	0.0481	0.0058	0.0003		
	6	0.9988	0.9487	0.8610	0.7217	0.3743	0.1189	0.0203	0.0014	0.0000	
	7	0.9998	0.9837	0.9431	0.8593	0.5634	0.2403	0.0576	0.0061	0.0002	
	8	1.0000	0.9957	0.9807	0.9404	0.7368	0.4073	0.1347	0.0210	0.0009	
	9		0.9991	0.9946	0.9790	0.8653	0.5927	0.2632	0.0596	0.0043	0.0000
	10		0.9998	0.9988	0.9939	0.9424	0.7597	0.4366	0.1407	0.0163	0.0002
	11		1.0000	0.9998	0.9986	0.9797	0.8811	0.6257	0.2783	0.0513	0.0012
	12			1.0000	0.9997	0.9942	0.9519	0.7912	0.4656	0.1329	0.0064
	13				1.0000	0.9987	0.9846	0.9058	0.6673	0.2836	0.0282
	14					0.9998	0.9962	0.9672	0.8354	0.4990	0.0982
	15					1.0000	0.9993	0.9918	0.9400	0.7287	0.2662
	16						0.9999	0.9987	0.9858	0.9009	0.5497
	17						1.0000	0.9999	0.9984	0.9820	0.8499
	18							1.0000	1.0000	1.0000	1.0000
19	0	0.1351	0.0144	0.0042	0.0011	0.0001					
	1	0.4203	0.0829	0.0310	0.0104	0.0008	0.0000				

		p									
n	*r*	0.10	0.20	0.25	0.30	0.40	0.50	0.60	0.70	0.80	0.90
	2	0.7054	0.2369	0.1113	0.0462	0.0055	0.0004	0.0000			
	3	0.8850	0.4551	0.2631	0.1332	0.2030	0.0022	0.0001			
	4	0.9648	0.6733	0.4654	0.2822	0.0696	0.0096	0.0006	0.0000		
	5	0.9914	0.8369	0.6678	0.4739	0.4629	0.0318	0.0031	0.0001		
	6	0.9983	0.9324	0.8251	0.6655	0.3081	0.0835	0.0116	0.0006		
	7	0.9997	0.9767	0.9225	0.8180	0.4878	0.1796	0.0352	0.0028	0.0000	
	8	1.0000	0.9933	0.9713	0.9161	0.6675	0.3228	0.0885	0.0105	0.0003	
	9		0.9984	0.9911	0.9674	0.8139	0.5000	0.1861	0.0326	0.0016	
	10		0.9997	0.9977	0.9895	0.9115	0.6762	0.3325	0.0839	0.0067	0.0000
	11		0.9999	0.9995	0.9972	0.9648	0.8204	0.5122	0.1820	0.0233	0.0003
	12		1.0000	0.9999	0.9994	0.9884	0.9165	0.6919	0.3345	0.0676	0.0017
	13			1.0000	0.9999	0.9969	0.9682	0.8371	0.5261	0.1631	0.0086
	14				1.0000	0.9994	0.9904	0.9304	0.7178	0.3267	0.0352
	15					0.9999	0.9978	0.9770	0.8668	0.5449	0.1150
	16					1.0000	0.9996	0.9945	0.9538	0.7631	0.2946
	17						1.0000	0.9992	0.9896	0.9171	0.5797
	18							0.9999	0.9989	0.9856	0.8649
	19							1.0000	1.0000	1.0000	1.0000
20	0	0.1216	0.0115	0.0032	0.0008	0.0000					
	1	0.3917	0.0692	0.0243	0.0076	0.0005	0.0000				
	2	0.6769	0.2061	0.0913	0.0355	0.0036	0.0002	0.0000			
	3	0.8670	0.4114	0.2252	0.1071	0.0160	0.0013	0.0001			
	4	0.9568	0.6296	0.4148	0.2375	0.0510	0.0059	0.0003			
	5	0.9887	0.8042	0.6172	0.4164	0.1256	0.0207	0.0016	0.0000		
	6	0.9976	0.9133	0.7858	0.6080	0.2500	0.0577	0.0065	0.0003		
	7	0.9996	0.9679	0.8982	0.7723	0.4159	0.1316	0.0210	0.0013	0.0000	
	8	0.9999	0.9900	0.9591	0.8867	0.5956	0.2517	0.0565	0.0051	0.0001	
	9	1.0000	0.9974	0.9861	0.9520	0.7553	0.4119	0.1275	0.0171	0.0006	
	10		0.9994	0.9961	0.9829	0.8725	0.5881	0.2447	0.0480	0.0026	0.0000
	11		0.9999	0.9991	0.9949	0.9435	0.7483	0.4044	0.1133	0.0100	0.0001
	12		1.0000	0.9998	0.9987	0.9790	0.8684	0.5841	0.2277	0.0321	0.0004
	13			1.0000	0.9997	0.9935	0.9423	0.7500	0.3920	0.0867	0.0024
	14				1.0000	0.9984	0.9793	0.8744	0.5836	0.1958	0.0113
	15					0.9997	0.9941	0.9490	0.7625	0.3704	0.0432
	16					1.0000	0.9987	0.9840	0.8929	0.5886	0.1330

(Contd.)

n	r	p 0.10	0.20	0.25	0.30	0.40	0.50	0.60	0.70	0.80	0.90
	17						0.9998	0.9964	0.9645	0.7939	0.3231
	18						1.0000	0.9995	0.9924	0.9308	0.6083
	19							1.0000	0.9992	0.9885	0.8784
	20								1.0000	1.0000	1.0000

A.2 Poisson Distribution Function

Table A.2 Poisson probability sums: $F(x, \lambda) = \sum_{k=0}^{x} e^{-\lambda} \frac{\lambda^k}{k!}$

λ \ x	0	1	2	3	4	5	6	7	8	9
0.02	0.980	1.000								
0.04	0.961	0.999	1.000							
0.06	0.942	0.998	1.000							
0.08	0.923	0.997	1.000							
0.10	0.905	0.995	1.000							
0.15	0.861	0.990	0.999	1.000						
0.20	0.819	0.982	0.999	1.000						
0.25	0.779	0.974	0.998	1.000						
0.30	0.741	0.963	0.996	1.000						
0.35	0.705	0.951	0.994	1.000						
0.40	0.670	0.938	0.992	0.999	1.000					
0.45	0.638	0.925	0.989	0.999	1.000					
0.50	0.607	0.910	0.986	0.998	1.000					
0.55	0.577	0.894	0.982	0.998	1.000					
0.60	0.549	0.878	0.977	0.997	1.000					
0.65	0.522	0.861	0.972	0.996	0.999	1.000				
0.70	0.497	0.844	0.966	0.994	0.999	1.000				
0.75	0.472	0.827	0.959	0.993	0.999	1.000				
0.80	0.449	0.809	0.953	0.991	0.999	1.000				
0.85	0.427	0.791	0.945	0.989	0.998	1.000				
0.90	0.407	0.772	0.937	0.987	0.998	1.000				
0.95	0.387	0.754	0.929	0.984	0.997	1.000				
1.00	0.368	0.736	0.920	0.981	0.996	0.999	1.000			

λ \ x	0	1	2	3	4	5	6	7	8	9
1.1	0.333	0.699	0.900	0.974	0.995	0.999	1.000			
1.2	0.301	0.663	0.879	0.966	0.992	0.998	1.000			
1.3	0.273	0.627	0.857	0.957	0.989	0.998	1.000			
1.4	0.247	0.592	0.833	0.946	0.986	0.997	0.999	1.000		
1.5	0.233	0.558	0.809	0.934	0.981	0.996	0.999	1.000		
1.6	0.202	0.525	0.783	0.921	0.976	0.994	0.999	1.000		
1.7	0.183	0.493	0.757	0.907	0.970	0.992	0.998	1.000		
1.8	0.165	0.463	0.731	0.891	0.964	0.990	0.997	0.999	1.000	
1.9	0.150	0.434	0.704	0.875	0.956	0.987	0.997	0.999	1.000	
2.0	0.135	0.406	0.677	0.857	0.947	0.983	0.995	0.999	1.000	
λ \ x	0	1	2	3	4	5	6	7	8	9
2.2	0.111	0.355	0.623	0.819	0.928	0.975	0.993	0.998	1.000	
2.4	0.091	0.308	0.570	0.779	0.904	0.964	0.988	0.997	0.999	1.000
2.6	0.074	0.267	0.518	0.736	0.877	0.951	0.983	0.995	0.999	1.000
2.8	0.061	0.231	0.469	0.692	0.848	0.935	0.976	0.992	0.998	0.999
3.0	0.050	0.199	0.423	0.647	0.815	0.916	0.966	0.988	0.966	1.000
3.2	0.041	0.171	0.380	0.603	0.781	0.895	0.955	0.983	0.994	0.998
3.4	0.033	0.147	0.340	0.558	0.744	0.871	0.942	0.977	0.992	0.997
3.6	0.027	0.126	0.303	0.515	0.706	0.844	0.927	0.969	0.988	0.996
3.8	0.022	0.107	0.269	0.473	0.668	0.816	0.909	0.960	0.984	0.994
4.0	0.018	0.092	0.238	0.433	0.629	0.785	0.889	0.949	0.979	0.002
4.2	0.015	0.078	0.210	0.395	0.590	0.753	0.867	0.936	0.972	0.989
4.4	0.012	0.066	0.185	0.359	0.551	0.720	0.844	0.921	0.964	0.985
4.6	0.010	0.056	0.163	0.326	0.513	0.686	0.818	0.905	0.955	0.980
4.8	0.008	0.048	0.143	0.294	0.476	0.651	0.791	0.887	0.944	0.975
5.0	0.007	0.040	0.125	0.265	0.440	0.616	0.762	0.867	0.932	0.968
5.2	0.006	0.034	0.109	0.238	0.406	0.581	0.732	0.845	0.918	0.960
5.4	0.005	0.029	0.095	0.213	0.373	0.546	0.702	0.822	0.903	0.951
5.6	0.004	0.024	0.082	0.191	0.342	0.512	0.670	0.797	0.886	0.941
5.8	0.003	0.021	0.072	0.170	0.313	0.478	0.638	0.771	0.867	0.929
6.0	0.02	0.017	0.062	0.151	0.285	0.446	0.606	0.744	0.847	0.916

(*Contd.*)

	10	11	12	13	14	15	16
2.8	1.000						
3.0	1.000						
3.2	1.000						
3.4	0.999	1.000					
3.6	0.999	1.000					
3.8	0.998	0.999	1.000				
4.0	0.997	0.999	1.0000				
4.2	0.996	0.999	1.000				
4.4	0.994	0.998	0.999	1.000			
4.6	0.992	0.997	0.999	1.000			
4.8	0.990	0.996	0.999	1.000			
5.0	0.986	0.995	0.998	0.999	1.000		
5.2	0.982	0.993	0.997	0.999	1.000		
5.4	0.977	0.990	0.996	0.999	1.000		
5.6	0.972	0.988	0.995	0.998	0.999	1.000	
5.8	0.965	0.984	0.993	0.997	0.999	1.000	
6.0	0.957	0.980	0.991	0.996	0.999	0.999	1.000

λ \ x	0	1	2	3	4	5	6	7	8	9
6.2	0.002	0.015	0.054	0.134	0.259	0.014	0.574	0.716	0.826	0.902
6.4	0.002	0.012	0.046	0.119	0.235	0.384	0.542	0.687	0.803	0.886
6.6	0.001	0.010	0.040	0.105	0.213	0.355	0.511	0.658	0.780	0.869
6.8	0.001	0.009	0.034	0.093	0.192	0.327	0.480	0.628	0.755	0.850
7.0	0.001	0.007	0.030	0.082	0.173	0.301	0.450	0.599	0.729	0.830
7.2	0.001	0.006	0.025	0.072	0.156	0.276	0.420	0.569	0.703	0.810
7.4	0.001	0.005	0.022	0.063	0.140	0.253	0.392	0.539	0.676	0.788
7.6	0.001	0.004	0.019	0.055	0.125	0.231	0.365	0.510	0.648	0.765
7.8	0.000	0.004	0.016	0.048	0.112	0.210	0.338	0.481	0.620	0.741
8.0	0.000	0.003	0.014	0.042	0.100	0.191	0.313	0.453	0.593	0.717
8.5	0.000	0.002	0.009	0.030	0.074	0.150	0.256	0.386	0.523	0.653
9.0	0.000	0.001	0.006	0.021	0.055	0.116	0.207	0.324	0.456	0.587
9.5	0.000	0.000	0.004	0.015	0.040	0.089	0.165	0.269	0.392	0.522
10.0	0.000	0.000	0.003	0.010	0.029	0.067	0.130	0.220	0.333	0.458

	10	11	12	13	14	15	16	17	18	19
6.2	0.949	0.975	0.989	0.995	0.998	0.999	1.000			
6.4	0.939	0.969	0.986	0.994	0.997	0.999	1.000			
6.6	0.927	0.963	0.982	0.992	0.997	0.999	0.999	1.000		
6.8	0.915	0.955	0.978	0.990	0.996	0.998	0.999	1.000		
7.0	0.901	0.947	0.973	0.987	0.994	0.998	0.999	1.000		
7.2	0.887	0.937	0.967	0.984	0.993	0.997	0.999	1.000	1.000	
7.4	0.871	0.926	0.961	0.980	0.991	0.996	0.998	0.999	1.000	
7.6	0.854	0.915	0.954	0.976	0.989	0.995	0.998	0.999	1.000	
7.8	0.835	0.902	0.945	0.971	0.986	0.993	0.997	0.999	1.000	
8.0	0.816	0.888	0.936	0.966	0.983	0.992	0.996	0.998	0.999	1.000
8.5	0.763	0.849	0.909	0.949	0.973	0.986	0.993	0.997	0.999	0.999
9.0	0.706	0.803	0.876	0.926	0.959	0.978	0.989	0.995	0.998	0.999
9.5	0.645	0.752	0.836	0.898	0.940	0.967	0.982	0.991	0.996	0.998
10.0	0.583	0.697	0.792	0.864	0.917	0.951	0.973	0.986	0.993	0.997

	20	21	22
8.5	1.000		
9.0	1.000		
9.5	0.999	1.000	
10.0	0.998	0.999	1.000

λ \ x	0	1	2	3	4	5	6	7	8	9
10.5	0.000	0.000	0.002	0.007	0.021	0.050	0.102	0.179	0.279	0.397
11.0	0.000	0.000	0.001	0.005	0.015	0.038	0.079	0.143	0.232	0.341
11.5	0.000	0.000	0.001	0.003	0.011	0.028	0.060	0.114	0.191	0.289
12.0	0.000	0.000	0.001	0.002	0.008	0.020	0.046	0.090	0.155	0.242
12.5	0.000	0.000	0.000	0.002	0.005	0.015	0.035	0.070	0.125	0.201
13.0	0.000	0.000	0.000	0.001	0.004	0.011	0.026	0.054	0.100	0.166
13.5	0.000	0.000	0.000	0.001	0.003	0.008	0.019	0.041	0.079	0.135
14.0	0.000	0.000	0.000	0.000	0.002	0.006	0.014	0.032	0.062	0.109
14.5	0.000	0.000	0.000	0.000	0.001	0.004	0.010	0.024	0.048	0.088
15.0	0.000	0.000	0.000	0.000	0.001	0.003	0.008	0.018	0.037	0.070

	10	11	12	13	14	15	16	17	18	19
10.5	0.521	0.639	0.742	0.825	0.888	0.932	0.960	0.978	0.988	0.994
11.0	0.460	0.579	0.689	0.781	0.854	0.907	0.944	0.968	0.982	0.991

(*Contd.*)

11.5	0.402	0.520	0.633	0.733	0.815	0.878	0.924	0.954	0.974	0.986
12.0	0.347	0.462	0.576	0.682	0.772	0.844	0.899	0.937	0.963	0.979
12.5	0.297	0.406	0.519	0.628	0.725	0.806	0.869	0.916	0.948	0.969
13.0	0.252	0.353	0.463	0.573	0.675	0.764	0.835	0.890	0.930	0.957
13.5	0.211	0.304	0.409	0.518	0.623	0.718	0.798	0.861	0.908	0.942
14.0	0.176	0.260	0.358	0.464	0.570	0.669	0.756	0.827	0.883	0.923
14.5	0.145	0.220	0.311	0.413	0.518	0.619	0.711	0.790	0.853	0.901
15.0	0.118	0.185	0.268	0.363	0.466	0.568	0.664	0.749	0.819	0.875
	20	21	22	23	24	25	26	27	28	29
10.5	0.997	0.999	0.999	1.000						
11.0	0.995	0.998	0.999	1.000						
11.5	0.992	0.996	0.998	0.999	1.000					
12.0	0.988	0.994	0.997	0.999	0.999	1.000				
12.5	0.983	0.991	0.995	0.998	0.999	0.999	1.000			
13.0	0.975	0.986	0.992	0.996	0.998	0.999	1.000			
13.5	0.965	0.980	0.989	0.994	0.997	0.998	0.999	1.000		
14.0	0.952	0.971	0.983	0.991	0.995	0.997	0.999	0.999	1.000	
14.5	0.936	0.960	0.976	0.986	0.992	0.996	0.998	0.999	0.999	1.000
15.0	0.917	0.947	0.967	0.981	0.989	0.994	0.997	0.998	0.999	1.000

A.3 AREA UNDER THE STANDARD NORMAL CURVE FROM O TO Z

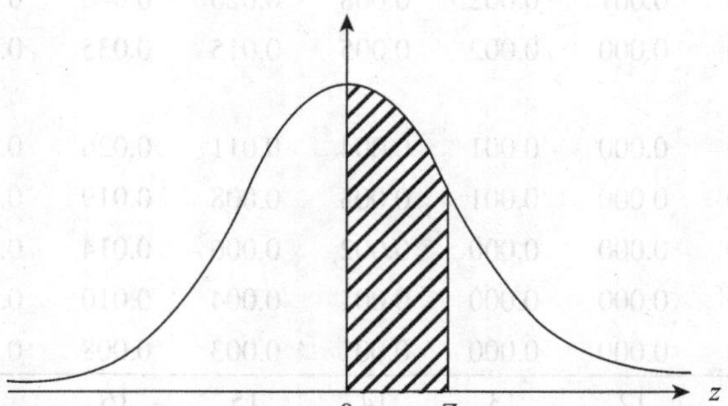

Figure A.1 Area under the standard normal curve from 0 to Z.

Table A.3 Normal tables

Z	0	1	2	3	4	5	6	7	8	9
0.0	0.0000	0.0040	0.0080	0.0120	0.0160	0.0199	0.0239	0.0279	0.0319	0.0359
0.1	0.0398	0.0438	0.0478	0.0517	0.0557	0.0596	0.0636	0.0675	0.0714	0.0754
0.2	0.0793	0.0832	0.0871	0.0910	0.0948	0.0987	0.1026	0.1064	0.1103	0.1141
0.3	0.1179	0.1217	0.1256	0.1293	0.1331	0.1368	0.1406	0.1443	0.1480	0.1517
0.4	0.1554	0.1591	0.1628	0.1664	0.1700	0.1736	0.1772	0.1808	0.1844	0.1879
0.5	0.1916	0.1950	0.1985	0.2019	0.2054	0.2088	0.2123	0.2157	0.2190	0.2224
0.6	0.2258	0.2291	0.2324	0.2357	0.2389	0.2422	0.2454	0.2486	0.2518	0.2649
0.7	0.2580	0.2612	0.2642	0.2673	0.2704	0.2734	0.2764	0.2794	0.2828	0.2852
0.8	0.2881	0.2910	0.2939	0.2967	0.2996	0.3023	0.3051	0.3078	0.3106	0.3133
0.9	0.3159	0.3186	0.3212	0.3238	0.3264	0.3289	0.3315	0.3340	0.3365	0.3389
1.0	0.3413	0.3438	0.3461	0.3485	0.3508	0.3531	0.3554	0.3577	0.3599	0.3621
1.1	0.3643	0.3665	0.3686	0.3708	0.3729	0.3749	0.3770	0.3790	0.3810	0.3830
1.2	0.3849	0.3869	0.3888	0.3907	0.3925	0.3944	0.3962	0.3980	0.3997	0.4015
1.3	0.4032	0.4049	0.4066	0.4082	0.4099	0.4115	0.4131	0.4147	0.4162	0.4177
1.4	0.4192	0.4207	0.4222	0.4236	0.4251	0.4265	0.4279	0.4292	0.4306	0.4319
1.5	0.4332	0.4345	0.4357	0.4370	0.4382	0.4394	0.4406	0.4418	0.4429	0.4441
1.6	0.4452	0.4463	0.4474	0.4484	0.4495	0.4505	0.4515	0.4525	0.4535	0.4545
1.7	0.4654	0.4564	0.4573	0.4582	0.4591	0.4599	0.4608	0.4616	0.4625	0.4633
1.8	0.4641	0.4649	0.4656	0.4664	0.4671	0.4678	0.4686	0.4693	0.4699	0.4706
1.9	0.4713	0.4719	0.4726	0.4732	0.4738	0.4744	0.4750	0.4756	0.4761	0.4767
2.0	0.4772	0.4778	0.4783	0.4788	0.4793	0.4798	0.4803	0.4808	0.4812	0.4817
2.1	0.4821	0.4826	0.4830	0.4834	0.4838	0.4842	0.4846	0.4850	0.4854	0.4857
2.2	0.4861	0.4864	0.4868	0.4871	0.4875	0.4878	0.4881	0.4884	0.4887	0.4890
2.3	0.4893	0.4896	0.4898	0.4901	0.4904	0.4906	0.4909	0.4911	0.4913	0.4916
2.4	0.4918	0.4920	0.4922	0.4925	0.4927	0.4929	0.4931	0.4932	0.4934	0.4936
2.5	0.4938	0.4940	0.4941	0.4943	0.4945	0.4946	0.4948	0.4949	0.4951	0.4952
2.6	0.4953	0.4955	0.4956	0.4957	0.4959	0.4960	0.4961	0.4962	0.4963	0.4964
2.7	0.4965	0.4966	0.4967	0.4968	0.4969	0.4970	0.4971	0.4972	0.4973	0.4974
2.8	0.4974	0.4975	0.4976	0.4977	0.4979	0.4978	0.4979	0.4979	0.4980	0.4981
2.9	0.4981	0.4982	0.4982	0.4983	0.4984	0.4984	0.4985	0.4985	0.4986	0.4986
3.0	0.4987	0.4987	0.4987	0.4988	0.4988	0.4989	0.4989	0.4989	0.4990	0.4990
3.1	0.4990	0.4991	0.4991	0.4991	0.4992	0.4992	0.4992	0.4992	0.4993	0.4993

(*Contd.*)

3.2	0.4993	0.4993	0.4994	0.4994	0.4994	0.4994	0.4994	0.4995	0.4955	0.4955
3.3	0.4995	0.4995	0.4995	0.4996	0.4996	0.4996	0.4996	0.4996	0.4996	0.4997
3.4	0.4997	0.4997	0.4997	0.4997	0.4997	0.4997	0.4997	0.4997	0.4997	0.4998
3.5	0.4998	0.4998	0.4998	0.4998	0.4998	0.4998	0.4098	0.4998	0.4998	0.4998
3.6	0.4998	0.4998	0.4999	0.4999	0.4999	0.4999	0.4999	0.4999	0.4999	0.4999
3.7	0.4999	0.4999	0.4999	0.4999	0.4999	0.4999	0.4999	0.4999	0.4999	0.4999
3.8	0.4999	0.4999	0.4999	0.4999	0.4999	0.4999	0.4999	0.4999	0.4999	0.4999
3.9	0.5000	0.5000	0.5000	0.5000	0.5000	0.5000	0.5000	0.5000	0.5000	0.5000

A.4 Critical Values of the *t*-Distribution

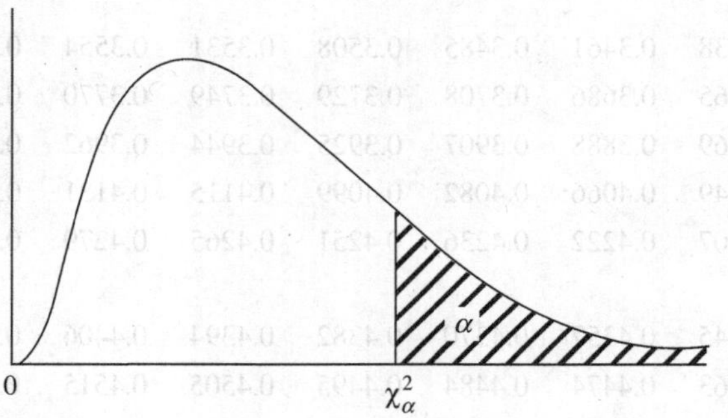

Figure A.2 Critical values of *t*-distribution.

Table A.4 Critical values of the *t*-Distribution (t_α)

	α						
ν	0.40	0.30	0.20	0.15	0.10	0.05	0.025
1	0.325	0.727	1.376	1.963	3.078	6.314	12.706
2	0.289	0.617	1.061	1.386	1.886	2.920	4.303
3	0.277	0.584	0.978	1.250	1.638	2.353	3.182
4	0.271	0.569	0.941	1.190	1.533	2.132	2.776
5	0.267	0.559	0.920	1.156	1.476	2.015	2.571
6	0.265	0.553	0.906	1.134	1.440	1.943	2.447
7	0.263	0.549	0.896	1.119	1.415	1.895	2.365
8	0.262	0.546	0.889	1.108	1.397	1.860	2.306
9	0.261	0.543	0.883	1.100	1.383	1.833	2.262
10	0.260	0.542	0.879	1.093	1.372	1.812	2.228

11	0.260	0.540	0.876	1.088	1.363	1.796	2.201
12	0.259	0.539	0.873	1.083	1.356	1.782	2.179
13	0.259	0.537	0.870	1.079	1.350	1.771	2.160
14	0.258	0.537	0.868	1.076	1.345	1.761	2.145
15	0.258	0.536	0.866	1.074	1.341	1.753	2.131
16	0.258	0.535	0.865	1.071	1.337	1.746	2.120
17	0.257	0.534	0.863	1.069	1.333	1.740	2.110
18	0.257	0.534	0.862	1.067	1.330	1.734	1.101
19	0.257	0.533	0.861	1.066	1.328	1.729	2.093
20	0.257	0.533	0.860	1.064	1.325	1.725	2.086
21	0.257	0.532	0.859	1.063	1.323	1.721	2.080
22	0.256	0.532	0.858	1.061	1.321	1.717	2.074
23	0.256	0.532	0.858	1.060	1.319	1.714	2.069
24	0.256	0.531	0.857	1.059	1.318	1.711	2.064
25	0.256	0.531	0.856	1.058	1.316	1.708	2.060
26	0.256	0.531	0.856	1.058	1.315	1.706	2.056
27	0.256	0.531	0.855	1.057	1.314	1.703	2.052
28	0.256	0.530	0.855	1.056	1.313	1.701	2.048
29	0.256	0.530	0.854	1.055	1.311	1.699	2.045
30	0.256	0.530	0.854	1.055	1.310	1.697	2.042
40	0.255	0.529	0.851	1.050	1.303	1.684	2.021
60	0.254	0.527	0.848	1.045	1.296	1.671	2.000
120	0.254	0.526	0.845	1.041	1.289	1.658	1.980
∞	0.253	0.524	0.842	1.036	1.282	1.645	1.960

	α						
v	0.02	0.015	0.01	0.0075	0.005	0.0025	0.0005
1	15.895	21.205	31.821	43.434	63.657	127.322	636.590
2	4.849	5.643	6.965	8.013	9.925	14.089	31.598
3	3.482	3.896	4.541	5.047	5.841	7.453	12.924
4	2.999	3.298	3.747	4.088	4.604	5.598	8.610
5	2.757	3.003	3.365	3.634	4.032	4.773	6.869

(*Contd.*)

6	2.612	2.829	3.143	3.372	3.707	4.317	5.959
7	2.517	2.715	2.998	3.203	3.499	4.029	5.408
8	2.449	2.634	2.896	3.085	3.355	3.833	5.041
9	2.398	2.574	2.821	2.998	3.250	3.690	4.781
10	2.359	2.527	2.764	2.932	3.169	3.581	4.587
11	2.328	2.491	2.718	2.879	3.106	3.497	4.437
12	2.303	2.461	2.681	2.836	3.055	3.428	4.318
13	2.282	2.436	2.650	2.801	3.012	3.372	4.221
14	2.264	2.415	2.624	2.771	2.977	3.326	4.410
15	2.249	2.397	2.602	2.746	2.947	3.286	4.073
16	2.235	2.382	2.583	2.724	2.921	3.252	4.015
17	2.224	2.368	2.567	2.706	2.898	3.222	3.965
18	2.214	2.356	2.552	2.689	2.878	3.197	3.922
19	2.205	2.346	2.539	2.674	2.861	3.174	3.883
20	2.197	2.336	2.528	2.661	2.845	3.153	3.849
21	2.189	2.328	2.518	2.649	2.831	3.135	3.819
22	2.183	2.320	2.508	2.639	2.819	3.119	3.792
23	2.177	2.313	2.500	2.629	2.807	3.104	3.768
24	2.172	2.307	2.492	2.620	2.797	3.091	3.745
25	2.167	2.301	2.485	2.612	2.787	3.078	3.725
26	2.162	2.296	2.479	2.605	2.779	3.067	3.707
27	2.158	2.291	2.473	2.598	2.771	3.057	3.690
28	2.154	2.286	2.467	2.592	2.763	3.047	3.674
29	2.150	2.282	2.462	2.586	2.756	3.038	3.659
30	2.147	2.278	2.457	2.581	2.750	3.030	3.646
40	2.125	2.250	2.423	2.542	2.704	2.971	3.551
60	2.099	2.223	2.390	2.504	2.660	2.915	3.460
120	2.076	2.196	2.358	2.468	2.617	2.860	3.373
∞	2.054	2.170	2.326	2.432	2.576	2.807	3.291

A.5 CRITICAL VALUES OF THE CHI-SQUARED DISTRIBUTION

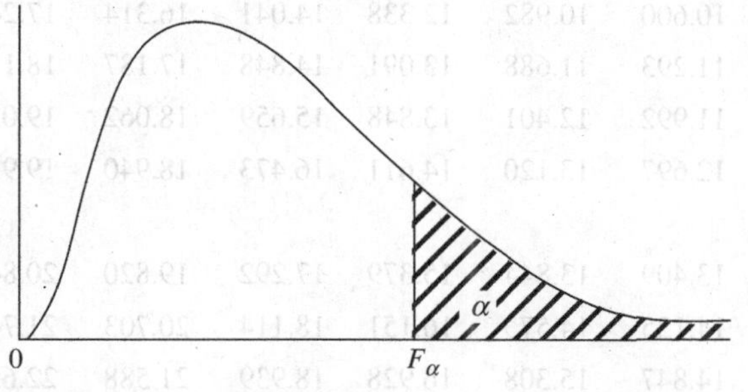

Figure A.3 Critical values of Chi-squared distribution.

Table A.5 Critical values of the Chi-squared Distribution (χ^2_α)

	α									
ν	0.995	0.99	0.98	0.975	0.95	0.90	0.80	0.75	0.75	0.50
1	0.04393	0.03157	0.03628	0.03982	0.00393	0.0158	0.0642	0.102	0.148	0.455
2	0.0100	0.0201	0.0404	0.0506	0.103	0.211	0.446	0.575	0.713	1.386
3	0.0717	0.115	0.185	0.216	0.352	0.584	1.005	1.213	1.424	2.366
4	0.207	0.297	0.429	0.484	0.711	1.064	1.649	1.923	2.195	3.357
5	0.412	0.554	0.752	0.831	1.145	1.610	2.343	2.675	3.000	4.351
6	0.676	0.872	1.134	1.237	1.635	2.204	3.070	3.455	3.828	5.348
7	0.989	1.239	1.564	1.690	2.167	2.833	3.822	4.255	4.671	6.346
8	1.344	1.646	2.032	2.180	2.733	3.490	4.594	5.071	5.527	7.344
9	1.735	2.088	2.532	2.700	3.325	4.168	5.380	5.899	6.393	8.343
10	2.156	2.558	3.059	3.247	3.940	4.865	6.179	6.737	7.267	9.342
11	2.603	3.053	3.609	3.816	4.575	5.578	6.989	7.584	8.148	10.341
12	3.074	3.571	4.178	4.404	5.226	6.304	7.807	8.438	9.034	11.340
13	3.565	4.107	4.765	5.009	5.892	7.042	8.634	9.299	9.926	12.340
14	4.075	4.660	5.368	5.629	6.571	7.790	9.467	10.165	10.821	13.339
15	4.601	5.229	5.985	6.262	7.261	8.547	10.307	11.036	11.721	14.339
16	5.142	5.812	6.614	6.908	7.962	9.312	11.152	11.912	12.624	15.338
17	5.697	6.408	7.255	7.564	8.672	10.085	12.002	12.792	13.531	16.338
18	6.265	7.015	7.906	8.231	9.390	10.865	12.857	13.675	14.440	17.338
19	6.844	7.633	8.567	8.907	10.117	11.651	13.716	14.562	15.352	18.338
20	7.434	8.260	9.237	9.591	10.851	12.443	14.578	15.452	16.266	19.337

(Contd.)

21	8.034	8.897	9.915	10.283	11.591	13.240	15.445	16.344	17.182	20.337
22	8.643	9.542	10.600	10.982	12.338	14.041	16.314	17.240	18.101	21.337
23	9.260	10.196	11.293	11.688	13.091	14.848	17.187	18.137	19.021	22.337
24	9.886	10.856	11.992	12.401	13.848	15.659	18.062	19.037	19.943	23.337
25	10.520	11.524	12.697	13.120	14.611	16.473	18.940	19.939	20.867	24.337
26	11.160	12.198	13.409	13.844	15.379	17.292	19.820	20.843	21.792	25.336
27	11.808	12.879	14.125	14.573	16.151	18.114	20.703	21.749	22.719	26.336
28	12.461	13.565	14.847	15.308	16.928	18.939	21.588	22.657	23.647	27.336
29	13.121	14.256	15.574	16.047	17.708	19.768	22.475	23.567	24.577	28.336
30	13.787	14.953	16.306	16.791	18.493	20.599	23.364	24.478	25.508	29.336

	α									
ν	0.30	0.25	0.20	0.10	0.05	0.025	0.02	0.01	0.005	0.001
1	1.074	1.323	1.642	2.706	3.841	5.024	5.412	6.635	7.879	10.827
2	2.408	2.773	3.219	4.605	5.991	7.378	7.824	9.210	10.597	13.815
3	3.665	4.108	4.642	6.251	7.815	9.348	9.837	11.345	12.838	16.268
4	4.878	5.385	5.989	7.779	9.488	11.143	11.668	13.277	14.860	18.465
5	6.064	6.626	7.289	9.236	11.070	12.832	13.388	15.086	16.750	20.517
6	7.231	7.841	8.558	10.645	12.592	14.449	15.033	16.812	18.548	22.457
7	8.383	9.037	9.803	12.017	14.067	16.013	16.622	18.475	20.278	24.322
8	9.524	10.219	11.030	13.362	15.507	17.535	18.168	20.090	21.955	26.125
9	10.656	11.389	12.242	14.684	16.919	19.023	19.679	21.666	23.589	27.877
10	11.781	12.549	13.442	15.987	18.307	20.483	21.161	23.209	25.188	29.588
11	12.899	13.701	14.631	17.275	19.675	21.920	22.618	24.725	26.757	31.264
12	14.011	14.845	15.812	18.549	21.026	23.337	24.054	26.217	28.300	32.909
13	15.119	15.984	16.985	19.812	22.362	24.736	25.472	27.688	29.819	34.528
14	16.222	17.117	18.151	21.064	23.685	26.119	26.873	29.141	31.319	36.123
15	17.322	18.245	19.311	22.307	24.996	27.488	28.259	30.578	32.801	37.697
16	18.418	19.369	20.465	23.542	26.296	28.845	29.633	32.000	34.267	39.252
17	19.511	20.489	21.615	24.769	27.587	30.191	30.995	33.409	35.718	40.790
18	20.601	21.605	22.760	25.989	28.869	31.526	32.346	34.805	37.156	42.312
19	21.689	22.718	23.900	27.204	30.144	32.852	33.687	36.191	38.582	43.820
20	22.775	23.828	25.038	28.412	31.410	34.170	35.020	37.566	39.997	45.315

21	23.858	24.935	26.171	29.615	32.671	35.479	36.343	38.932	41.401	46.797
22	24.939	26.039	27.301	30.813	33.924	36.781	37.659	40.289	42.796	48.268
23	26.018	27.141	28.429	32.007	35.172	38.076	38.968	41.638	44.181	49.728
24	27.096	28.241	29.553	33.196	36.415	39.365	40.270	42.980	45.558	51.179
25	28.172	29.339	30.675	34.382	37.652	40.646	41.566	44.314	46.928	52.620
26	29.246	30.434	31.795	35.563	38.885	41.923	42.856	45.642	48.290	54.052
27	30.319	31.528	32.912	36.741	40.113	43.194	44.140	46.963	49.645	55.476
28	31.391	32.620	34.027	37.916	41.337	44.461	45.419	48.278	50.993	56.893
29	32.461	33.711	35.139	39.087	42.557	45.722	46.693	49.588	52.336	58.302
30	33.530	34.800	36.250	40.256	43.773	46.979	47.962	50.892	53.672	59.703

A.6 Critical Values of the *F*-Distribution

Table A.6 Critical values of $F_{0.01}(\upsilon 1, \upsilon 2)$

	ν_1								
ν_2	1	2	3	4	5	6	7	8	9
1	161.4	199.5	215.7	224.6	230.2	234.0	236.8	238.9	240.5
2	18.51	19.00	19.16	19.25	19.30	19.33	19.35	19.37	19.38
3	10.13	9.55	9.28	9.12	9.01	8.94	8.89	8.85	8.81
4	7.71	6.94	6.59	6.39	6.26	6.16	6.09	6.04	6.00
5	6.61	5.79	5.41	5.19	5.05	4.95	4.88	4.82	4.77
6	5.99	5.14	4.76	4.53	4.39	4.28	4.21	4.15	4.10
7	5.59	4.74	4.35	4.12	3.97	3.87	3.79	3.73	3.68
8	5.32	4.46	4.07	3.84	3.69	3.58	3.50	3.44	3.39
9	5.12	4.26	3.86	3.63	3.48	3.37	3.29	3.23	3.18
10	4.96	4.10	3.71	3.48	3.33	3.22	3.14	3.07	3.02
11	4.84	3.98	3.59	3.36	3.20	3.09	3.01	2.95	2.90
12	4.75	3.89	3.49	3.26	3.11	3.00	2.91	2.85	2.80
13	4.67	3.81	3.41	3.18	3.03	2.92	2.83	2.77	2.71
14	4.60	3.74	3.34	3.11	2.96	2.85	2.76	2.70	2.65
15	4.54	3.68	3.29	3.06	2.90	2.79	2.71	2.64	2.59
16	4.49	3.63	3.24	3.01	2.85	2.74	2.66	2.59	2.54
17	4.45	3.59	3.20	2.96	2.81	2.70	2.61	2.55	2.49
18	4.41	3.55	3.16	2.93	2.77	2.66	2.58	2.51	2.46

(Contd.)

19	4.38	3.52	3.13	2.90	2.74	2.63	2.54	2.48	2.42
20	4.35	3.49	3.10	2.87	2.71	2.60	2.51	2.45	2.39
21	4.32	3.47	3.07	2.84	2.68	2.57	2.49	2.42	2.37
22	4.30	3.44	3.05	2.82	2.66	2.55	2.46	2.40	2.34
23	4.28	3.42	3.03	2.80	2.64	2.53	2.44	2.37	2.32
24	4.26	3.40	3.01	2.78	2.62	2.51	2.42	2.36	2.30
25	4.24	3.39	2.99	2.76	2.60	2.49	2.40	2.34	2.28
26	4.23	3.37	2.98	2.74	2.59	2.47	2.39	2.32	2.27
27	4.21	3.35	2.96	2.73	2.57	2.46	2.37	2.31	2.25
28	4.20	3.34	2.95	2.71	2.56	2.45	2.36	2.29	2.24
29	4.18	3.33	2.92	2.69	2.53	2.42	2.33	2.27	2.21
30	4.17	3.32	2.92	2.69	2.53	2.42	2.33	2.27	2.21
40	4.08	3.23	2.84	2.61	2.45	2.34	2.25	2.18	2.12
60	4.00	3.15	2.76	2.53	2.37	2.25	2.17	2.10	2.04
120	3.92	3.07	2.68	2.45	2.29	2.17	2.09	2.02	1.96
∞	3.84	3.00	2.60	2.37	2.21	2.10	2.01	1.94	1.88

	ν_1									
ν_2	10	12	15	20	24	30	40	60	120	
1	241.9	243.9	245.9	248.0	249.1	250.1	251.1	252.2	253.3	254.3
2	19.40	19.41	19.43	19.45	19.45	19.46	19.47	19.48	19.49	19.50
3	8.79	8.74	8.70	8.66	8.64	8.62	8.59	8.57	8.55	8.53
4	5.96	5.91	5.86	5.80	5.77	5.75	5.72	5.69	5.66	5.63
5	4.74	4.68	4.62	4.56	4.53	4.50	4.46	4.43	4.40	4.36
6	4.06	4.00	3.94	3.87	3.84	3.81	3.77	3.74	3.70	3.67
7	3.64	3.57	3.51	3.44	3.41	3.38	3.34	3.30	3.27	3.23
8	3.35	3.28	3.22	3.15	3.12	3.08	3.04	3.01	2.97	2.93
9	3.14	3.07	3.01	2.94	2.90	2.86	2.83	2.79	2.75	2.71
10	2.98	2.91	2.85	2.77	2.74	2.70	2.66	2.62	2.58	2.54
11	2.85	2.79	2.72	2.65	2.61	2.57	2.53	2.49	2.45	2.40
12	2.75	2.69	2.62	2.54	2.51	2.47	2.43	2.38	2.34	2.30
13	2.67	2.60	2.53	2.46	2.42	2.38	2.34	2.30	2.25	2.21

14	2.60	2.53	2.46	2.39	2.35	2.31	2.27	2.22	2.18	2.13
15	2.54	2.48	2.40	2.33	2.29	2.25	2.20	2.16	2.11	2.07
16	2.49	2.42	2.35	2.28	2.24	2.19	2.15	2.11	2.06	2.01
17	2.45	2.38	2.31	2.23	2.19	2.15	2.10	2.06	2.01	1.96
18	2.41	2.34	2.27	2.19	2.15	2.11	2.06	2.02	1.97	1.92
19	2.38	2.31	2.23	2.16	2.11	2.07	2.03	1.98	1.93	1.88
20	2.35	2.28	2.20	2.12	2.08	2.04	1.99	1.95	1.90	1.84
21	2.32	2.25	2.18	2.10	2.05	2.01	1.96	1.92	1.87	1.81
22	2.30	2.23	2.15	2.07	2.03	1.98	1.94	1.89	1.84	1.78
23	2.27	2.20	2.13	2.05	2.01	1.96	1.91	1.86	1.81	1.76
24	2.25	2.18	2.11	2.03	1.98	1.94	1.89	1.84	1.79	1.73
25	2.24	2.16	2.09	2.01	1.96	1.92	1.87	1.82	1.77	1.71
26	2.22	2.15	2.07	1.99	1.95	1.90	1.85	1.80	1.75	1.69
27	2.20	2.13	2.06	1.97	1.93	1.88	1.84	1.79	1.73	1.67
28	2.19	2.12	2.04	1.96	1.91	1.87	1.82	1.77	1.71	1.65
29	2.18	2.10	2.03	1.94	1.90	1.85	1.81	1.75	1.70	1.64
30	2.16	2.09	2.01	1.93	1.89	1.84	1.79	14.75	1.68	1.62
40	2.08	2.00	1.92	1.84	1.79	1.74	1.69	1.64	1.58	1.51
60	1.99	1.92	1.84	1.75	1.70	1.65	1.59	1.53	1.47	1.39
120	1.91	1.83	1.75	1.66	1.61	1.55	1.50	1.43	1.35	1.25
∞	1.83	1.75	1.67	1.57	1.52	1.46	1.39	1.32	1.22	1.00

Table A.7 Critical values of $F_{0.05}(v_1, v_2)$

	v_1								
v_2	1	2	3	4	5	6	7	8	9
1	4052	4999.5	5403	5625	5764	5859	5928	5981	6022
2	98.50	99.00	99.17	99.25	99.30	99.33	99.36	99.37	99.39
3	34.12	30.82	29.46	28.71	28.24	27.91	27.67	27.49	27.35
4	21.20	18.00	16.69	15.98	15.52	15.21	14.98	14.80	14.66
5	16.26	13.27	12.06	11.39	10.97	10.67	10.46	10.29	10.16
6	13.75	10.92	9.78	9.15	8.75	8.47	8.26	8.10	7.98
7	12.25	9.55	8.45	7.85	7.46	7.19	6.99	6.84	6.72
8	11.26	8.65	7.59	7.01	6.63	6.37	6.18	6.03	5.91

(Contd.)

9	10.56	8.02	6.99	6.42	6.06	5.80	5.61	5.47	5.35
10	10.04	7.56	6.55	5.99	5.64	5.39	5.20	5.06	4.94
11	9.65	7.21	6.22	5.67	5.32	5.07	4.89	4.74	4.63
12	9.33	6.93	5.95	5.41	5.06	4.82	4.64	4.50	4.39
13	9.07	6.70	5.74	5.21	4.86	4.62	4.44	4.30	4.19
14	8.86	6.51	5.56	5.04	4.69	4.46	4.28	4.14	4.03
15	8.68	6.36	5.42	4.89	4.56	4.32	4.14	4.00	3.89
16	8.53	6.23	5.29	4.77	4.44	4.20	4.03	3.89	3.78
17	8.40	6.11	5.18	4.67	4.34	4.10	3.93	3.79	3.68
18	8.29	6.01	5.09	4.58	4.25	4.01	3.84	3.71	3.60
19	8.18	5.93	5.01	4.50	4.17	3.94	3.77	3.63	3.52
20	8.10	5.85	4.94	4.43	4.10	3.87	3.70	3.56	3.46
21	8.02	5.78	4.87	4.37	4.04	3.81	3.64	3.51	3.40
22	7.95	5.72	4.82	4.31	3.99	3.76	3.59	3.45	3.35
23	7.88	5.66	4.76	4.26	3.94	3.71	3.54	3.41	3.30
24	7.82	5.61	4.72	4.22	3.90	3.67	3.50	3.36	3.26
25	7.77	5.57	4.68	4.18	3.85	3.63	3.46	3.32	3.22
26	7.72	5.53	4.64	4.14	3.82	3.59	3.42	3.29	3.18
27	7.68	5.49	4.60	4.11	3.78	3.56	3.39	3.26	3.15
28	7.64	5.45	4.57	4.07	3.75	3.53	3.36	3.23	3.12
29	7.60	5.42	4.54	4.04	3.73	3.50	3.33	3.20	3.09
30	7.56	5.39	4.51	4.02	3.70	3.47	3.30	3.17	3.07
40	7.31	5.18	4.31	3.83	3.51	3.29	3.12	2.99	2.89
60	7.08	4.98	4.13	3.65	3.34	3.12	2.95	2.82	2.72
120	6.85	4.79	3.95	3.48	3.17	2.96	3.79	2.66	2.56
∞	6.63	4.61	3.78	3.32	3.02	2.80	2.64	2.51	2.41

	ν_1									
ν_2	10	12	15	20	24	30	40	60	120	
1	6056	6106	6157	6209	6235	6261	6287	6313	6339	6366
2	99.40	99.42	99.43	99.45	99.46	99.47	99.47	99.48	99.49	99.50
3	27.23	27.05	26.87	26.69	26.60	26.50	26.41	26.32	26.22	26.13

4	14.55	14.37	14.20	14.02	13.93	13.84	13.75	13.65	13.56	13.46
5	10.05	9.89	9.72	9.55	9.47	9.38	9.29	9.20	9.11	9.02
6	7.87	7.72	7.56	7.40	7.31	7.23	7.14	7.06	6.97	6.88
7	6.62	6.47	6.31	6.16	6.07	5.99	5.91	5.82	5.74	5.65
8	5.81	5.67	5.52	5.36	5.28	5.20	5.12	5.03	4.95	4.86
9	5.26	5.11	4.96	4.81	4.73	4.65	4.57	4.48	4.40	4.31
10	4.85	4.71	4.56	4.41	4.33	4.25	4.17	4.08	4.00	3.91
11	4.54	4.40	4.25	4.10	4.02	3.94	3.86	3.78	3.69	3.60
12	4.30	4.16	4.01	3.86	3.78	3.70	3.62	3.54	3.45	3.36
13	4.10	3.96	3.82	3.66	3.59	3.51	3.43	3.34	3.25	3.17
14	3.94	3.80	3.66	3.51	3.43	3.35	3.27	3.18	3.09	3.00
15	3.80	3.67	3.52	3.37	3.29	3.21	3.13	3.05	2.96	2.87
16	3.69	3.55	3.41	3.26	3.18	3.10	3.02	2.93	2.84	2.75
17	3.59	3.46	3.31	3.16	3.08	3.00	2.92	2.83	2.75	2.65
18	3.51	3.37	3.23	3.08	3.00	2.92	2.84	2.75	2.66	2.57
19	3.43	3.30	3.15	3.00	2.92	2.84	2.76	2.67	2.58	2.49
20	3.37	3.23	3.09	2.94	2.86	2.78	2.69	2.61	2.52	2.42
21	3.31	3.17	3.03	2.88	2.80	2.72	2.64	2.55	2.46	2.36
22	3.26	3.12	2.98	2.83	2.75	2.67	2.58	2.50	2.40	2.31
23	3.21	3.07	2.93	2.78	2.70	2.62	2.54	2.45	2.35	2.26
24	3.17	3.03	2.89	2.74	2.66	2.58	2.49	2.40	2.31	2.21
25	3.13	2.99	2.85	2.70	2.62	2.54	2.45	2.36	2.27	2.17
26	3.09	2.96	2.81	2.66	2.58	2.50	2.42	2.33	2.23	2.13
27	3.06	2.93	2.78	2.63	2.55	2.47	2.38	2.29	2.20	2.10
28	3.03	2.90	2.75	2.60	2.52	2.44	2.35	2.26	2.17	2.06
29	3.00	2.87	2.73	2.57	2.49	2.41	2.33	2.23	2.14	2.03
30	2.98	2.84	2.70	2.55	2.47	2.39	2.30	2.21	2.11	2.01
40	2.80	2.66	2.52	2.37	2.29	2.20	2.11	2.02	1.92	1.80
60	2.63	2.50	2.35	2.20	2.12	2.03	1.94	1.84	1.73	1.60
120	2.47	2.34	2.19	2.03	1.95	1.86	1.76	1.66	1.53	1.38
∞	2.32	2.18	2.04	1.88	1.79	1.70	1.59	1.47	1.32	1.00

(Contd.)

A.7 FISHER'S Z-TRANSFORMATION

Table A.8 Critical values of $Z = \frac{1}{2}\ln\left(\frac{1+r}{1-r}\right)$

r	0.00	0.01	0.02	0.03	0.04	0.05	0.06	0.07	0.08	0.09
0.0	0.000	0.010	0.020	0.030	0.040	0.050	0.060	0.070	0.080	0.090
0.1	0.100	0.110	0.121	0.131	0.141	0.151	0.161	0.172	0.182	0.192
0.2	0.203	0.213	0.224	0.234	0.245	0.255	0.266	0.277	0.288	0.299
0.3	0.310	0.321	0.332	0.343	0.354	0.365	0.377	0.388	0.400	0.412
0.4	0.424	0.436	0.448	0.460	0.472	0.485	0.497	0.510	0.523	0.536
0.5	0.549	0.563	0.576	0.590	0.604	0.618	0.633	0.648	0.662	0.678
0.6	0.693	0.709	0.725	0.741	0.758	0.775	0.793	0.811	0.829	0.848
0.7	0.867	0.887	0.908	0.929	0.950	0.973	0.996	1.020	1.045	1.071
0.8	1.099	1.127	1.157	1.188	1.221	1.256	1.293	1.333	1.376	1.422
0.9	1.472	1.528	1.589	1.658	1.738	1.832	1.946	2.092	2.298	2.647

Note: For negative values of r, put a minus sign in front of the corresponding Z's, and vice versa.

Appendix B: Basic Results

B.1 Exponential Function e^x

$$e = 2.71828\ 18284 \text{ (Figure B.1)}$$

$$e^x\, e^y = e^{x+y}$$

$$\frac{e^x}{e^y} - e^{x-y}$$

$$(e^x)^y = e^{xy}$$

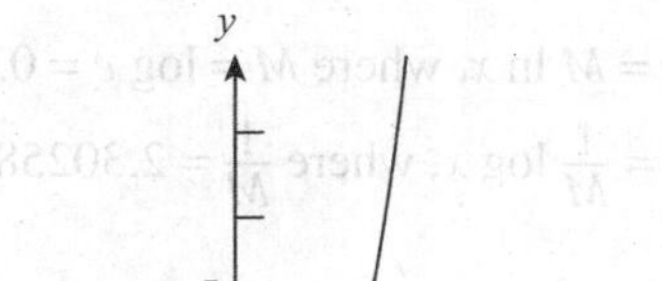

Figure B.1 Exponential function e^x.

B.2 Logarithm

B.2.1 Natural Logarithm

$\ln x$ is the inverse of e^x and has base e (Figure B.2) and

$$e^{\ln x} = x$$

$$e^{-\ln x} = e^{\ln \frac{1}{x}} = \frac{1}{x}$$

$$\ln(xy) = \ln x + \ln y$$

$$\ln\left(\frac{1}{x}\right) = \ln x - \ln y$$

$$\ln(x^a) = a \ln x$$

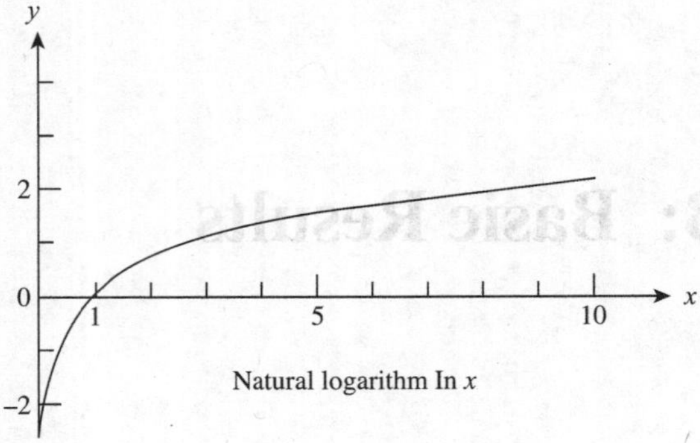

Figure B.2 Natural logarithm (ln *x*).

B.2.2 Common Logarithm: $\log_{10} x$ or Simply $\log x$

$\log x$ is the inverse of 10^x and

$$10^{\log x} = x$$

$$10^{-\log x} = \frac{1}{x}$$

$$\log x = M \ln x, \text{ where } M = \log e = 0.43429$$

$$\ln x = \frac{1}{M} \log x, \text{ where } \frac{1}{M} = 2.30258$$

B.3 Trigonometric Functions

B.3.1 Sine and Cosine Functions

$\sin x$ is odd.

$$\sin(-x) = -\sin x$$

$$\cos(-x) = \cos x$$

Note Angles are measured in real numbers corresponding to radians in calculus, so that $\sin x$ (Figure B.3) and $\cos x$ (Figure B.4) have period 2π.

$$\sin^2 x + \cos^2 x = 1$$

$$\sin(x \pm y) = \sin x \cos y \pm \cos x \sin y$$

$$\cos(x \pm y) = \cos x \cos y \mp \sin x \sin y$$

$$\sin 2x = 2 \sin x \cos x = 2 \tan x/(1 + \tan^2 x)$$

$$\cos 2x = \cos^2 x - \sin^2 x = 1 - 2\sin^2 x$$

$$= 2\cos^2 x - 1 = \frac{1 - \tan^2 x}{1 + \tan^2 x}$$

$$\sin x = \cos\left(x - \frac{\pi}{2}\right) = \cos\left(\frac{\pi}{2} - x\right)$$

$$\cos x = \sin\left(x + \frac{\pi}{2}\right) = \sin\left(\frac{\pi}{2} - x\right)$$

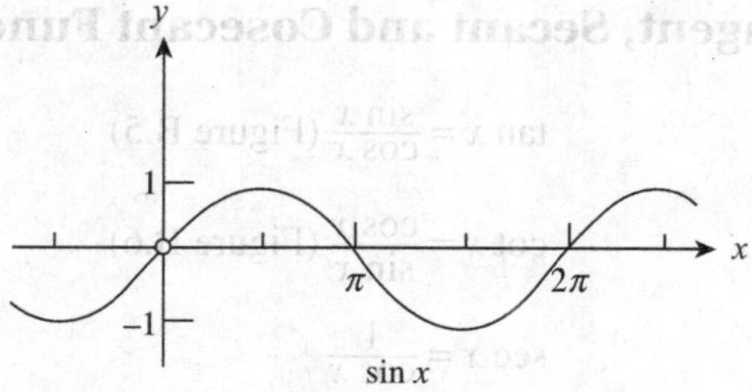

Figure B.3 Sin x.

$$\sin(\pi - x) = \sin x, \ \cos(\pi - x) = -\cos x$$

$$\cos^2 x = \frac{1}{2}(1 + \cos 2x)$$

$$\sin^2 x = \frac{1}{2}(1 - \cos 2x)$$

$$\sin x \sin y = \frac{1}{2}[-\cos(x+y) + \cos(x-y)]$$

$$\cos x \cos y = \frac{1}{2}[\cos(x+y) + \cos(x-y)]$$

$$\sin x \cos y = \frac{1}{2}[\sin(x+y) + \sin(x-y)]$$

$$\cos x \sin y = \frac{1}{2}[\sin(x+y) - \sin(x-y)]$$

$$\sin u + \sin v = 2\sin\frac{u+v}{2}\cos\frac{u-v}{2}$$

$$\sin u - \sin v = 2\cos\frac{u+v}{2}\sin\frac{u-v}{2}$$

$$\cos u + \cos v = 2\cos\frac{u+v}{2}\cos\frac{u-v}{2}$$

$$\cos u - \cos v = -2\sin\frac{u+v}{2}\sin\frac{u-v}{2}$$

$$A\cos x + B\sin x = \sqrt{A^2 + B^2}\cos(x \pm \delta)$$

$$\tan\delta = \frac{\sin\delta}{\cos\delta} \pm \frac{B}{A}$$

$$A\cos x + B\sin x = \sqrt{A^2 + B^2}\sin(x \pm \delta)$$

$$\tan\delta = \frac{\sin\delta}{\cos\delta} \pm \frac{A}{B}$$

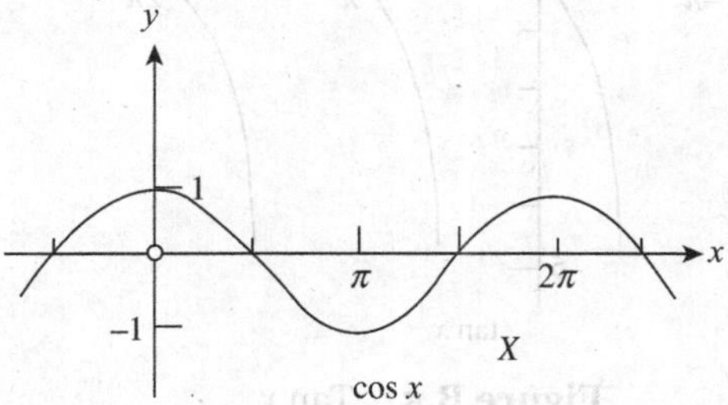

Figure B.4 Cos x.

B.3.2 Tangent, Cotangent, Secant and Cosecant Functions

$$\tan x = \frac{\sin x}{\cos x} \text{ (Figure B.5)}$$

$$\cot x = \frac{\cos x}{\sin x} \text{ (Figure B.6)}$$

$$\sec x = \frac{1}{\cos x}$$

$$\csc x = \frac{1}{\sin x}$$

$$\tan (x \pm y) = \frac{\tan x \pm \tan y}{1 \pm \tan x \tan y}$$

$$\tan 2x = \frac{2 \tan x}{1 - \tan^2 x}$$

Any trigonometric ratio of $n \times 90° \pm \theta$

$= \pm$ same ratio of θ when n is even

$= \pm$ co-ratio of θ when n is odd

The sign + or – is to be decided from the quadrant in which $n \times 90° \pm \theta$ lies.

Examples

$$\sin 570° = \sin (6 \times 90° + 30°) = -\sin 30° = \frac{1}{2}$$

$$\tan 315° = \tan (3 \times 90° + 45°) = -\cot 45°$$

In any ΔABC, $\frac{a}{\sin A} = \frac{b}{\sin B} = \frac{c}{\sin C}$ and $\cos C = \frac{a^2 + b^2 - c^2}{2ab}$.

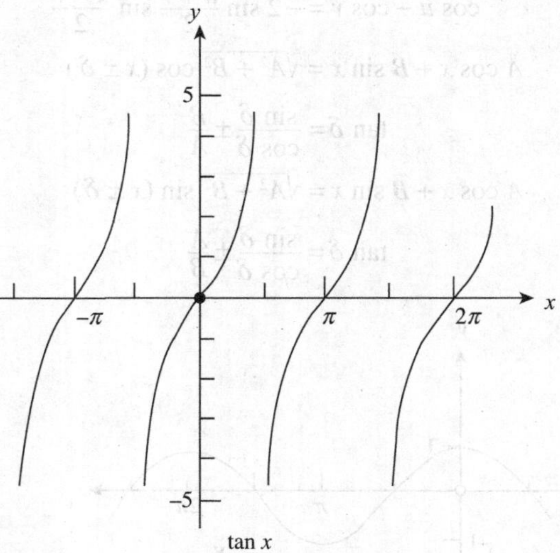

Figure B.5 Tan *x*.

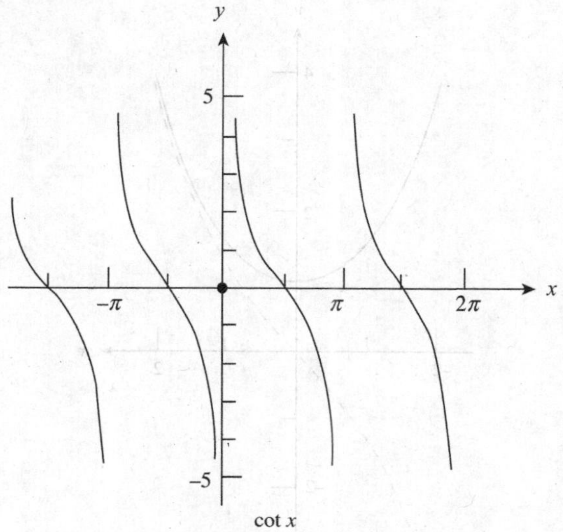

Figure B.6 Cot x.

B.4 Hyperbolic Functions

$$\sinh x = \frac{1}{2}\,(e^x - e^{-x})$$

$$\cosh x = \frac{1}{2}\,(e^x - e^{-x})$$

$$\tanh x = \frac{\sinh x}{\cosh x}$$

$$\coth x = \frac{\cosh x}{\sinh x}$$

$$\cosh x + \sinh x = e^x$$

$$\cosh x - \sinh x = e^{-x}$$

$$\cosh^2 x - \sinh^2 x = 1$$

$$\sinh^2 x = \frac{1}{2}\,(\cosh 2x - 1)$$

$$\cosh^2 x = \frac{1}{2}\,(\cosh 2x + 1)$$

$$\begin{cases} \sinh\,(x \pm y) = \sinh x \cosh y \pm \cosh x \sinh y \\ \cosh\,(x \pm y) = \cosh x \cosh y \pm \sinh x \sinh y \end{cases}$$

$$\tanh\,(x \pm y) = \frac{\tanh x \pm \tanh y}{1 \pm \tanh x \tanh y}$$

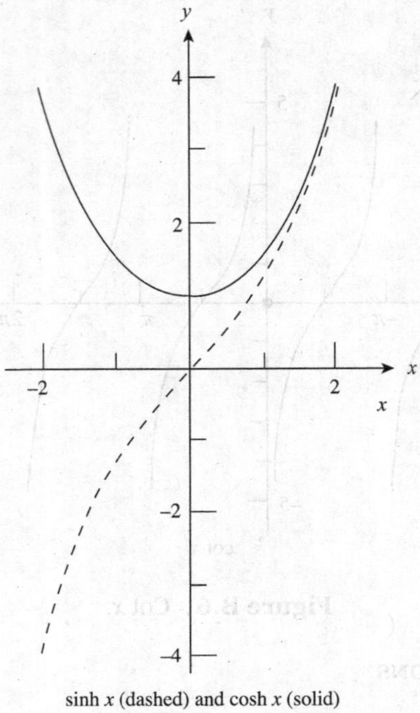

Figure B.7 Sinh x (dashed) and cosh x (solid).

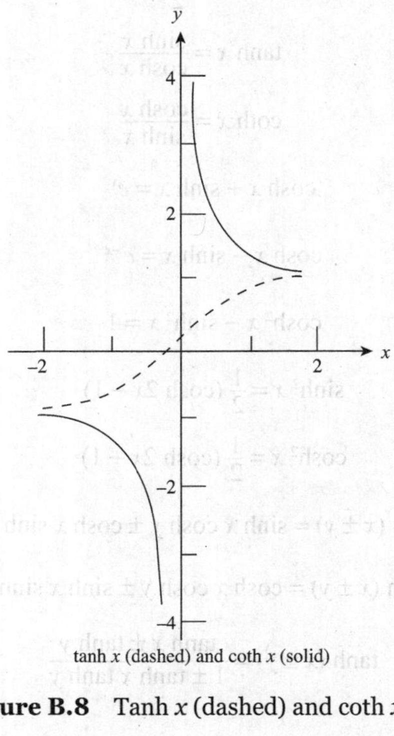

Figure B.8 Tanh x (dashed) and coth x (solid).

B.5 Differentiation

$$\frac{d}{dx}(uv) = u\frac{dv}{dx} + v\frac{du}{dx} \qquad \frac{d}{dx}\left(\frac{u}{v}\right) = \frac{vdu/dx - udv/dx}{v^2}$$

$$\frac{du}{dx} = \frac{du}{dy} \times \frac{dy}{dx} \text{ (Chain rule)} \qquad \frac{d}{dx}(ax+b)^n = n(ax+b)^{n-1}.a$$

$$\frac{d}{dx}(e^x) = e^x \qquad \frac{d}{dx}(a^x) = a^x \log_e a$$

$$\frac{d}{dx}(\log_e x) = \frac{1}{x} \qquad \frac{d}{dx}(\log_a x) = \frac{1}{x}\log_a$$

$$\frac{d}{dx}(\sin x) = \cos x \qquad \frac{d}{dx}(\cos x) = -\sin x$$

$$\frac{d}{dx}(\tan x) = \sec^2 x \qquad \frac{d}{dx}(\cot x) = -\operatorname{cosec}^2 x$$

$$\frac{d}{dx}(\sec x) = \sec x \tan x \qquad \frac{d}{dx}(\operatorname{cosec} x) = -\operatorname{cosec} x \cot x$$

$$\frac{d}{dx}(\sin^{-1} x) = \frac{1}{\sqrt{(1-x^2)}} \qquad \frac{d}{dx}(\cos^{-1} x) = -\frac{-1}{\sqrt{(1-x)^2}}$$

$$\frac{d}{dx}(\sec^{-1} x) = \frac{1}{1+x^2} \qquad \frac{d}{dx}(\cot^{-1} x) = -\frac{-1}{(1-x)^2}$$

$$\frac{d}{dx}(\sec^{-1} x) = \frac{1}{\sqrt{(x^2-1)}} \qquad \frac{d}{dx}(\operatorname{cosec}^{-1} x) = -\frac{-1}{x\sqrt{(x^2-1)}}$$

$$\frac{d}{dx}(\sinh x) = \cosh x \qquad \frac{d}{dx}(\cosh x) = \sinh x$$

$$D^n(e^{mx}) \qquad D^n(a^{mx}) = m^n(\log a)^n . a^{mx}$$

$$D^n(ax+b)^n = m(m-1)(m-2)\ldots(m-n+1)(ax+b)^{m-n}$$

$$D^n \log(ax+b) = (-1)^{n-1}(n-1)!\, a^n/(ax+b)^n$$

$$D^n \sin(ax+b) = a^n \sin(ax+b+n\pi/2)$$

$$D^n \cos(ax+b) = a^n \cos(ax+b+n\pi/2)$$

$$D^n[e^{ax}\sin(bx+c)] = (a^2+b^2)^{n/2} e^{ax}$$
$$\sin(bx+c+n\tan^{-1} b/a)$$

$$D^n[e^{ax}\cos(bx+c)] = (a^2+b^2)^{n/2} e^{ax}$$
$$\cos(bx+c+n\tan^{-1} b/a)$$

$$D^n(uv) = u_n + nC_1 u_{n-1}v + n\,C_2 u_{n-2}v_2$$
$$+\cdots+ n\,C_r u_{n-r} v_r + \cdots + nC_n v_n$$

B.6 Integration

$$\int x^n\,dx = \frac{x^{n+1}}{n+1}\ (n \neq 1) \qquad \int \frac{1}{x}\,dx = \log_e x$$

$$\int e^x\,dx = e^x \qquad \int a^x dx = a^x/\log_e a$$

$$\int \sin x\,dx = -\cos x \qquad \int \cos x\,dx = \sin x$$

$$\int \tan x \, dx = -\log \cos x \qquad \int \cot x \, dx = \log \sin x$$

$$\int \sec x \, dx = \log (\sec x + \tan x) \qquad \int \operatorname{cosec} x \, dx = \log (\operatorname{cosec} x - \cot x)$$

$$\int \sec^2 x dx = \tan x \qquad \int \operatorname{cosec}^2 x dx = -\cot x$$

$$\int \sinh x \, dx = \cosh x \qquad \int \cosh x \, dx = \sinh x$$

$$\int \frac{dx}{a^2 + x^2} = \frac{1}{a} \tan^{-1} \frac{x}{a} \qquad \int \frac{dx}{a^2 - x^2} = \sin^{-1} \frac{x}{a}$$

$$\int \frac{dx}{a^2 - x^2} = \frac{1}{2a} \log \frac{a + x}{a - x} \qquad \int \frac{dx}{a^2 + x^2} = \sinh^{-1} \frac{x}{a}$$

$$\int \frac{dx}{x^2 - a^2} = \frac{1}{2a} \log \frac{x - a}{x + a} \qquad \int \frac{dx}{x^2 - a^2} = \cosh^{-1} \frac{x}{a}$$

$$\int \sqrt{(a^2 - x^2)} \, dx = \frac{x\sqrt{(a^2 - x^2)}}{2} + \frac{a^2}{2} \sin^{-1} \frac{x}{a}$$

$$\int \sqrt{(a^2 + x^2)} \, dx = \frac{x\sqrt{(a^2 + x^2)}}{2} + \frac{a^2}{2} \sinh^{-1} \frac{x}{a}$$

$$\int \sqrt{(x^2 - a^2)} \, dx = \frac{x\sqrt{(x^2 - a^2)}}{2} - \frac{a^2}{2} \cosh^{-1} \frac{x}{a}$$

$$\int e^{ax} \sin bx dx = \frac{e^{ax}}{a^2 + b^2} (a \sin bx - b \cos bx)$$

$$\int_0^\infty e^{-ax} \sin bx dx = \frac{b}{a^2 + b^2}$$

$$\int e^{ax} \cos bx dx = \frac{e^{ax}}{a^2 + b^2} (a \cos bx - b \sin bx)$$

$$\int_0^\infty e^{-ax} \cos bx dx = \frac{b}{a^2 + b^2}$$

$$\int_{-\infty}^{\infty} e^{-x^2} dx = \sqrt{\pi}$$

$$\int_0^\infty \frac{e^{-ax}}{x} \sin bx dx = \tan^{-1} \frac{b}{c}, \; c > 0, b > 0$$

$$\int_0^\infty \sin \frac{a^x}{x} \, dx = \frac{\pi}{2} \text{ if } a > 0$$

$$\int_0^\infty \frac{e^{ax} - e^{-ax}}{e^{-\pi x} + e^{-\pi x}} \, dx = \frac{1}{2} \tan \frac{a}{2}$$

$$\int_0^\infty \frac{e^{ax} - e^{-ax}}{e^{-\pi x} + e^{-\pi x}} \, dx = \frac{1}{2} \sec \frac{a}{2}$$

$$\int_0^{\pi/2} \sin^n x dx = \int_0^{\pi/2} \cos^n x\, dx$$

$$= \frac{(n-1)(n-3)(n-5)...}{n(n-2)(n-4)...}$$

Note RHS is multiplied by $\pi/2$ when n is even.

$$\int_0^{\pi/2} \sin^m x dx \cos^n x dx = \frac{(m-1)(m-3)... \times (n-1)(n-3)...}{(m+n)(m+n-2)(m+n-4)...}$$

Note RHS is multiplied by $\pi/2$ when both n and m are even.

$$\int_{-a}^{a} f(x)dx = \begin{cases} 2\int_0^a f(x)dx, & \text{if } f(x) \text{ is an even function} \\ 0, & \text{if } f(x) \text{ is an odd function.} \end{cases}$$

$$\int_0^{2a} f(x)dx = \begin{cases} 2\int_0^a f(x)dx, & \text{if } f(2a-x) = f(x) \\ 0, & \text{if } f(2a-x) = -f(x). \end{cases}$$

Leibnitz general rule of integration by parts is

$$\int u\, dv = uv - \int v\, du$$

$$\int u(x)\, v(x)dx = uv_1 - u'v_2 + u''v_3 - u'''v_4 + ...$$

Note Superscript $'$ denotes differentiation, i.e., u'' denotes of differentiation of u twice. Subscript number denotes number of times integration of v, i.e., v_3 denotes integration of v thrice.

B.7 Series

B.7.1 Exponential Series

$$e^x = 1 + \frac{x}{1!} + \frac{x^2}{2!} + \frac{x^3}{3!} + ...$$

B.7.2 Sin, Cos, Sinh and Cosh Series

$$\sin x = x - \frac{x^3}{3!} + \frac{x^5}{5!} - ...$$

$$\cos x = 1 - \frac{x^2}{2!} + \frac{x^4}{x!} - \dots$$

$$\sinh x = x + \frac{x^3}{3}! + \frac{x^5}{5!} + \dots$$

$$\cosh x = 1 + \frac{x^2}{2!} + \frac{x^4}{4!} + \dots$$

B.7.3 Log Series

$$\log (1 + x) = x - \frac{x^2}{2} + \frac{x^3}{3} + \dots$$

$$\log (1 - x) = - x + \frac{x^2}{2} + \frac{x^3}{3} + \dots$$

B.7.4 Gregory Series

$$\tan^{-1} x = x - \frac{x^3}{3} + \frac{x^5}{5} - \dots$$

$$\tanh^{-1} x = \frac{1}{2} \log \frac{1 + x}{1 - x} = x + \frac{x^3}{3} + \frac{x^5}{5} + \dots$$

B.7.5 Binomial Series

$$(1 + x)^n = 1 + nx + \frac{n\,(n-1)}{1 \times 2}\, x^2 + \frac{n(n-1)(n-2)}{1 \times 2 \times 3}\, x^3 + \cdots$$

$$(1 + x)^{-1} = 1 - nx + \frac{n\,(n+1)}{1 \times 2}\, x^2 + \frac{n(n+1)(n+2)}{1 \times 2 \times 3}\, x^3 + \cdots$$

$$(1 + x)^{-n} = 1 + nx + \frac{n\,(n+1)}{1 \times 2}\, x^2 + \frac{n\,(n+1)(n+2)}{1 \times 2 \times 3}\, x^3 + \cdots$$

B.8 Progressions

1. Numbers a, $a + d$, $a + 2d$, … are said to be in arithmetic progression (AP). Its nth term $T_n = a + \overline{n - 1}d$ and sum $S_n = \frac{n}{2}(2a + \overline{n - 1}d)$.
2. Numbers a, ar, ar^2, … are said to be in geometric progression (GP). Its nth term $T_n = ar^{n-1}$ and sum $S_n = \frac{a(1 - r^n)}{1 - r}$, $\lim_{n\to\infty} S_n = \frac{a}{1 - r}$ when $|r| < 1$.
3. Numbers a_1, a_2, a_3, … are said to be in harmonic progression (HP) if $1/a_1$, $1/a_2$, $1/a_3$, … are in AP.
4. For any two numbers a and b, their

   Arithmetic mean $= \frac{1}{2}(a + b)$

Geometric mean = $\sqrt{ab}$

Harmonic mean = $\dfrac{2ab}{(a+b)}$

5. For the first n natural numbers 1, 2, 3, …, n,

$$\sum n = \frac{n(n+1)}{2}$$

$$\sum n^2 = \frac{n(n+1)(2n+1)}{6}$$

$$\sum n^3 = \frac{n^2(n+1)^2}{2^2} = \left[\frac{n(n+1)}{2}\right]^2$$

6. **Stirling's approximation:** When n is large, $n! \sim \sqrt{2\pi n} \times n^n e^{-n}$.

B.9 Permutations and Combinations

$$P(n, r) = {}^nP_r = \frac{n!}{(n-r)!}$$

$$C(n, r) = {}^nC_r = \frac{n!}{r!(n-r)!} = \frac{{}^nP_r}{r!}$$

$$nC_{n-r} = nC_r$$

$$nC_0 = 1 = nC_n$$

B.10 Matrices

$$A^{-1} = \frac{1}{|A|}\,\text{adj}A$$

$$(AB)^{-1} = B^{-1}A^{-1}$$

$$(AB)^T = B^TA^T$$

$$(AT)^{-1} = (A^{-1})^T$$

B.11 Ordinary Differential Equations

B.11.1 First Order Linear Differential Equation

$$y' + p(x)y = q(x)$$

IF = $e^{\int p(x)dx}$, where IF is the integrating factor

By general solution, we get

$$y(\text{IF}) = \int (\text{IF})\, q(x)dx$$

B.11.2 Bessel Equation

$$x^2y'' + xy' + (x^2 - v^2)y = 0$$

B.11.3 Legendre Equation

$$(1 - x)y'' - 2xy' + \lambda y = 0$$

Additional Solved Problems

Example 1

A 10-digit number is formed using the digit from 0 to 9, every digit being used only once. Find the probability that the number is divisible by 4.

Solution The 10 digits can be arranged in 10! ways. Of these 9! will begin with the digit 0. The total number of 10 digit numbers formed is 10! – 9! = 36,28,800 – 3,62,880 = 32,65,920.

A number will be divisible by 4 if the last two digit number is divisible by 4, i.e. if it 04, 08, 12, 16, 20, 24, 28, 32, 36, 40, 48, 52, 56, 60, 64, 68, 72, 76, 80, 84, 92 or 96. Of 10 digit numbers ending with

04 are 8! = 40,320
12 are 8! – 7! = 35,280 (zero is part of 8 digit numbers)
20 are 8! = 40,320
24 are 8! – 7! = 35,280 (zero is part of 8 digit numbers)
32 are 8! – 7! = 35,280 (zero is part of 8 digit numbers)
40 are 8! = 40,320 and so on.

∴ The total number of 10-digit numbers divisible by 4 is

$$= 6 \times 8! + 16(8! - 7!)$$

$$= 6 \times 40{,}320 + 16 \times 35{,}280$$

$$= 2{,}41{,}920 + 5{,}64{,}480 = 8{,}06{,}400$$

The required probability $= \dfrac{8{,}06{,}400}{32{,}65{,}920} - 0.2469.$

Example 2

A and B throw alternatively a pair of dice. One who first throws a total of 9 wins. What are their respective chances of winning if A starts the game?

Solution Sum 9 occurs in the following four cases: (4, 5), (5, 4), (3, 6) and (6, 3)

Total number of trials = 36

The probability p that a sum of 9 occurs is

$$p = \frac{4}{36} = \frac{1}{9}$$

$$\therefore q = 1 - p = 1 - \tfrac{1}{9} = \frac{8}{9}$$

A wins in the following sequence of events:

A gets 9; (or) A loses, B loses and A wins; (or) A loses, B loses, A loses, B loses and A wins and so on.

$$= P(\text{A}) + P(\overline{\text{A}})P(\overline{\text{B}})P(\text{A}) + P(\overline{\text{A}})P(\overline{\text{B}})P(\overline{\text{A}})P(\overline{\text{B}})P(\text{A}) + \ldots$$

$$= \frac{1}{9} + \frac{8}{9} \times \frac{8}{9} \times \frac{1}{9} + \frac{8}{9} \times \frac{8}{9} \times \frac{8}{9} \times \frac{8}{9} \times \frac{1}{9} + \ldots$$

$$= \frac{1}{9}\left[1 + \left(\frac{8}{9}\right)^2 + \left(\frac{8}{9}\right)^4 + \ldots\right] = \frac{1}{9} \times \frac{1}{1 - \left(\frac{8}{9}\right)^2} = \frac{1}{9} \times \frac{1}{1 - \frac{64}{81}} = \frac{1}{9} \times \frac{81}{81 - 64}$$

$$= \frac{9}{17} = 0.5294$$

Example 3

A and B throw alternatively a pair of dice. A wins if he throws 6 before B throws 7; and B wins if he throws 7 before A throws 6. If A wins, show that his chance of winning is 30/61.

Solution Probability of sum 6 occurring with a pair of dice $= \frac{5}{36}$

Number of trials = 36 and favorable = {(1, 5) (2, 4) (3, 3) (4, 2) (5, 1)} = 5

Probability of sum 7 occurring with a pair of dice $= \frac{6}{36}$

Number of trials = 36 and favorable = {(1, 6) (2, 5) (3, 4) (4, 3) (5, 2) (6, 1)} = 6

Probability of A winning = P((A getting 6) or (A not getting 6 and B not getting 7 and A getting) or…)

$$= P(\text{A getting 6}) + P(\text{A not getting 6}) \times P(\text{B not getting 7}) \times P(\text{A getting 6}) + \ldots$$

$$= \frac{5}{36} + \left(1 - \frac{5}{36}\right)\left(1 - \frac{6}{36}\right)\frac{5}{36} + \frac{31}{36} \times \frac{30}{36} \times \frac{31}{36} \times \frac{30}{36} \times \frac{5}{36} + \ldots$$

$$= \frac{5}{36}\,1 + \frac{31 \times 30}{36^2} + \left(\frac{31 \times 30}{36^2}\right)^2 + \ldots = \frac{5}{36} \times \frac{1}{1 - \frac{930}{1296}} = \frac{5}{36} \times \frac{1296}{366} = \frac{30}{61}$$

Example 4

Suppose 5 men out of 100 and 25 women out of 10,000 are colour blind. A colour blind person is chosen at random. What is the probability of the person being a male? Assume that male and female persons to be equal in numbers.

Solution Here $P(M) = 0.5$ and $P(F) = 0.5$. Let X be a colour blind person.

$$P(X|M) = \frac{5}{100} = 0.05 \text{ and } P(X|F) = \frac{25}{100} = 0.0025$$

$$P(X) = P(M) \times P(X|M) + P(F) \times P(X|F)$$

$$= 0.5 \times 0.05 + 0.5 \times 0.0025$$

$$= 0.02625$$

$$P(M|X) = \frac{P(M) \times P(X|M)}{P(X)} = \frac{0.5 \times 0.05}{0.02625} = \frac{0.025}{0.02625} = \frac{2,500}{2,600} = \frac{20}{21} = 0.9524$$

Example 5

Cards are dealt one by one from a well-shuffled pack until an ace appears. Find the probability that exactly n cards are dealt before the ace appears.

Solution

$$\text{Probability of an ace} = \frac{4}{52} = \frac{1}{13}$$

$$P(X = 1) = \frac{1}{13}$$

$$P(X = 2) = \left(1 - \frac{1}{13}\right)\frac{1}{13} = \frac{12}{13}\frac{1}{13}$$

$$P(X = 3) = \left(1 - \frac{1}{13}\right)\left(1 - \frac{1}{13}\right)\frac{1}{13} = \left(\frac{12}{13}\right)^2 \frac{1}{13}$$

$$P(X = n) = \left(\frac{12}{13}\right)^{n-1} \frac{1}{13}$$

Example 6

In a factory, machine A produce 40% of the output and machine B produces 60%. On the average, 9 items in 1000 produced by A are defective and 1 item in 250 produced by B is defective. An item drawn at random from a day's output is defective. What is the probability that it is produced by A or B?

Solution Here $P(A) = 0.4$, $P(B) = 0.6$. Hence $P(X/A) = 0.009$ and $P(X/B) = 0.004$.
Let X denote the event that the product is defective.

$$P(X) = 0.006,\ P(A/X) = \frac{0.4(0.009)}{0.006} = 0.6$$

$$P(B/X) = 1 - 0.6 = 0.4$$

Example 7

A player tosses 3 fair coins. He wins Rs 800 if 3 tails occur, Rs 500 if 2 tails occur, Rs 300 if one tail occurs; on the other hand, he loses Rs 1000 if 3 heads occur. Find the value of the game to the player. Is it favorable to him?

Solution

$$P(TTT) = \frac{1}{8},\ P(HHH) = \frac{1}{8}$$

$$P(2 \text{ tails}) = P(HTT) + P(THT) + P(TTH) = \frac{3}{8}$$

$$P(1 \text{ tail}) = P(THH) + P(HTH) + P(HHT) = \frac{3}{8}$$

Discrete probability distribution

X (number of tails)	0	1	2	3
P(X)	$\frac{1}{8}$	$\frac{3}{8}$	$\frac{3}{8}$	$\frac{1}{8}$

Value $E = \frac{1}{8} \times 800 + \frac{3}{8} \times 500 + \frac{3}{8} \times 300 - \frac{1}{8} \times 1000 = 300$

Since $E > 0$, the game is favorable to the player.

Example 8

A pair of fair dice is tossed. Let X denote the maximum of the numbers appearing, i.e. $X(a, b) = \max(a, b)$ and Y denote the sum of the numbers appearing, i.e. $Y(a, b) = a + b$. Find the variance and standard deviation of X on Y.

Solution

$$X(a, b) = \max(a, b)$$

$$X\{(1, 1)\} = 1 \Rightarrow P(X = 1) = \frac{1}{36}$$

$$X\{(1, 2), (2, 2), (2, 1)\} = 2 \Rightarrow P(X = 2) = \frac{3}{36}$$

$$X\{(1, 3), (2, 3), (3, 3), (3, 2), (3, 1)\} = 3 \Rightarrow P(X = 3) = \frac{5}{36} \text{ and so on.}$$

The discrete probability distribution is

X	1	2	3	4	5	6
P(X)	$\frac{1}{36}$	$\frac{3}{36}$	$\frac{5}{36}$	$\frac{7}{36}$	$\frac{9}{36}$	$\frac{11}{36}$

$$E(X) = \sum x_i P_i = 4.472$$

$$E(X^2) = 21.972$$

$$\sigma_X^2 = 21.972 - (4.472)^2 = 1.973$$

$$\sigma_X = 1.405$$

$$Y(a, b) = a + b$$

$$Y\{(1, 1)\} = 2 \Rightarrow P(Y = 2) = \frac{1}{36} \qquad \because (1, 1) \text{ occurs once}$$

$$Y\{(1, 2), (2, 1)\} = 3 \Rightarrow P(Y = 3) = \frac{2}{36}$$

$$Y\{(1, 3), (3, 1), (2, 2)\} = 4 \Rightarrow P(Y = 4) = \frac{3}{36} \text{ and so on.}$$

Y	2	3	4	5	6	7	8	9	10	11	12
P(Y)	$\frac{1}{36}$	$\frac{2}{26}$	$\frac{3}{36}$	$\frac{4}{36}$	$\frac{5}{36}$	$\frac{6}{36}$	$\frac{5}{36}$	$\frac{4}{36}$	$\frac{3}{36}$	$\frac{2}{36}$	$\frac{1}{36}$

$E(Y) = 7$

$E(Y^2) = 54.833$

$\sigma_Y^2 = 54.333 - 72 = 5.8333$

$\sigma_Y = 2.415$

Example 9

If the chances that 1 of the 10 telephone lines is busy at an instant in 0.3:

(a) What is the chance that 5 of the lines are busy?
(b) What is the most probable number of busy lines and what is the probability of this number?
(c) What is the probability that all the lines are busy?

Solution Let X denote the number of busy lines, the number of lines $n = 10$ and the probability that the line is busy $p = 0.3$ and $q = 1 - p = 1 - 0.3 = 0.7$.

(a) $P(X = 5) = \binom{10}{3} (0.3)^5 (0.7)^5 = 0.1029$

(b) Mean number of busy lines $np = 10 \times 0.3 = 3$

$P(X = 3) = \binom{10}{3}(0.3)^3 (0.7)^7 = 0.2688$

(d) $P(X = 10) = \binom{10}{10}(0.3)^{10} (0.7)^0 = (0.3)^{10}$

Example 10

Assume that 60% of all engineering students are good in English. Determine the probability that among 18 engineering students (a) exactly 10, (b) at least 10, (c) at most 8, (d) at least 2 and (e) at most 9 are good in English.

Solution Probability that an engineering student is good in English $p = 0.6$ and $n = 18$. Let X be the number of students good in English.

X: $b(x; n, p) \Rightarrow P(X = x) = \binom{n}{x}(0.6)^{10} (0.4)^{n-x}$

$P(X = 10) = \binom{18}{10} (0.6)^{10} (0.4)^8 = 0.1734$

$P(X \geq 10) = \sum_{x=0}^{18}(18)(0.6)^x (0.4)^{18-x} = 1 - \sum_{x=0}^{9}\binom{18}{x}(0.6)^x (0.4)^{18-x}$

$P(X \leq 8) = \sum_{x=0}^{10}\binom{18}{x}(0.6)^x (0.4)^{10-x}$

$P(2 \leq X \leq 9) = \sum_{x=2}^{9}\binom{18}{x}(0.6)^x (0.4)^{18-x}$

Example 11

The probability that an entering student will graduate is 0.4. Determine the probability that out of 5 students (a) none, (b) one and (c) at least one will graduate.

Solution Here $n = 5, p = 0.4$ and $q = 1 - p = 1 - 0.4 = 0.6$

Let X be the number of students graduating, then

$$P(X = 0) = (0.6)^5 = 0.07776$$

$$P(X = 1) = \binom{5}{1}(0.4)(0.6)^4 = 0.2592$$

$$P(X \geq 1) = 1 - P(X = 0) = 1 - (0.6)^5 = 092224.$$

Example 12

A population random variable X has a mean 100 and a standard deviation 16. What are the mean and standard deviation of the sample mean for the random samples of size 4 drawn with replacement?

Solution Since the sampling is done with replacement, the population may be considered as infinite.

$$E(\overline{X}) = \mu = 100$$

$$\sigma(\overline{X}) = \frac{\sigma}{\sqrt{n}} = \frac{10}{\sqrt{4}} = 8$$

Example 13

Find the maximum difference that we can expect with probability 0.95 between the means of sizes 10 and 12 from a normal population if their standard deviations are found to be 2 and 3 respectively.

Solution Here x_1 and LOS $= \alpha = 5\%$

$$s_p^2 = \frac{10(2)^2 + 2(3)^2}{10 + 12 - 2} = \frac{148}{20} = 7.4 \Rightarrow \therefore s_p = 2.72$$

Here the sample is small. So, we use Student's t-distribution. The statistics is

$$t = \frac{|\overline{X}_1 - \overline{X}_2| - (\mu_1 - \mu_2)}{s_p\sqrt{\frac{1}{x_1} + \frac{1}{x_2}}}$$

From this, we obtain

$$|\overline{X}_1 - \overline{X}_2| = t_{\alpha/2}\, s_p\sqrt{\frac{1}{x_1} + \frac{1}{x_2}} \text{ with degrees of freedom (dof)} = 20$$

$t_{0.025}$ for dof is 2.086.

Hence $$|\overline{X}_1 - \overline{X}_2| = (2.086) \times (2.72) \times \sqrt{\frac{1}{10} + \frac{1}{12}} = 2.429$$

Example 14

A sample of size 3 is selected at random from a box containing 12 items of which 3 are defective. Let X denote the number of defective items in the sample. Find the expected number $E(X)$ of defective items.

Solution

$$P_0 = P(X=0) = \frac{\binom{3}{0}\binom{9}{3}}{\binom{12}{3}} = \frac{9\times 8\times 7}{12\times 11\times 10} = \frac{21}{35} = 0.3818$$

$$P_1 = P(X=1) = \frac{\binom{3}{0}\binom{9}{3}}{\binom{12}{9}} = \frac{3\times 9\times 8\times 3}{12\times 11\times 10} = \frac{27}{55} = 0.4909$$

$$P_2 = P(X=2) = \frac{\binom{3}{1}\binom{9}{3}}{\binom{12}{3}} = \frac{3\times 9\times 2\times 3}{12\times 11\times 10} = 0.1227$$

$$P_3 = P(X=3) = \frac{\binom{3}{0}\binom{9}{3}}{\binom{12}{3}} = \frac{1}{220} = 0.0045$$

Expectation $E(X) = \sum X_i P_i$

$$= 0(0.3818) + 1(0.4909) + 2(0.1227) + 3(0.0045)$$

$$= 0.7498$$

Example 15

If the probability of a bad reaction from a certain injection is 0.001, determine the chance that out of 3000 individuals more than two will get a bad reaction.

Solution Probability of bad reaction due to injection $p = 0.001$, $x = 3000$ and $\lambda = xp = 3000 \times 0.001$.

$$P(X=x)\ \frac{e^{-\lambda}\lambda^x}{x!} = \frac{e^{13}\, 3^x}{x!} \qquad \text{for } x = 0, 1, 2, \ldots$$

If X is the number of persons who get bad reaction due to injection, then

$$P(X>2) = 1 - P(X \le 2)$$

$$= 1 - P(X=0) - P(X=1) - P(X=2)$$

$$= 1 - e^{-3}\left(1 + \frac{3}{1!} + \frac{3^2}{2!}\right)$$

$$= 1 - e^{-3}(8.5) = 0.5768$$

Example 16

A burglar alarm system has 6 fail-safe components. The probability of each failing is 0.05. Find the probability that (a) exactly 3 will fail, (b) fewer than 2 will fail, (c) none will fail and (d) compare the answers for (a), (b) and (c) and explain why the results are reasonable.

Solution Here $p = 0.05$ and $q = 1 - p = 1 - 0.05 = 0.95$.

Let X be the number of alarm system components, i.e. $X = 6$.

$$\therefore P(X=x) = \binom{6}{x}(0.05)^x(0.95)^{6-x}$$

(a) $P(X=3) = \binom{6}{3}(0.05)^3(0.95)^3 = 0.00214$

(b) $P(X<2) = P(X=0) + P(X=1) = (0.95)^6 + \binom{6}{1}(0.05)(0.95)^5 = 0.9672$

(c) $P(X=0) = (0.95)^6 = 0.7351$

(d) The question is incomplete.

Example 17

Each month, a Hyderabad household generates an average of 28 lb of newspaper for garbage or recycling. Assume the standard deviation is 2 lb of a household is selected at random, find the probability of its generating

(a) Between 27 and 31 lb

(b) More than 30.2 lb/month

Assume the variable is approximately normally distributed.

Solution Let X denote the amount of garbage.
Here $\mu = 28$ and $\sigma = 2$.

$$Z = \frac{X-\mu}{\sigma} = \frac{X-28}{2}$$

(a) $P(27 < X < 31) = P(-0.5 \le 1.5) = P(0 < Z < 0.5) + P(0 < Z < 1.5)$

$$= 0.1916 + 0.4332 = 0.6248$$

(b) $P(X > 30.2) = P(Z > 1.1) = 0.5 - 0.3643 = 0.1357$

Example 18

Let $s = \{1, 5, 6, 8\}$. Find the probability distribution of the sample mean for random samples of size 2 drawn without replacement.

Solution

$$\text{Population mean } \mu = \frac{1+5+6+8}{4} = 5$$

The number of samples of size Z is $\binom{4}{2} = 6$

The samples are (1, 5), (1, 6), (1, 8), (5, 6), (5, 8) and (6, 8)
The corresponding means of the samples are 3, 3.5, 4.5, 5.5, 6.5 and 7
The mean of the sample means

$$\mu_{\bar{X}} = \frac{3\times 1 + 3\times 51 + 4\times 51 + 5\times 51 + 6\times 51 + 7\times 1}{6} = \frac{30}{6} = 5$$

Example 19

Fit a binomial distribution for the following data and compare the theoretical frequencies with the actual ones:

x_i	0	1	2	3	4	5
f_i	2	14	20	34	22	8

Solution Here $x = 5$

$$xp = \overline{X} = \frac{\sum x_i f_i}{\sum f_i} = \frac{0 + 1(14) + 2(20) + 3(34) + 4(22) + 5(8)}{2 + 14 + 20 + 34 + 22 + 8} = \frac{284}{100} = 2.84$$

$$p = \frac{2.84}{5} = 0.560 \text{ and } q = 1 - p = 1 - 0.568 = 0.432$$

$$N = \sum f_i = 100$$

Binomial distribution is

$$P(X = i) = \binom{x}{i} p^i \, 8^{x-i} \text{ for } I = 0, 1, 2, 3, 4 \text{ and } 5$$

$$P(X = 0) = (0.432)^2$$

$$P(X = 1) = 5(0.568)(0.432)^4$$

$$P(X = 2) = 10(0.568)^2(0.432)^3$$

$$P(X = 3) = 10(0.568)^3(0.432)^2$$

$$P(X = 4) = 5(0.568)^4(0.432)$$

$$P(X = 5) = (0.568)^5$$

Comparison

x_i	0	1	2	3	4	5
f_i	2	14	20	34	22	8
Expected frequencies	2	10	26	34	22	6

Example 20

If two independent random samples of $x_1 = 9$ and $x_2 = 16$ are taken from a normal population, what is the probability that the variance of the first sample will be at least 4 times as large as the variance of the second sample.

Solution From the table of F-distribution,

$F_{0.01} = 4$ for $v_1 = x_1 - 1 = 9 - 1 = 8$ dof

and $v_2 = x_2 - 1 = 16 - 1 = 15$ dof

$\therefore$ The required probability = 0.01.

Example 21

Two random samples of sizes $x_1 = 15$ and $x_2 = 25$ are taken from N. Find the probability that the ration of the sample variances does not exceed 2.28.

Solution Here $x_1 = 15$ and $x_2 = 25$. From the table of F-distribution,

we find $F_{0.05}(14, 24) = 2.15(x_1 - 1 = 14, x_2 - 1 = 24)$

So, the required probability = 0.05 approximately.

Example 22

If two independent random samples of sizes $x_1 = 9$ and $x_2 = 16$ are taken from a normal population, what is the probability that the variance of the first sample will be at least 4 times as large as the variance of the second sample?

Solution From the F tables, we have

$F_{0.01} = 4$ for $v_1 = x_1 - 1 = 9 - 1$ dof

and $v_2 = x_2 - 1 = 16 - 1 = 15$ dof

Thus, the required probability = 0.01.

Example 23

Determine the probability that the sample mean area covered by a sample of 40 of 1-L paint boxes will be between 510 and 520 sq. ft given that 1 L of such paint box covers on the average 513.3 sq. ft with standard deviation of 31.5 sq. ft.

Solution Here sample size $x = 40$, standard deviation $\sigma = 31.5$ and mean $\mu = 513.3$.

$$Z = \frac{\overline{X} - \mu}{\sigma} = \frac{\overline{X} - 513.3}{31.5}$$

$$\therefore P(510 < \overline{X} < 520) = P(-0.662 < Z < 1.345)$$

$$= 0.2454 + 0.4115 = 0.6569$$

Example 24

Two independent random samples of sizes 8 and 7 gave variances 4.2 and 3.9 respectively. Do you think that such a difference has probability less than 0.05. Justify your answer.

Solution

Null hypothesis $H_0 : \sigma_1^2 = \sigma_2^2$

Alternative hypothesis $H_1 : \sigma_1^2 \neq \sigma_2^2$

Variances $s_1^2 = 4.2$ and $s_2^2 = 3.9$

$$\therefore F = \frac{s_1^2}{s_2^2} = \frac{4.2}{3.9} = 1.076$$

For $\alpha = 0.05$, $F_{0.05} = 4.21$ with

$v_1 = 8 - 1 = 7$, $v_2 = 7 - 1 = 6$ dof

Accept null hypothesis since

$$F_{cal} = 1.076 < F_{0.05} = 4.21$$

Hence there is no difference in variances at the level of significance = 0.05.

Example 25

Let X be equal to the weight of an unbranded 200 fish fry at a local restaurant in East Greenbrush, New York. Assume that the distribution of X is $N(11, a_2)$. A random sample of $n = 24$ weights are 4.4, 3.8, 5.1, 4.6, 4.5, 4.5, 4.8, 4.1, 3.9, 4,2, 4.4, 4.9, 5.0, 4.3,4.4, 3.6, 5.2, 4.8, 4.4, 4.6, 4.6, 5.0, 4.0 and 4.5.

(a) Find paint estimate of μ, σ^2 and σ.

(b) Find a 95% one-sided confidence interval for μ which gives a lower bound for μ.

Solution

(a) Here $n = 24$, $\overline{X} = 4.4833$, $s^2 = 0.1719$ and $s = 0.4146$.
The paint estimates of μ, σ^2 and σ, are 4.4833, 0.1719 and 0.4146 respectively.

(b) A 95% one-sided confidence interval for μ is

$$\left(\overline{X} - 1.64\left(\frac{\sigma}{\sqrt{n}}\right), \infty\right)$$

$$= \left(4.483 - 1.64\left(\frac{0.4146}{\sqrt{n}}\right), \infty\right) = (4.43445, \infty)$$

Example 26

Independent random samples of the heights of adult males living in two countries yielded the following results:

$n = 12$, $\overline{X} = 65.7$ in. and $s_x = 4$ in.

$m = 15$, $\overline{Y} = 68.2$ in. and $s_y = 3$ in.

Find an approximate 98% confidence interval for the difference $\mu_x - \mu_y$ of the means of the population of heights. Assume $\sigma_x^2 - \sigma_y^2$.

Solution Here

$n = 12$, $\overline{X} = 65.7$ and $s_x = 4$

$m = 15$, $\overline{Y} = 68.2$ and $s_y = 3$

Pooled variance $s_p^2 = \dfrac{12 \times 42 + 15 \times 32}{12 + 15 - 2} = 13.08 \Rightarrow s_p = 3.617$

For $\alpha = 0.02$, $t_{\alpha/2} = 2.485$ with 25 dof.

Then 98% confidence interval for $\mu_x - \mu_y = (\overline{X} - \overline{Y}) \pm t_{\alpha/2}\, s_p \sqrt{\dfrac{1}{n} + \dfrac{1}{m}}$

$$\Rightarrow -5.98 \le \mu_x - \mu_y \le 4.46$$

Example 27

Consider the butterfat production (in pounds) for a cow during a 305-day milk production period following the birth of a calf. Let X and Y be the butterfat production for such cows on a farm in Wisconsin and a farm in Michigan respectively. Twelve observations of X are 649, 657, 714, 877, 975, 468, 567, 849, 721, 791, 874 and 405. Sixteen observations of Y are 699, 891, 632, 815, 589, 764, 524, 727, 597, 868, 652, 978, 479, 733, 549 and 790.

(a) Assuming that X is an $N(\mu_x, \sigma_x^2)$ and Y is an $N(\mu_y, \sigma_y^2)$. Find a 95% confidence interval for $\mu_x - \mu_y$.

(b) Construct box-and-whisker diagrams for these two sets of data on the same graph.

(c) Does there seem to be a significant difference in butterfat production for cows on these two farms?

Solution

(a) Here $n_1 = 12$ and $n_2 = 16$. From the data,

$\overline{X} = 712.25,\ s_x = 173.083$

$\overline{Y} = 705.44,\ s_y = 141.71$

$$\text{Pooled variance } s_p^2 = \frac{11(173.083)^2 + 15(141.71)^2}{12 + 16 - 2} = 24260.032$$

For $\alpha = 0.05$, $t_{0.025} = 2.056$ at 26 dof.

Then 95% confidence interval for $\mu_x - \mu_y$ is $\overline{X} - \overline{Y} \pm t_{\alpha/2} \cdot s_p \sqrt{\frac{1}{x_1} + \frac{1}{x_2}}$ i.e. (−115.48, 129.1)

(b) Out of syllabus.

(c) Null hypothesis H_0: $\mu_x = \mu_y$

Alternative hypothesis H_1: $\mu_x \neq \mu_y$

$$\textbf{Test statistic: } t_{cal} = \frac{\overline{X}\,\overline{Y}}{s_p\sqrt{\frac{1}{x_1} + \frac{1}{x_2}}} = \frac{0.01}{155.76\sqrt{\frac{1}{12} + \frac{1}{16}}} = 0.1145$$

Conclusion: Since $t_{cal} = 0.1145 < t_{0.025} = 2.056$ accept H_0. There is no significant difference in butterfat production.

Example 28

An interior automatic supplier is considering its electrical wire harness to save money. The idea is to replace a current 20-gauge wire with a 22-gauge wire. Since not all wires in the harness can be changed, the new wire must work with the current wire splice process. To determine if the new wire is compatible, random samples were selected and measured with a pull test. The minimum pull force required by the customer is 20 lb. Twenty observations of the forces needed for the current wire are 28.8, 24.4, 30.1, 25.6, 26.4, 23.9, 22.1, 22.5, 27.6, 28.1, 20.8, 27.7, 24.4, 25.1, 24.6, 26.3, 28.2, 22.2, 26.3 and 24.4.

Twenty observations of the forces needed for the new wire are 14.1, 12.2, 14.0, 14.6, 8.5, 12.6, 13.7, 14.8, 14.1, 13.2, 12.1, 11.4, 10.1, 14.2, 13.6, 13.1, 11.9, 14.8, 11.1 and 13.5.

(a) Does the current wire meet the customer's specifications?

(b) Find a 90% confidence interval for the differences of the means for these two sets of wire.

(c) Construct box-and-whisker diagrams of the two sets of data on the same figure.

(d) What is your recommendation for this company?

Solution From the given data, we have

Sample size $= x_1 = 20$, mean $\overline{X}_1 = 25.475$ and standard deviation $s_1 = 2.4935$

Sample size $= x_2 = 20$, mean $\overline{X}_2 = 12.88$ and standard deviation $s_2 = 1.6608$

(a) The minimum pull force required is 20 lb and since the current wire has pull force equal to $\overline{X}_1 = 25.475$, the current wire meets the customer's specifications.

$$s_p^2 = \frac{(x_1 - 1)s_1^2 + (x_2 - 1)s_2^2}{x_1 + x_2 - 2}$$

(b) $\dfrac{19(2.4935)^2 + 19(1.6608)^2}{38} = 4.4879$

$\therefore s_p = 2.1184$

For $\alpha = 0.1$, $t_{\alpha/2} = t_{0.05} = 1.68$ with 38 dof.

So, 90% confidence interval for $\mu_1 - \mu_2$ is

$$\overline{X}_1 - \overline{X}_2 \pm t_{\alpha/2} \cdot s_p\sqrt{\frac{1}{x_1} + \frac{1}{x_2}} = (25.475 - 12.88) \pm (1.68)(2.1184)\sqrt{\frac{1}{20} + \frac{1}{20}}$$

(c) Out of syllabus.

(d) The company is recommended to remain with the use of current wire.

Example 29

A manufacturer of electronic equipment subjects samples of two competing brands of transistors to accelerated performance test of 45 of 180 transistors of the first kind and 34 of 120 transistors of the second kind fails the test. What can be conclude at the level of significance $\alpha = 0.05$ about the difference between the corresponding sample size?

Solution

(I) Sample proportion and sample size

$P_1 = \dfrac{45}{180} = 0.25$ and $x_1 = 180$

(II) Sample proportion and sample size

$P_2 = \dfrac{34}{120} = 0.2833$ and $x_2 = 120$

(1) Null hypothesis H_0: $P_1 = P_2$

(2) Alternative hypothesis H_1: $P_1 \neq P_2$

$$\hat{P} = \frac{X_1P_1 + X_2P_2}{X_1 + X_2} = \frac{180(0.25) + 120(0.2833)}{180 + 120} = 0.263$$

$$\hat{q} = 1 - \hat{p} = 0.737$$

(3) **Test statistic:** $Z_{cal} = \dfrac{P_1 - P_2}{\hat{P}\hat{q}\sqrt{\dfrac{1}{x_1} + \dfrac{1}{x_2}}} = \dfrac{0.25 - 0.283}{0.263 \times 0.737\sqrt{\dfrac{1}{180} + \dfrac{1}{120}}}$

$= -1.458$

(4) **Conclusion:** For $\alpha = 0.05$, $Z_{cal} = -1.458 < Z_{\alpha/2} = Z_{0.025} = 1.96$, accept the null hypothesis H_0.

Example 30

On the basis of their total scores, 200 candidates of a civil services examination are divided into two groups: (1) the upper 30% and (2) the remaining 70% consider the first question of the examination. In the first group, 40 had written the correct answer, whereas in the second group, 80 candidates had written the correct answer. On the basis of these results, can one conclude that the first question is no good at the discriminating ability of the type being examined here?

Solution

Sample size $x_1 = 60$ and proportion $P_1 = \dfrac{40}{60} = 0.6667$

Sample size $x_2 = 140$ and proportion $P_2 = \dfrac{80}{140} = 0.5714$

(1) Null hypothesis H_0: $P_1 = P_2$

(2) Alternative hypothesis H_1: $P_1 \neq P_2$

$$\hat{P} = \frac{x_1P_1 + x_2P_2}{x_1 + x_2} = \frac{60\left(\frac{40}{60}\right) + 140\left(\frac{80}{140}\right)}{60 + 140} = \frac{120}{200} = 0.6$$

$\hat{q} = 1 - \hat{P} = 0.4$

(3) **Test statistic:** $Z_{\text{cal}} = \dfrac{P_1 - P_2}{\hat{P}\,\hat{q}\sqrt{\dfrac{1}{x_1} + \dfrac{1}{x_2}}} = \dfrac{0.6667 - 0.5714}{0.6 \times 0.4\sqrt{\dfrac{1}{60} + \dfrac{1}{40}}} = 1.9657$

(4) **Conclusion:** At 5% LOS, $Z_{\alpha/2} = 1.96$. Since calculated value of Z $(= 1.9657) > Z_{\alpha/2} = 1.96$, reject the null hypothesis H_0. The equation is good at determining the ability.

Example 31

In an air population study, the following amounts of suspended benzene soluble organic matter (in micrograms/cubic meter) were obtained at an experiment station for 8 different samples of air: 2.2, 1.8, 3.1, 2.0, 2.4, 2.0, 2.1 and 1.2. Construct a 0.95 confidence interval for the corresponding time mean.

Solution For $\alpha = 0.05$, we have from tables $Z_{\alpha/2} = 2.447$ with $\nu = 6$ dof

Also, $\bar{X} = 2.1$ and $s^2 = \dfrac{1}{x - 1}\sum (x_i - \bar{X})^2$

$\Rightarrow s = 0.537$

Confidence interval is

$\bar{X} \pm t_{\alpha/2} = \left(\dfrac{s}{\sqrt{x}}\right) = 2.1 \pm 2.447\left(\dfrac{0.537}{\sqrt{8}}\right)$

$\Rightarrow 1.625 < \mu < 2.564$

Example 32

It is desired to test the hypothesis $\mu_0 = 40$ against the alternative hypothesis $\mu_1 = 42$ on the basis of a random sample from a normal population with the standard deviation $\sigma = 4$. If the probability of a Type I error is to be 0.05 and the probability of a Type 2 error is to be 0.24, find the required size of the sample.

Solution Here $\mu_0 = 40$, $\mu_1 = 42$, $\sigma = 4$, $\alpha = 0.05$ and $\beta = 0.24$

So, $Z_\beta = Z_{0.24} = 0.645$ and $Z_{\alpha/2} = Z_{0.025} = 1.96$

(1) Null hypothesis H_0: $\mu = \mu_0 = 40$

(2) Alternative hypothesis H_1: $\mu > \mu_0$

$$\sigma^2 \frac{(Z_{\alpha/2} + Z_\beta)^2}{(\mu_1 - \mu_0)^2}$$

$$\text{Sample size} = x = \frac{16(1.96 + 0.645)^2}{(42 - 40)^2}$$

$$= 4 \times 6.886025 = 27.544$$

$$= 28 \text{ (rounding to higher integer)}$$

Example 33

An oceanographer wants to check whether the average depth of the ocean in a certain region is 57.4 fathoms, as had previously been recorded. What can be concluded at the level of significance $\alpha = 0.05$ if soundings taken at 40 random locations in the given region yielded a mean of 59.1 fathoms with a standard deviation of 5.2 fathoms?

x	1	2	3	4	5	6
y	2.98	4.26	5.21	6.10	6.80	7.50

Solution From the given data, we have sample size $n = 40$, mean $\overline{X} = 59.1$ and standard deviation $s = 5.2$.

x	y	$X = \log x$	$Y = \log y$	X_2	XY
1	2.98	0	1.0919	0	0
2	4.26	0.6931	1.4492	0.4804	1.0044
3	5.21	1.0986	1.6506	1.2069	1.8133
4	6.10	1.3863	1.8083	1.9218	2.5068
5	6.80	1.6094	1.9169	2.5902	3.0851
6	7.50	1.7918	2.0149	3.2105	3.6103
		$\sum X = 6.5792$	$\sum Y = 9.9318$	$\sum X^2 = 9.4098$	$\sum XY = 12.0199$

(1) Null hypothesis H_0: $\mu = 57.4$

(2) Alternative hypothesis H1: $\mu \neq 57.4$

(3) Level of significance $\alpha = 0.05$

(4) **Test statistic:** $Z_{\text{cal}} = \dfrac{\overline{X}-\mu}{\sigma/\sqrt{n}} = \dfrac{59.1-57.4}{5.2/\sqrt{40}}$

(5) **Conclusion:** Since $Z_{\text{cal}} = 2.067 > 1.96 = Z_{\alpha/2}$ for a two-tailed test, reject the null hypothesis. So, we conclude that the average depth of the ocean is not 57.4.

Example 34

Predict y at $x = 3.75$ by fitting a power curve to the following data:

x	1	2	3	4
y	7	11	17	27

Solution Let $y = ax^b$ be the given power curve. Then $\log y = \log a + b \log x$.
Put $\log y = Y$, $\log a = A$, $\log x = X$.

x	y	$Y = \log y$	x^2	XY
1	7	1.9459	1	1.9459
2	11	2.3979	4	4.7958
3	17	2.8332	9	8.4996
4	27	3.2958	16	13.1832
$\sum x = 10$		$\sum Y = 10.4728$	$\sum x^2 = 30$	$\sum XY = 28.4245$

Normal equations are

$$nA + b\sum X = \sum Y \Rightarrow 6A + 6.579b = 9.932 \tag{1}$$

$$A\sum X + b\sum X^2 = \sum XY \Rightarrow 6.579A + 9.410b = 12.020 \tag{2}$$

$$\Delta = \begin{pmatrix} 6 & 6.579 \\ 6.579 & 9.41 \end{pmatrix}$$

$$= 13.1767$$

$$\Delta_A = \begin{pmatrix} 9.932 & 6.579 \\ 12.020 & 9.41 \end{pmatrix} = 14.38$$

$$\Delta_B = \begin{pmatrix} 6 & 9.932 \\ 6.579 & 12.02 \end{pmatrix} = 6.777$$

$$A = \frac{14.38}{13.1767} = 1.06132 \Rightarrow a = e^A = 2.9782$$

$$b = \frac{6.777}{13.1767} = 0.5143$$

The required curve is $y = 2.978\, x^{0.5143} = 2.978 \times (3.75)^{0.5143} = 5.8769.$

Example 35

Fit an exponential curve of the form $y = ae^{bx}$ for the following data:

x	63	50	55	65	55	70	64	70	58	68	52	60
y	87	74	76	90	85	87	92	98	82	91	77	78

Solution Let $y = ab^b$ be the given power curve. Then $\log y = \log a + b \log x$.
Put $\log y = Y$, $\log a = A$, $\log x = X$. Constructing the table of values:

Normal Equations are

$$nA + \left(\sum x\right)b = \sum Y \Rightarrow 4A + 10b = 10.4728 \quad (1)$$

$$\left(\sum x\right)A + \left(\sum x^2\right)b = \sum XY \Rightarrow 10A + 30b = 28.4245 \quad (2)$$

$$\Delta = \begin{pmatrix} 4 & 10 \\ 10 & 30 \end{pmatrix} = 20$$

$$\Delta_A = \begin{pmatrix} 10.4728 & 10 \\ 28.4245 & 30 \end{pmatrix} = 29.939$$

$$\Delta = \begin{pmatrix} 4 & 10.4728 \\ 10 & 28.4245 \end{pmatrix} = 8.79$$

$A = 1.49695 \Rightarrow a = 4.468$

$b = 0.4485$

The required curve is $y = 4.468x^{0.4485}$.

Example 36

Calculate the correlation coefficient r for the following data:

x	12	10	14	11	12	9
y	18	17	23	19	20	15

Solution

Number of data points $N = 12$

Sum of x values $\sum x = 63 + 50 + 55 + 65 + 55 + 70 + 64 + 70 + 58 + 68 + 52 + 60 = 730$

Sum of squares of x values $= \sum x^2 = 63^2 + 50^2 + 55^2 + 65^2 + 55^2 + 70^2 + 64^2 + 70^2 + 64^2 + 58^2 + 68^2 + 52^2 + 60^2$

Sum of y values $\sum y = 87 + 74 + 76 + 90 + 85 + 87 + 92 + 98 + 82 + 91 + 77 + 78 = 1,017$

Sum of squares of y values $= \sum y^2 = 87^2 + 74^2 + 76^2 + 90^2 + 85^2 + 87^2 + 92^2 + 98^2 + 82^2 + 91^2 + 77^2 + 78^2 = 86801$

Sum of the product of x and y values $\sum xy = 63.87 + 50.74 + 55.76 + 65.90 + 55.85 + 70.87 + 64.92 + 70.98 + 58.82 + 68.91 + 52.77 + 60.78 = 62,352$

The coefficient of correlation r is given by

$$r = \frac{n\sum xy - \sum x \sum y}{\left[n\sum x^2 - \left(\sum x\right)^2\right]\left[n\sum y^2 - \left(\sum y\right)^2\right]}$$

$$= \frac{12(62{,}352) - (730)(1{,}017)}{\sqrt{[12(44{,}932) - (730)^2]\,[12(8{,}680) - (1{,}017)^2]}} = 0.857$$

Example 37

Determine the least squares regression line of (a) y on x, (b) x on y, (c) find r using the regression coefficients, (d) find $y(8)$ and (e) find $x(16)$.

x	21	23	30	54	57	58	72	78	87	90
y	60	71	72	83	110	84	100	92	113	135

Solution Here $N = 6$, $\sum x = 68$, $\sum x^2 = 786$, $\sum y = 112$, $\sum y^2 = 2128$ and $\sum xy = 1292$.

(a) Let $y = a + bx$ be the least squares regression line of y on x.

Normal equations are

$Na + b\sum x = \sum y \Rightarrow 6a + 68b = 112$

$a\sum x + b\sum x^2 = \sum xy \Rightarrow 68a + 786b = 1292$

Solving, we get $a = 1.913$ and $b = 1.478$

$\therefore$ The regression line is $y = 1.913 + 1.478x$.

(b) Let $y = c + dx$ be the least squares regression line of x on y. Normal equations are

$Nc + d\sum y = \sum x \Rightarrow 6c + 112d = 68$

$c\sum y + d\sum y^2 = \sum xy \Rightarrow 112c + 2128d = 1292$

Solving, we get $c = 0$ and $d = 0.607$

The regression line is $y = 0.607x$.

(c) $r^2 = bd \Rightarrow r^2 = (1.4708)(0.607) = 0.8928 \Rightarrow r = 0.9449$

(d) $y(8) = 1.913 + 1.478 \times 8 = 13.737$

(e) $x(16) = 0.607 \times 16 = 9.712$

Example 38

Use

$$r = \frac{\sigma_x^2 + \sigma_y^2 - \sigma_{x-y}^2}{2\,\sigma_x \sigma_y}$$

to find r for the following data:

Solution

$$\sigma_x^2 = \frac{\sum x^2}{N} - \left(\frac{\sum x}{N}\right)^2 = 584.6 \Rightarrow \sigma_x = 24.178$$

$$\sigma_y^2 = \frac{\sum y^2}{N} - \left(\frac{\sum y}{N}\right)^2 = 468.8 \Rightarrow \sigma_y = 21.652$$

$$\sigma_{x-y}^2 = \frac{\sum (x-y)^2}{N} - \left(\frac{\sum (x-y)^2}{N}\right) = 134.6 \Rightarrow \sigma_{x-y} = 11.601$$

$$r = \frac{584.6 + 468.8 - 134.6}{2(24.178)(21652)} = 0.877$$

Index

U

V